Springer Proceedings in Mathematics

Volume 9

For further volumes:
http://www.springer.com/series/8806

Springer Proceedings in Mathematics

The book series will feature volumes of selected contributions from workshops and conferences in all areas of current research activity in mathematics. Besides an overall evaluation, at the hands of the publisher, of the interest, scientific quality, and timeliness of each proposal, every individual contribution will be refereed to standards comparable to those of leading mathematics journals. It is hoped that this series will thus propose to the research community well-edited and authoritative reports on newest developments in the most interesting and promising areas of mathematical research today.

Valentin Blomer • Preda Mihăilescu
Editors

Contributions in Analytic and Algebraic Number Theory

Festschrift for S. J. Patterson

Springer

Editors
Valentin Blomer
Mathematisches Institut
Universität Göttingen
Bunsenstr. 3-5
37073 Göttingen, Germany
blomer@uni-math.gwdg.de

Preda Mihăilescu
Mathematisches Institut
Universität Göttingen
Bunsenstr. 3-5
37073 Göttingen, Germany
preda@uni-math.gwdg.de

ISSN 2190-5614 e-ISSN 2190-5622
ISBN 978-1-4614-1218-2 e-ISBN 978-1-4614-1219-9
DOI 10.1007/978-1-4614-1219-9
Springer New York Dordrecht Heidelberg London

Library of Congress Control Number: 2011941120

Printed on acid-free paper

Springer is part of Springer Science+Business Media (www.springer.com)

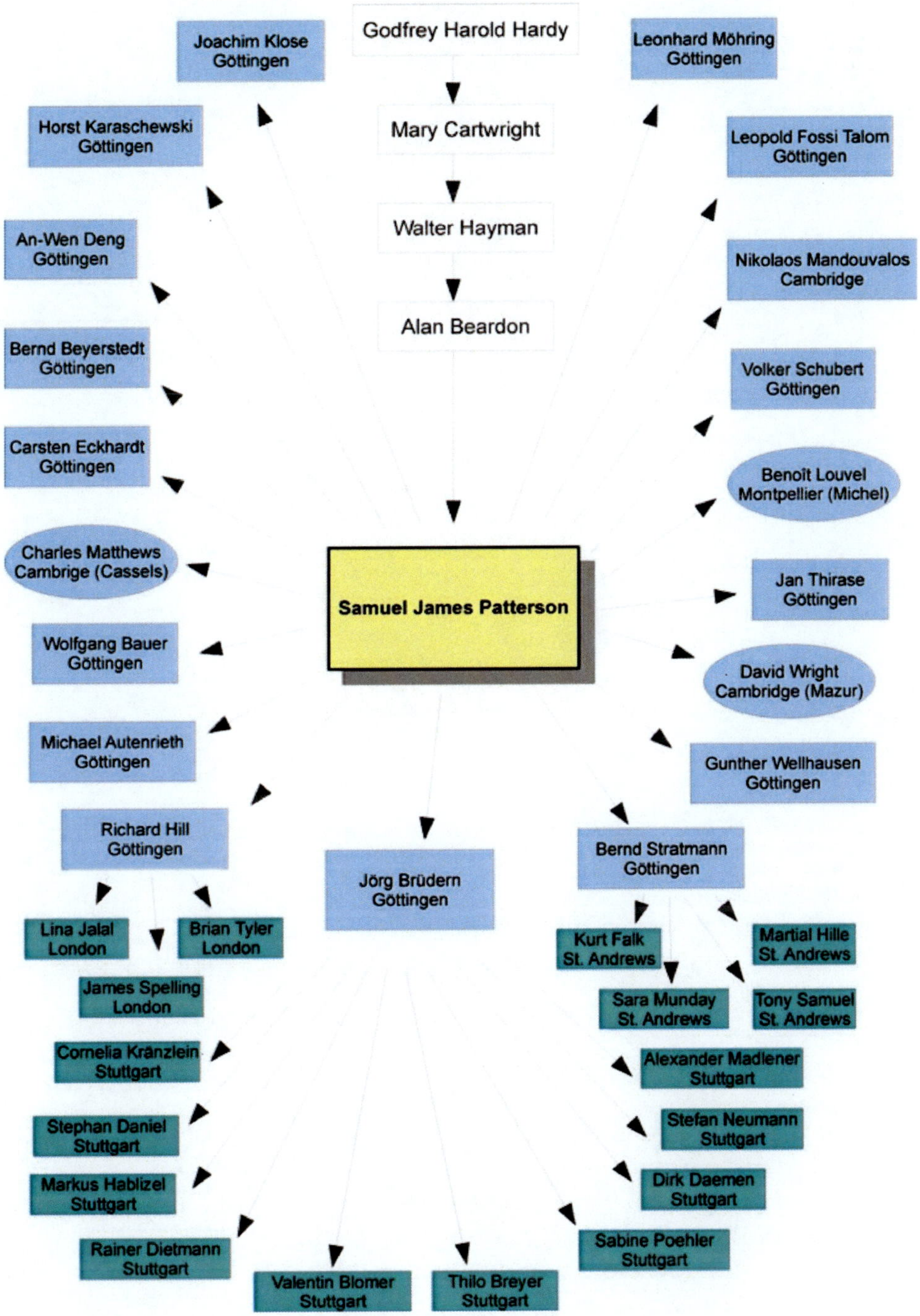
Godfrey Harold Hardy
Mary Cartwright
Walter Hayman
Alan Beardon
Samuel James Patterson
Joachim Klose Göttingen
Horst Karaschewski Göttingen
An-Wen Deng Göttingen
Bernd Beyerstedt Göttingen
Carsten Eckhardt Göttingen
Charles Matthews Cambrige (Cassels)
Wolfgang Bauer Göttingen
Michael Autenrieth Göttingen
Richard Hill Göttingen
Lina Jalal London
Brian Tyler London
James Spelling London
Cornelia Kränzlein Stuttgart
Stephan Daniel Stuttgart
Markus Hablizel Stuttgart
Rainer Dietmann Stuttgart
Valentin Blomer Stuttgart
Thilo Breyer Stuttgart
Jörg Brüdern Göttingen
Sabine Poehler Stuttgart
Dirk Daemen Stuttgart
Stefan Neumann Stuttgart
Alexander Madlener Stuttgart
Bernd Stratmann Göttingen
Kurt Falk St. Andrews
Sara Munday St. Andrews
Martial Hille St. Andrews
Tony Samuel St. Andrews
Gunther Wellhausen Göttingen
David Wright Cambridge (Mazur)
Jan Thirase Göttingen
Benoît Louvel Montpellier (Michel)
Volker Schubert Göttingen
Nikolaos Mandouvalos Cambridge
Leopold Fossi Talom Göttingen
Leonhard Möhring Göttingen

Foreword

Few mathematicians in Germany in the last 30 years have influenced analytic number theory as much and as deeply as Samuel Patterson. His impressive genealogy is only one of the results of his active mathematical career.

Many of the articles in this volume have been presented at the conference "Patterson 60++" at the University of Göttingen in July 2009 on the occasion of his 60th birthday. The conference featured four generations of mathematicians including Patterson's children and grandchildren as well as his own advisor Prof. Alan F. Beardon who did not miss the opportunity to contribute an enjoyable and lively talk to the many mathematical birthday presents.

The articles in this volume reflect Patterson's manifold interests ranging from classical analytic number theory and exponential sums via automorphic forms on metaplectic groups to measure-theoretic aspects of Fuchsian groups. The combination of measure theory and spectral theory, envisioned by Patterson 30 years ago, is still a most fruitful instance of interdisciplinary mathematics. Besides purely mathematical topics, Samuel Patterson is also well known for his interest in the history of mathematics; a place like Göttingen where he has been teaching and researching for the last 30 years seems to be an appropriate location in this respect. We hope that this book gives a lively image of all of these aspects.

We would like to thank the staff of Springer for their cooperation, Michaela Wasmuth for preparing the genealogy, Stefan Baur for excellent and competent typesetting, the numerous referees for their valuable time and expertise, and first and foremost the authors for contributing to this volume. Some colourful and very personal recollections of Samuel Patterson have been compiled on the following pages.

Göttingen, Germany
Summer 2011

Valentin Blomer
Preda Mihăilescu

Encounters with Samuel J. Patterson

I came to Göttingen in October 1982, as a first year student in mathematics. Samuel J. Patterson was lecturing on linear algebra. The course was very intense, never loosing a beat in the flow of ideas. His style was passionate, getting in touch with the subject even literally: he would wipe the board with his arm, or his right hand, and when it was over, after ninety minutes of hard work, there was chalk all over the place, including his pants and, sometimes, even his impressive beard. Needless to say, the girls liked it very much.

At that time, I did not know much about mathematics and contemporary research, but it was clear to the audience that an enthusiastic scientist was at work here, and it was a significant encouragement to more than just a few to become serious about mathematics.

Later, when I became his research student, he had gathered a larger group of younger people in the SFB 170. Those days, his efforts in number theory concentrated on Gauss sums and metaplectic forms, so it seemed. My reading list, however, reflected his heritage as a more classically oriented analytic number theorist, and it would send me off into a different direction. His approach to supervision is to guide the student toward independence as early as is possible. He would push the project as long as guidance is necessary, but get out of the way when there is enough momentum. Also, in the Oberseminar, we got a weekly treat of what he thought we should know, and that was quite a bit.

It is not a surprise that his students work in different fields of mathematics. At heart, his love for the theory of numbers and classical mathematics is always present. Now he is soon to retire, whatever this means. Certainly, he will not retire as a mathematician, and we hope that he also continues to encourage the next generation.

Jörg Brüdern

I remember June 2001 in Oberwolfach, the conference commemorating the bicentennary of Carl Friedrich Gauss's *Disquisitiones Arithmeticae*. We had brought together for the occasion mathematicians with historians and philosophers of mathematics. Fairly explicit mathematics dominated the first talks: composition of binary quadratic forms, irreducibility questions in Gauss's Seventh Section on cyclotomy, nineteenth century reports on the state of the theory of numbers, etc. This changed when it was Patterson's turn to talk. His subject was Gauss's determination of the sign of (certain) quadratic Gauss sums, as they are called today. Not that he banned explicit mathematics from his talk – as the lecture went on, he indicated explicitly several proofs of Gauss's relation, and generalisations thereof. But for some 20 min, he just talked about Gauss's well-known letter to Olbers of 3 September 1805. This is the letter, where Gauss describes how long he had struggled with the proof – proof of a result, as Patterson duly pointed out, which Gauss had announced in the *Disquisitiones* on the sole basis of experimental evidence – and how suddenly, with no apparent connection to the ideas he had pursued before – *wie der Blitz einschlägt* – the crucial idea had finally struck him.

A specialist of German literature commenting on this letter would be expected to point to the ambient discourse on *genius* at the time – *der Geniebegriff bei Goethe (und anderen)*. At the Oberwolfach conference, it was precisely the most active research mathematician among all those present at the conference who struck this note. He explained how to read this letter as a self-portrait of Gauss working on his image, on the ethos reflected in it. Beethoven's Eroica was mentioned as a point of reference for the document, as was Tolstoi's *War and Peace*.

Unlike the romantic writer, the poet of Roman antiquity was supposed, not only just to capture an immediate impression and conjure up its deeper meaning with a stroke – ... *die Vöglein schweigen im Walde; warte nur, balde* .... – but also to prove himself a *poeta doctus*, able to build global knowledge from geography and natural history into his polished narration. Samuel J. Patterson is for me a rare and challenging example of *mathematicus litterarius* – a species of which we could use many more, provided they are as capable as he is to use the literary, the humanistic vantage point for detaching themselves from naive identifications with mathematical heroes of the past.

Cf. The expanded writeup of the talk: S.J. Patterson, Gauss sums. In: The Shaping of Arithmetic after C.F. Gauss's Disquisitiones Arithmeticae (C. Goldstein, N. Schappacher, J. Schwermer, eds.), Springer, 2007; pp. 505–528.

Norbert Schappacher

Es gibt wenige Personen im Leben, die dauerhaft im Gedächtnis haften bleiben. Samuel J. Patterson ist eine davon. Nicht nur als Mensch, sondern mehr noch durch seine Impulse und Ideen, die entscheidende Weichenstellungen geben, und zu glücklichen und wertvollen Entwicklungen führen.

Er nahm zum Wintersemester 1981 den Ruf auf die Nachfolge von Max Deuring an, die Antrittsvorlesung hielt er am 20. Januar 1982: "Gauß'sche Summen: ein Thema mit Variationen", ein Gebiet, dass ihn stets fasziniert hat. Es ist bezeichnend, dass er in seiner Antrittsvorlesung Gauß mit den Worten zitiert[1]: "... *nämlich die Bestimmung des Wurzelzeichens ist es gerade, was mich immer gequält hat. Dieser Mangel hat mir alles Übrige, was ich fand, verleidet; und seit vier Jahren wird selten eine Woche hingegangen sein, wo ich nicht einen oder den anderen vergeblichen Versuch, diesen Knoten zu lösen, gemacht hätte — besonders lebhaft wieder in der letzten Zeit. Aber alles Brüten, alles Suchen ist umsonst gewesen, traurig habe ich jedesmal die Feder weglegen müssen. Endlich vor ein paar Tagen ist's gelungen – aber nicht meinem mühsamen Suchen, sondern bloß durch die Gnade Gottes möchte ich sagen. Wie der Blitz einschlägt, hat sich das Rätsel gelöst;...*". An diese Sätze von Gauß werde ich stets erinnert, wenn ich Pattersons Arbeitsweise charakterisieren soll; ähnliche Bemerkungen von ihm selbst zu seiner Arbeitsweise belegen dies.

[1]Brief von Gauß an Olbers vom 3.9.1805.

Auf Grund seiner Arbeiten zu Fuchs'schen und Klein'schen Gruppen[2] wurde D. Sullivan dazu angeregt, über analoge Fragen für dynamische Systeme nachzudenken. Dies wiederum führte dazu, dass ich mit Patterson im SFB 170 im Jahr 1984 ein Teilprojekt zu geodätischen Strömungen beantragte, das erfolgreich über mehr als zehn Jahre lief. Natürlich war Patterson die treibende Kraft hinter diesem Projekt, bezeichnend hierbei seine Weitsicht und die Fähigkeit weitgesteckte Fragen zu formulieren. Bezeichnend ist auch die Begeisterung, mit der er Mitte der achtziger Jahre fraktale Geometrie erklärte. Auf der Konferenz über dynamische Systeme und Ergodentheorie in Oberwolfach konnte er stundenlang darüber erzählen und auch Laien dafür begeistern. Die Fragestellungen des SFB waren schwierig, wurden dann aber doch in Zusammenarbeit mit der Berliner Gruppe im SFB 288 (A. Juhl) beantwortet[3]: "Es war 1984 nicht klar, wie die beiden Zugänge (symbolische Dynamik und Selberg'sche Zetafunktionen) zu der analytischen Theorie der Kleinschen Gruppen zueinander passten. Es war auch nicht klar, ob ein Zugang dem anderen überlegen war. Ein Ziel war ein Verständnis der beiden Theorien und, wenn möglich, ihre Synthese. Etwas später hat Gromov eine sehr anschauliche, wenn nicht immer wörtlich korrekte Darstellung der Ripsschen Theorie der hyperbolischen Gruppen gegeben. Es war klar, dass die beiden Zugänge auch hier anwendbar waren. Da der kombinatorische Rahmen in mancher Hinsicht einfacher und in anderer komplizierter als der differentialgeometrische ist, lag es auch nahe, diese auch in die Betrachtungen einzubeziehen."

Der SFB 170 "Geometrie und Analysis" war der zweite Sonderforschungsbereich in reiner Mathematik, der durch die Deutsche Forschungsgemeinschaft gefördert wurde. Nicht zuletzt durch die treibende Kraft von Hans Grauert, Tammo tom Dieck und Samuel J. Patterson wurde dies ermöglicht. Der SFB 170 hatte 4 Teilprojekte: 1. Komplexe Analysis (Grauert, Flenner), 2. Topologie (tom Dieck, Smith) 3. Metaplektische Formen (Patterson, Christian) und 4. Geodätische Strömungen (Patterson, Denker). Er war nach heutigen Maßstäben recht klein, trotzdem wissenschaftlich wegweisend, für die damalige Fakultät lebenserhaltend und für den wissenschaftlichen Nachwuchs in den achtziger Jahren eine der wenigen Möglichkeiten, sich in Deutschland fortzubilden.[4] Der SFB ermöglichte zudem eine verbesserte Außendarstellung der Fakultät.

Am Ende der Laufzeit des SFB 1995 war klar, dass die Fakultät neu ausgerichtet werden musste. Auf Grund mehrerer Umstände war dieses Ziel kurzfristig nicht erreichbar. Verschiedene Versuche waren nicht erfolgreich, aber eine wegweisenden Neuausrichtung war für Paddy stets oberste Priorität. Als eine solche (unter mehreren) erwies sich die Möglichkeit für die Fakultät, über Zentren in Informatik und Statistik mehr Zusammenarbeit innerhalb der Universität zu generieren und so die Mathematik mehr ins Blickfeld zu rücken. Das Zentrum für Statistik wurde

[2] siehe mein Beitrag mit Bernd Stratmann in diesem Band.

[3] Abschlußbericht des SFB 170.

[4] Die Göttinger Fakultät hatte viele seiner Mittelbaustellen verloren; der SFB schaffte hier einen kleinen Ausgleich.

2001 als eine Einrichtung der Fakultät gegründet.[5] Patterson war sicherlich der fakultätsinterne Initiator dieser Initiative und verfolgte die Entwicklung des ZfS als Vorstandsmitglied mit Rat und Tat. Es soll hier nicht diskutiert werden, wie sehr diese entscheidende Idee von Patterson zur Entwicklung der Fakultät in den letzten Jahren beigetragen hat. Es zeigt aber, wie sehr auch Patterson der Gauß'schen Maxime zuneigt ist, dass angewandte und reine Mathematik eins sind. Auf jeden Fall kann seither eine positive Entwicklung der Fakultät beobachtet werden, nicht zuletzt durch Patterson's konstantes Drängen auf Qualität gepaart mit Sinn für das Machbare.

Schon in seiner Antrittsvorlesung legte Patterson großen Wert auf historische Zusammenhänge. Dies ist eine bemerkenswerte Eigenschaft seiner Arbeitsweise. Es gab über all die Jahre keinen besser qualifizierten Wissenschaftler in Göttingen, der die umfangreiche Modellsammlung des Mathematischen Instituts[6] und seine historisch wertvollen Schriften betreuen konnte. Es ist stets ein Vergnügen zu beobachten, mit welchem Enthusiasmus und welche tiefsinnigen Beschreibungen Patterson die Modelle zahlreichen Besuchern, auch Schulklassen, näher bringt. Er ist auch einer der universitätsweit anerkannten Fachleute der mathematischen Nachlässe, die die Universitätsbibliothek besitzt. Sicherlich hat Paddy hier sein Hobby mit seiner Arbeit verbunden. Anläßlich des Gauß Jahres 2005 beschäftigte sich Patterson mit dem mathematischen Werk von Gauß im verstärkten Maße, und das führte zu einem Beitrag in der Begleitband zur Gauß-Ausstellung 2005.[7] Es bleibt nur zu wünschen, dass Paddy weiterhin so aktiv und erfolgreich wirken kann.

Manfred Denker

Maybe you had a God-father, maybe you did not. But when one joins as a new professor the Institute of Mathematics in Göttingen, then one *must* have a God-father, it is the law. More precisely, it became recently the law. When I arrived in Göttingen, I had the chance to have one by his own initiative: Samuel J. Patterson, also called Paddy.

Believing it might be useful for someone new to the Institute – and who had in fact only recently rejoined Academia – to have a point of reference, he developed for us a regular meeting ceremony. A little bit like the fox and the Little Prince of Saint-Exupéry, we met regularly on Wednesday, around the *Café Hemer*. I soon discovered that Paddy was a fundamental link in the long line, which connected present to the *foundation times*; at our Institute, these are situated around the personalities of Hilbert and Klein. With Paddy, halfways mythical names such as Grauert, Kneser or Siegel and Hasse were personal, direct or indirect encounters from his early years in Göttingen, and then stories and associations brought imperceptibly the founders

[5] http://www.uni-protokolle.de/nachrichten/id/19562/.

[6] http://www.math.uni-goettingen.de/historisches/modellsammlung.html.

[7] S. J. Patterson, M. Denker: *Gauß – der geniale Mathematiker*, Göttinger Bibliotheksschriften 30.

on the scene. As a recent person in the Alma Mater, I had encountered few active mathematicians in my life – but talking about so many, as if somehow I had just missed them by a fraction, the number of *acquaintances* grew from one Wednesday to another. At times, the stories of people who missed in one or another way their duty or natural expectations put in them, would come to complete the picture – and an attentive look would check then how the message was received.

That was the God-father at work, recalling me if necessary, that ethics was there constantly as a choice, failure a most probable menace, yet not a rule – and the tiny gap between the two is where the light comes in. Ethics and mathematics going on two indistinguishably close, parallel roads, the counterpart was openly described in Paddy's favorite formula "a mathematician is a person who fails 99% of the time and knows it". With or without quantization, failure is a rule of statistics but not one of necessity, and we are working at its borderline, either as mathematicians, or simply as creatures trying to stay involved in the art of being human. In the two realms, our scepticisms were compensating each other. This friendly initiation lasted for less than two years, and afterwards no one ever cared to ask me, how I graduated from Paddy's God-fathership.

At this time, I learned to respect in Paddy one of the rare personalities I happened to encounter, who seemed to have offered all of his dedication in community tasks in the academic business, for the finality of inflicting the Institutions with the best of the spirit they were intended for. One who tried, modestly and with tenacity, *to change the system from within*, as some song might put it. I was amazed by his knowledge of the legal background, coupled by a venerable lack of (utilitarian) pragmatism, which accompanied all of his attitudes.

It happened that towards the end of the God-father apprenticeship, I was getting involved with understanding some things about Iwasawa theory, which I believed should be simple in their essence, but did not appear like that in any book or publication I encountered. Paddy had at that time a "Working-group seminar" on Tuesdays, and I asked him if I could become one of that group and use some time slots for talking about these questions. He accepted, and using the attention of my small auditory, I could start organizing my new area of research. As a result, there was a series of four manuscripts of about 30–70 pages in length, which expanded on one another, partially refuting previous claims and gaining focus. The attention of this auditory, and most of all, the one of Paddy, was of great importance, in an early spotting of some crude errors of wishful thinking. The subject was certainly remote from his domain of expertise, but certainly not from his mathematical instinct. Once again, the discussions we had in our Wednesday walks remained a memorable help, and the first versions of the *Snoqit*[8] manuscripts, which were the immediate notes for the seminar and have been more than once rewritten, remained a valuable means for recovering some of the initial guiding questions, if not insights. Some results drawn from that work are included in this volume.

[8]Seminar notes on open questions in Iwasawa theory.

Paddy will take a slightly anticipated retirement and intends to retire in nature, watching birds and performing laic pilgrimages to places of Celtic Christianism. I wish him all the peace and luck at this – can we believe, this might also mean retirement from mathematics?

Preda Mihăilescu

Contents

Contributors

Valentin Blomer Mathematisches Institut, Universität Göttingen, Bunsenstraße 3–5, D-37073 Göttingen, Germany, blomer@uni-math.gwdg.de

Alexander Braverman Department of Mathematics, Brown University, 151 Thayer St., Providence, RI 02912, USA, braval@math.brown.edu

Benjamin Brubaker Department of Mathematics, MIT, Cambridge, MA 02139-4307, USA, brubaker@math.mit.edu

Jörg Brüdern Mathematisches Institut, Universität Göttingen, Bunsenstraße 3–5, D-37073 Göttingen, Germany, bruedern@uni-math.gwdg.de

Daniel Bump Department of Mathematics, Stanford University, Stanford, CA 94305-2125, USA, bump@math.stanford.edu

Ulrich Bunke NWF I - Mathematik, Universität Regensburg, 93040 Regensburg, Germany, ulrich.bunke@mathematik.uni-regensburg.de

Gautam Chinta Department of Mathematics, The City College of CUNY, New York, NY 10031, USA, chinta@sci.ccny.cuny.edu

Manfred Denker Mathematics Department, Pennsylvania State University, State College, PA 16802, USA, denker@math.psu.edu

Adrian Diaconu University of Minnesota, 127 Vincent Hall, Church St. SE, Minneapolis, MN 55455, USA, cad@math.umn.edu

Michael Finkelberg IMU, IITP and State University Higher School of Economics, Department of Mathematics, 20 Myasnitskaya st, Moscow 101000 Russia, fnklberg@gmail.com

Solomon Friedberg Department of Mathematics, Boston College, Chestnut Hill, MA 02467-3806, USA, friedber@bc.edu

Paul Garrett University of Minnesota, 127 Vincent Hall, Church St. SE, Minneapolis, MN 55455, USA, garre001@umn.edu

The Density of Rational Points on a Certain Threefold

Valentin Blomer and Jörg Brüdern

Abstract The equation

$$x_1y_2y_3 + x_2y_1y_3 + x_3y_1y_2 = 0$$

defines a singular threefold in $\mathbb{P}^2 \times \mathbb{P}^2$. Let $N(B)$ be the number of rational points on this variety with non-zero coordinates of height at most B. It is proved that $N(B) \asymp B(\log B)^4$.

1 Introduction

The equation

$$x_1y_2y_3 + x_2y_1y_3 + x_3y_1y_2 = 0 \tag{1}$$

is a homogenous linear one in $\mathbf{x} = (x_1, x_2, x_3)$ and a quadric in $\mathbf{y}$. Thus, the equation defines a threefold in $\mathbb{P}^2 \times \mathbb{P}^2$. The purpose of this note is to study the distribution of its rational points.

There are three singularities on this variety. It is readily confirmed that these are given by the equations $y_i = x_i = y_j = x_j = 0$, for each of the three choices of i, j with $1 \leq i < j \leq 3$. Any solution of (1) with $x_1x_2x_3y_1y_2y_3 = 0$ we consider as a "trivial" solution. Note that the singular points are included here.

We shall count non-trivial rational points on (1), sorted according to their anticanonical height that we now define. Recall that a rational point in $\mathbb{P}^2$ has a representation by a primitive vector $\mathbf{x} \in \mathbb{Z}^3$ (that is, the coordinates x_1, x_2, x_3 are coprime), and that $\mathbf{x}$ is unique up to sign. Then $|\mathbf{x}| = \max |x_j|$ is the natural height of the rational point. A rational point on (1) may be represented by a pair $\mathbf{x}, \mathbf{y}$ of

---

V. Blomer (✉) • J. Brüdern
Mathematisches Institut, Universität Göttingen, Bunsenstrasse 3–5, 37073 Göttingen, Germany
e-mail: blomer@uni-math.gwdg.de; bruedern@uni-math.gwdg.de

V. Blomer and P. Mihăilescu (eds.), *Contributions in Analytic and Algebraic Number Theory*, Springer Proceedings in Mathematics 9, DOI 10.1007/978-1-4614-1219-9_1,

primitive vectors, and its anticanonical height is defined by $|\mathbf{x}|^2|\mathbf{y}|$. It is now natural to consider the counting function $N(B)$, defined as the number of all pairs $\mathbf{x},\mathbf{y}$ of primitive vectors in $\mathbb{Z}^3$ with

$$|\mathbf{x}|^2|\mathbf{y}| \le B \tag{2}$$

that satisfy (1) and

$$x_1x_2x_3y_1y_2y_3 \neq 0. \tag{3}$$

We shall determine the order of magnitude of $N(B)$.

Theorem 1. *One has*

$$B(\log B)^4 \ll N(B) \ll B(\log B)^4.$$

It should be stressed that there are many more trivial solutions. For example, we may choose $x_1 = 0$, $x_2 = x_3 = 1$ and y_1, y_2 coprime. Then $y_3 = -y_2$ solves (1), and there are $\gg B^2$ such solutions satisfying (2).

It is very likely that there is an asymptotic formula for $N(B)$ in which the leading term is $CB(\log B)^4$, with some positive constant C. Such a formula is predicted by a very general conjecture of Manin (see [1, 3]). The relevant height zeta function should have an analytic continuation with a pole at its abscissa of convergence, so that one would even expect an asymptotic expansion

$$N(B) = BP(\log B) + o(B) \quad (B \to \infty), \tag{4}$$

in which P is a certain real polynomial of degree 4. However, results of this type seem to be beyond the limitations of the methods in this paper.

The anticanonical height is a very natural measure for the size of a solution of a bihomogenous equation. In a sense, this measure puts equal weight on the linear and the quadratic contributions, in our special case. There are very few examples where this choice of height has been studied via methods from analytic number theory (but see Robbiani [6] and Spencer [7]). Alternatively, one may also consider (1) as a cubic in $\mathbb{P}^5$, and then count rational points according to their natural height $\max(|\mathbf{x}|, |\mathbf{y}|)$. Then one can say much more. In particular, the analogue of (4) can be established by an analytic method. For this as well as a more thorough discussion of Manin's conjecture and other recent work related to it, we refer the reader to our forthcoming memoir [2].

Our approach has some similarity with work of Heath-Brown [4] on the Caley cubic. It has become customary in problems of this type to count points on the universal torsor, rather than on the original variety, and this paper is no exception. The torsor is described in Sect. 2. Then we use a lattice point estimate from Heath-Brown [5] to establish the upper bound estimate for $N(B)$ in Sects. 3 and 4. In certain cases, one encounters an unconventional problem: a direct use of the geometry of numbers leads to estimates that are seemingly too weak by a power of $\log\log B$. In Sect. 4, we bypass this difficulty. Our argument uses an iterative scheme that appears to be new in this context. The lower bound for $N(B)$ can be established by direct methods, as we shall see in Sect. 5. Two technical bounds are required here, and the proofs for these are given in Sect. 6.

2 The Passage to the Torsor

The sole purpose of this section is to describe a parametrization of the threefold. The result is crucial for the derivation of both the upper and the lower bound in the theorem. Although no use of this fact is made later, it is perhaps of interest to observe that Lemma 1 yields a bijection between the threefold and its universal torsor.

Let $\mathscr{C}$ denote the set of pairs of primitive $\mathbf{x},\mathbf{y} \in \mathbb{Z}^3$ that satisfy (1) and (3). Let $\mathscr{A}$ denote the set of all $(\mathbf{d},\mathbf{z},\mathbf{a}) \in \mathbb{N}^3 \times \mathbb{Z}^3 \times \mathbb{Z}^3$ with $a_1a_2a_3z_1z_2z_3 \neq 0$ that satisfy the lattice equation

$$a_1d_1 + a_2d_2 + a_2d_3 = 0 \tag{5}$$

and the coprimality constraints

$$\begin{aligned}&(d_i;d_j) = (z_i;z_j) = (d_k;z_k) = 1 \ (1 \le i < j \le 3, 1 \le k \le 3),\\ &(a_1z_1;a_2z_2;a_3z_3) = 1.\end{aligned} \tag{6}$$

Lemma 1. *The mapping $\mathscr{A} \to \mathscr{C}$ defined by*

$$\begin{aligned} x_1 &= a_1z_1, \quad x_2 = a_2z_2, \quad x_3 = a_3z_3\\ y_1 &= d_2d_3z_1,\ y_2 = d_1d_3z_2,\ y_3 = d_1d_2z_3 \end{aligned} \tag{7}$$

is a bijection.

Proof. Given an element of $\mathscr{A}$, one checks from (6) that $\mathbf{x}$ and $\mathbf{y}$ are primitive, and from (1) that $(\mathbf{x},\mathbf{y}) \in \mathscr{C}$. To construct the inverse mapping, let $(\mathbf{x},\mathbf{y}) \in \mathscr{C}$ be given. Then put

$$d_1 = (y_2;y_3), \quad d_2 = (y_1;y_3), \quad d_3 = (y_1;y_2).$$

Since $\mathbf{y}$ is primitive, the d_j are coprime in pairs, as required in (6). By (3), we can now define non-zero integers z_j through the equations

$$y_1 = d_2d_3z_1, \quad y_2 = d_1d_3z_2, \quad y_3 = d_1d_2z_3,$$

as in 7, and by construction, the coprimality conditions in the first line of (6) all hold. We substitute for $\mathbf{y}$ in (1) and obtain

$$d_1x_1z_2z_3 + d_2x_2z_1z_3 + d_3x_3z_1z_2 = 0.$$

This yields $z_1 \mid d_1x_1z_2z_3$, and with (6) now in hand, we conclude that $z_1 \mid x_1$. Similarly, $z_j \mid x_j$ for $j = 2,3$, and we may define $a_j \in \mathbb{Z}$ through $x_j = a_jz_j$. The condition that $(a_1z_1; a_2z_2; a_3z_3) = 1$ is implied by the fact that $\mathbf{x}$ is primitive. This completes the proof.

3 A First Auxiliary Upper Bound

We embark on the estimation of $N(B)$ from above through a series of auxiliary bounds for the number, say $V(X,Y)$, of $(\mathbf{x},\mathbf{y}) \in \mathscr{C}$ within the box

$$|\mathbf{x}| \le X, \quad |\mathbf{y}| \le Y. \tag{8}$$

Our principal tool is the theory of successive minima from the geometry of numbers. All the necessary information is imported through the next lemma.

Lemma 2. *Let $\mathbf{v} \in \mathbb{Z}^3$ be primitive, let $H_i > 0$ $(1 \le i \le 3)$, and suppose that $|v_3| \le H_1H_2$. Then the number of primitive $\mathbf{u} \in \mathbb{Z}^3$ that satisfy*

$$u_1v_1 + u_2v_2 + u_3v_3 = 0,$$

and that lie in the box $|u_i| \le H_i$ $(1 \le i \le 3)$, is at most $50H_1H_2|v_3|^{-1}$.

Proof. This is contained in Heath-Brown [5], Lemma 3 (see also [4], Lemma 3). Even without the hypothesis that $|v_3| \le H_1H_2$, Heath-Brown shows that $4 + 40H_1H_2|v_3|^{-1}$ is an admissible upper bound for the number of $\mathbf{u}$ that are counted here.

As a first application of Lemma 2, we demonstrate an essentially best possible bound for $V(X,Y)$ when X is smaller than Y.

Lemma 3. *Let $4 \le X \le Y$. Then*

$$V(X,Y) \ll X^2Y(\log X)^3.$$

Proof. By Lemma 1, we see that $V(X,Y)$ equals the number of $(\mathbf{d},\mathbf{a},\mathbf{z}) \in \mathscr{A}$ with

$$|a_jz_j| \le X, \quad d_1d_2|z_3| \le Y, \quad d_1d_3|z_2| \le Y, \quad d_2d_3|z_1| \le Y.$$

Here, all z_j enter the conditions only through $|z_j|$, so that at the cost of a factor 8, it is enough to count $\mathbf{z}$ with all z_j positive. A similar observation applies to $\mathbf{a}$: by (5), the a_j may not all have the sign. Hence, there are exactly six possible distributions of signs among a_1, a_2, a_3. By symmetry, it then suffices to consider the case where $a_1 > 0, a_2 > 0, a_3 < 0$. In this situation, it is convenient to replace $-a_3$ by a_3, and accordingly to rewrite (5) as

$$a_1d_1 + a_2d_2 = a_3d_3 \tag{9}$$

in which all variables are now positive. Since any $(\mathbf{d},\mathbf{a},\mathbf{z}) \in \mathscr{A}$ consists of primitive $\mathbf{d},\mathbf{a},\mathbf{z}$, by (6), it now follows that

$$V(X,Y) \leq 48U, \tag{10}$$

where U is the number of triples $\mathbf{d},\mathbf{a},\mathbf{z} \in \mathbb{N}^3$ of primitive vectors that satisfy (9) and

$$a_j z_j \leq X, \quad d_1 d_2 z_3 \leq Y, \quad d_1 d_3 z_2 \leq Y, \quad d_2 d_3 z_1 \leq Y. \tag{11}$$

Now fix $\mathbf{a}$ and $\mathbf{z}$ in line with these conditions. We shall count the number of possibilities for $\mathbf{d}$ by Lemma 2. By (9), we have $a_1 d_1 < a_3 d_3$, whence by (11),

$$d_1 \leq Y/(d_3 z_2) \leq Y a_3/(z_2 a_1 d_1).$$

Consequently,

$$d_1 \leq (Y a_3)^{1/2} (z_2 a_1)^{-1/2}, \tag{12}$$

and similarly, one also finds that

$$d_2 \leq (Y a_3)^{1/2} (z_1 a_2)^{-1/2}. \tag{13}$$

We now apply Lemma 2, with $\mathbf{a}$ in the role of $\mathbf{v}$, and with

$$H_1 = (Y a_3)^{1/2} (z_2 a_1)^{-1/2}, \quad H_2 = (Y a_3)^{1/2} (z_1 a_2)^{-1/2}, \quad H_3 = Y.$$

The condition that

$$|a_3| \leq H_1 H_2 = Y a_3 (a_1 z_1 a_2 z_2)^{-1/2}$$

is satisfied, because by (11) and the hypothesis of the lemma to be proved, one has $a_1 z_1 a_2 z_2 \leq X^2 \leq Y^2$. Hence, the number of primitive $\mathbf{d}$ to be counted does not exceed $50Y(a_1 z_1 a_2 z_2)^{-1/2}$. It follows that

$$U \leq 50 \sum_{\substack{a_j z_j \leq X \\ j=1,2,3}} \frac{Y}{(a_1 z_1 a_2 z_2)^{1/2}}.$$

The familiar asymptotic formula

$$\sum_{az \leq X} 1 = X \log X + O(X)$$

and partial summation now suffice to conclude that

$$U \ll X^2 Y (\log X)^3,$$

and the lemma follows from (10).

4 Another Auxiliary Upper Bound

In this section, we complement the work of the previous section by examining the situation when Y is smaller than X.

Lemma 4. *Suppose that* $4 \leq Y \leq X$. *Then*

$$V(X,Y) \ll X^2 Y (\log Y)^3.$$

This bound has the same strength than the one exhibited in Lemma 3, but the proof is considerably more complex. The basic idea is the same, with the roles of $\mathbf{d}$ and $\mathbf{a}$ interchanged. We shall estimate $V(X,Y)$ through an iterative scheme rooted in the following bound.

Lemma 5. *Let* $4 \leq Z \leq Y \leq X$ *and* $4 \leq Y_3 \leq Y$. *Let* $T = T(X,Y,Y_3,Z)$ *denote the number of* $(\mathbf{x},\mathbf{y}) \in \mathscr{C}$ *with*

$$|\mathbf{x}| \leq X, \quad Y/Z \leq |y_1|, |y_2| \leq Y, \quad |y_3| \leq Y_3.$$

Then

$$T \ll X^2 Y_3 (\log Y)^3 (\log Z)^2.$$

Proof. By Lemma 1, T equals the number of $(\mathbf{a},\mathbf{d},\mathbf{z}) \in \mathscr{A}$ satisfying the inequalities

$$\begin{aligned} &|a_j z_j| \leq X \quad (1 \leq j \leq 3), \quad && d_1 d_2 |z_3| \leq Y_3, \\ &Y/Z \leq d_2 d_3 |z_1| \leq Y, && Y/Z \leq d_1 d_3 |z_2| \leq Y. \end{aligned} \tag{14}$$

As in the previous section, only primitive $\mathbf{a}$ and $\mathbf{d}$ are counted here. Now fix an admissible choice of $\mathbf{d}$ and $\mathbf{z}$. We apply Lemma 2 with $\mathbf{d}$ in the role of $\mathbf{v}$, and with $H_j = X/|z_j|$. Repeated use of (14) and the postulated inequality $Y \leq X$ suffices to confirm that

$$d_3 \leq Y/|z_1| \leq X/|z_1| \leq X^2/|z_1 z_2| = H_1 H_2,$$

as required. Lemma 2 now shows that no more than $50 H_1 H_2 d_3^{-1}$ primitive $\mathbf{a}$ correspond to the given pair $\mathbf{z},\mathbf{d}$. We sum this over $\mathbf{z}$ and $\mathbf{d}$, and observe that there are 8 possible distributions of sign among z_1, z_2, z_3. It follows that

$$T \leq 400 \sum_{\mathbf{z},\mathbf{d}} \frac{X^2}{z_1 z_2 d_3}$$

in which the sum is over $\mathbf{z} \in \mathbb{N}^3$, $\mathbf{d} \in \mathbb{N}^3$ subject to the conditions

$$d_1 d_2 z_3 \leq Y_3, \quad Y/Z \leq d_1 d_3 z_2 \leq Y, \quad Y/Z \leq d_2 d_3 z_1 \leq Y$$

and $z_j \leq X$ for $1 \leq j \leq 3$. We carry out the sum over z_3 to infer that

$$T \leq 400 \sum_{z_1,z_2 \leq Y} \sum_{\mathbf{d}} \frac{X^2 Y_3}{z_1 z_2 d_1 d_2 d_3},$$

where the sum over **d** is now restricted by the inequalities

$$Y/Z \leq d_1 d_3 z_2 \leq Y, \quad Y/Z \leq d_2 d_3 z_1 \leq Y.$$

Before proceeding further, we record here an elementary inequality. In fact, when ξ, ζ are real numbers with $1 \leq \zeta \leq \xi$, then

$$\sum_{\xi/\zeta \leq d \leq \xi} \frac{1}{d} \leq 2 + \int_{\xi/\zeta}^{\xi} \frac{\mathrm{d}t}{t} \leq 2 + \log \zeta,$$

as one readily confirms. We apply this with $\xi = Y/(d_3 z_j)$ and $\zeta = \min(Z, Y/(d_3 z_j))$ to bound the sums over d_1 and d_2 above, and deduce that

$$T \leq 400 \sum_{z_1,z_2 \leq Y} \sum_{d_3 \leq Y} \frac{X^2 Y_3 (2 + \log Z)^2}{z_1 z_2 d_3}.$$

The conclusion of Lemma 5 is now immediate.

Note that $V(X,Y) = T(X,Y,Y,Y)$, so that Lemma 5 gives $V(X,Y) \ll X^2 Y (\log Y)^5$. This is too weak when Y is large, but the conclusion of Lemma 4 is now established for $Y \leq e^4$. For larger Y, one can do rather better. Let $\log_k$ be the iterated logarithm, defined by $\log_1 = \log$ and $\log_{k+1} = \log \circ \log_k$. When $Y \geq e^4$, determine the integer K by the inequalities

$$4 \leq \log_K Y < e^4 \tag{15}$$

and note that $K \geq 1$. For $1 \leq k \leq K$, the number

$$Z_k = (\log_k Y)^3 \tag{16}$$

is a real number with $Z_k \geq 64$. Now let $W_0 = W_0(X,Y)$ be the number of all $\mathbf{x}, \mathbf{y}$ counted by $V(X,Y)$, where one has $|y_j| \leq Y Z_1^{-1}$ for at least one $j \in \{1,2,3\}$, and when $1 \leq k \leq K$, let $V_k = V_k(X,Y)$ be the number of those $\mathbf{x}, \mathbf{y}$ counted by $V(X,Y)$ where $Y Z_k^{-1} < |y_j| \leq Y$ holds for all $j \in \{1,2,3\}$. Then one has $V(X,Y) = W_0 + V_1$. If we define W_k as the number of solutions counted by $V_k(X,Y)$ that have $|y_j| \leq X Z_{k+1}^{-1}$ for at least one $j \in \{1,2,3\}$, then $V_k = W_k + V_{k+1}$, and by repeated application, one finds that

$$V(X,Y) = W_0 + W_1 + \cdots + W_{K-1} + V_K. \tag{17}$$

Observe that $V_K = T(X,Y,Y,Z_K)$, so that Lemma 5 and (15) imply the bound $V_K \ll XY^2(\log Y)^3$. A similar estimate is available for W_k when $0 \le k < K$. By symmetry in the indices $1,2,3$, one finds that

$$W_0 \le 3T(X,Y,YZ_1^{-1},Y)$$

and

$$W_k \le 3T(X,Y,YZ_{k+1}^{-1},Z_k) \quad (1 \le k < K).$$

Lemma 5 now yields

$$W_0 \ll X^2YZ_1^{-1}(\log Y)^5 \ll X^2Y(\log Y)^2$$

and

$$W_k \ll X^2YZ_{k+1}^{-1}(\log Y)^3(\log Z_k)^2,$$

uniformly for $1 \le k < K$. By (16),

$$Z_{k+1}^{-1}(\log Z_k)^2 = 3(\log_{k+1} Y)^{-1},$$

so that, on collecting together, we now deduce from (17) the inequality

$$V(X,Y) \ll X^2Y(\log Y)^3\Big(1+\sum_{k=1}^{K}(\log_k Y)^{-1}\Big).$$

By (15), we see $\log_{K-l} Y \ge \exp_l(0)$ for $0 \le l < K$; here $\exp_1 = \exp$ and $\exp_{l+1} = \exp \circ \exp_l$ is the iterated exponential. Now

$$\sum_{k=1}^{K}(\log_k Y)^{-1} \le 1 + \sum_{l=1}^{\infty} \exp_l(0)^{-1},$$

and the sum on the right converges. This completes the proof of Lemma 4.

The upper bound for $N(B)$ is now available. Any pair $\mathbf{x},\mathbf{y}$ with $|\mathbf{x}|^2|\mathbf{y}| \le B$ satisfies $|\mathbf{x}| \le 4^j$, $|\mathbf{y}| \le 4^{2-2j}B$ for some j with $1 \le j \ll \log B$. Hence,

$$N(B) \le \sum_{1 \le j \ll \log B} V(4^j, 4^{2-2j}B), \tag{18}$$

and we may apply Lemmas 3 and 4 to conclude that $N(B) \ll B(\log B)^4$, as required.

5 The Lower Bound Method

The work in the previous sections involved a number of enveloping processes. For the upper bounds it sufficed to ignore the coprimality constraints (6) whenever they would otherwise complicate the argument, and also (18) is derived by a non-disjoint cover of the hyperbolic condition $|\mathbf{x}|^2|\mathbf{y}| \le B$. For the lower bound, we have to reverse the procedure, and we therefore now consider the number $V^*(X,Y)$ of all $(\mathbf{x},\mathbf{y}) \in \mathscr{C}$ with

$$|\mathbf{x}| \le X, \quad \frac{1}{4}Y < |\mathbf{y}| \le Y.$$

The crucial lower bound then is the following result.

Lemma 6. *Let Y be sufficiently large. Then, uniformly for X and Y with $X \ge Y^3$, one has*

$$V^*(X,Y) \gg X^2 Y (\log Y)^3.$$

Once this is established, we can mimic the strategy leading to (18), and obtain

$$N(B) \ge \sum_{4^j \le B^{1/7}} V^*(2^{-j}B^{1/2}, 4^j) \gg B(\log B)^4,$$

which completes the proof of the theorem.

In the remainder of this section, we shall derive Lemma 6. Again, the counting process will be performed on the torsor. Three technical estimates are required, the first of which is given in the next lemma.

Lemma 7. *Let A_1, A_2 be real numbers with $A_j \ge 2$. Let $z_1, z_2, d_3, b \in \mathbb{N}$ with $(z_1; z_2) = (b; d_3) = 1$. Let $S(A_1, A_2, z_1, z_2, d_3, b) = S$ denote the number of pairs a_1, a_2 with $(a_1; z_2) = (a_2; z_1) = (a_1; a_2) = 1$ and*

$$a_1 \le A_1, \quad a_2 \le A_2, \quad a_2 \equiv b a_1 \bmod d_3.$$

Then uniformly for A_1, A_2, z_1, z_2, d, b in the indicated ranges, one has

$$S \gg A_1 A_2 \frac{\varphi(z_2 d_3)\varphi(z_1)}{d_3^2 z_2 z_1} + O((\tau(z_1)A_1 + \tau(z_2)A_2)\log(A_1 z_1)).$$

We postpone the proof of this as well as the one for the following lemma to the final section.

Lemma 8. *Let $D \ge 1$ and $c \in \mathbb{N}$. Then*

$$\sum_{\substack{d \le D \\ (c;d)=1}} 1 = \frac{\varphi(c)}{c} D + O(\tau(c)). \tag{19}$$

In addition, let $D_2 \geq D_1 \geq 1$. Then

$$\sum_{\substack{D_1 \leq d \leq D_2 \\ (c;d)=1}} \frac{\varphi(d)^4}{d^5} \gg \frac{\varphi(c)}{c} \log \frac{D_2}{D_1} + O\left(\frac{\tau(c)(1+\log D_1)^4}{D_1}\right). \tag{20}$$

Proof (of Lemma 6). As a first step, we will only count those $(\mathbf{x},\mathbf{y}) \in \mathscr{C}$, where $(x_1;x_2) = 1$, before passing to the torsor. By Lemma 1, it follows that $V^*(X,Y)$ is bounded below by the number of $(\mathbf{d},\mathbf{a},\mathbf{z}) \in \mathbb{N}^3 \times \mathbb{Z}^3 \times \mathbb{N}^3$, where

$$(d_i;d_j) = (z_i;z_j) = (d_k;z_k) = 1 \tag{21}$$

hold for $1 \leq i < j \leq 3$, $1 \leq k \leq 3$, where

$$(a_1z_1;a_2z_2) = 1, \tag{22}$$

and where

$$1 \leq |a_j|z_j \leq X, \quad \frac{1}{4}Y < d_1d_2z_3, d_2d_3z_1, d_1d_3z_2 \leq Y. \tag{23}$$

Note that a_3 is now absent from the coprimality constraints.

Let V^{**} denote the number of $(\mathbf{d},\mathbf{a},\mathbf{z})$ counted above that also satisfy the inequalities

$$d_2d_3z_1 \geq \frac{1}{2}Y, \quad \frac{1}{2}d_3z_2 \leq d_2z_3 \leq d_3z_2 \leq Y^{2/5}. \tag{24}$$

Then

$$V^*(X,Y) \geq V^{**}. \tag{25}$$

In preparation for an application of Lemma 7, fix a choice of $\mathbf{z},\mathbf{d}$ in accordance with (21)–(24).

Now choose pairs $(a_1,a_2) \in \mathbb{N}^2$ with

$$a_1 \leq X/(4z_1), \quad a_2 \leq X/(2z_2), \quad (a_1z_1,a_2z_2) = 1, \quad d_1a_1 \equiv -a_2d_2 \bmod d_3.$$

In view of (21) and Lemma 7, there are $S(X/(4z_1),X/(2z_2),z_1,z_2,d_3,b)$ such choices, where b is defined by $bd_2 \equiv -d_1 \bmod d_3$, and we have

$$S \gg X^2 \frac{\varphi(z_1)\varphi(z_2d_3)}{(z_1z_2d_3)^2} - O(X(\log X)(z_1z_2)^{\varepsilon}) \gg X^2 \frac{\varphi(z_1)\varphi(z_2d_3)}{(z_1z_2d_3)^2}$$

in the current situation. For each such pair (a_1,a_2), there is an integer a_3 such that (6) holds, and we have

$$|a_3| = \frac{a_1d_1 + a_2d_2}{d_3} \leq \frac{X}{d_3}\left(\frac{d_1}{4z_1} + \frac{d_2}{2z_2}\right) \leq \frac{X}{d_3}\left(\frac{Y}{4z_1d_2z_3} + \frac{d_3}{2z_3}\right) \leq \frac{X}{z_3},$$

as one readily checks from (23) and (24). Hence, the triple $(\mathbf{d},\mathbf{a},\mathbf{z})$ is indeed counted by V^{**}. This shows that

$$V^{**} \gg X^2 \sum_{\mathbf{d},\mathbf{z}} \frac{\varphi(z_1)\varphi(z_2d_3)}{(z_1z_2d_3)^2},$$

in which the sum is restricted by (21), (24) and the second set of conditions in (23).

Next, observe that any d_1 with

$$Y/(4d_2z_3) \le d_1 \le Y/(2d_2z_3) \tag{26}$$

satisfies the previous size constraints, and that $Y/(d_2z_3) \ge Y^{3/5}$. The coprimality constraints on d_1 add up to $(d_1;d_2d_3z_1) = 1$. Hence, we may sum over all d_1 satisfying (26) and apply (19) with $c = d_2d_3z_1$. Then

$$V^{**} \gg X^2Y \sum_{\substack{\frac{1}{2}Y<d_2d_3z_1\le Y \\ \frac{1}{2}d_3z_2\le d_2z_3\le d_3z_2\le Y^{2/5}}} \frac{\varphi(z_1)^2\varphi(z_2)\varphi(d_2)\varphi(d_3)^2}{z_1^3z_2^2z_3d_2^2d_3^3},$$

where z_1,z_2,z_3,d_2,d_3 are still subject to all applicable conditions in (21). The next sum that we carry out is the one over z_1. The interval for z_1 is of length $Y/(2d_2d_3) \gg Y^{1/5}$, and the coprimality condition on z_1 is $(z_1;z_2z_3) = 1$. Since $\varphi(z_1)^2z_1^{-3} \ge \varphi(z_1)^3z_1^{-4}$, we may apply (20) with $c = z_2z_3$, $D_2/D_1 = 2$ and $D_1 \gg Y^{1/5}$ to find that

$$V^{**} \gg X^2Y \sum_{\frac{1}{2}d_3z_2\le d_2z_3\le d_3z_2\le Y^{2/5}} \frac{\varphi(z_2)^2\varphi(z_3)\varphi(d_2)\varphi(d_3)^2}{z_2^3z_3^2d_2^2d_3^3},$$

with the sum now subject to the coprimality conditions $(d_2;d_3) = (z_2;z_3) = (d_2;z_2) = (d_3;z_3) = 1$. We obtain again a lower bound if we sum for d_3,z_2,z_3 over the interval $[Y^{1/6},Y^{1/5}]$ only. If these three variables are in that range, the conditions on d_2 are $(d_2;d_3z_2) = 1$ and $d_3z_2/(2z_3) \le d_2 \le d_3z_2/z_3$, in which $d_3z_2/z_3 \ge Y^{1/5}$. We may sum over d_2 with the aid of (20), and then find that

$$V^{**} \gg X^2Y \sum_{\substack{Y^{1/6}\le d_3,z_2,z_3\le Y^{1/5} \\ (z_2;d_3z_3)=(z_3;d_3)=1}} \frac{\varphi(z_2)^3\varphi(z_3)\varphi(d_3)^3}{z_1^4z_3^2d_3^4}.$$

We now sum over z_2 first, using (20) with $c = d_3z_3$, and then over d_3 with $c = z_3$ and finally over z_3 with $c = 1$. In this way, we deduce that $V^{**} \gg X^2Y(\log Y)^3$. Lemma 6 now follows from (25).

6 Two Exercises

It remains to establish the technical estimates reported as Lemmas 7 and 8. Proofs are fairly standard and pedestrian, though also somewhat elaborate.

We begin with Lemma 7. In the notation introduced in the statement of that lemma, we have

$$S \geq \sum_{\substack{a_1 \leq A_1 \\ (a_1;d_3z_2)=1}} \sum_{\substack{a_2 \leq A_2 \\ (a_2;a_1z_1)=1 \\ a_2 \equiv ba_1 \bmod d_3}} 1 = \sum_{\substack{a_1 \leq A_1 \\ (a_1;d_3z_2)=1}} \sum_{f \mid a_1z_1} \mu(f) \sum_{\substack{a_2 \leq A_2 \\ a_2 \equiv 0 \bmod f \\ a_2 \equiv ba_1 \bmod d_3}} 1.$$

Under the hypotheses of Lemma 7, the inner sum is empty unless $(f;d_3) = 1$. In the latter case, the congruence conditions combine to one modulo d_3f, and we infer that

$$S \geq \sum_{\substack{a_1 \leq A \\ (a_1;d_3z_2)=1}} \sum_{\substack{f \mid a_1z_1 \\ (f;d_3)=1}} \mu(f)\left(\frac{A_2}{d_3f} + O(1)\right).$$

Here, the error term $O(1)$ sums up to at most

$$\sum_{a_1 \leq A_1} \tau(a_1z_1) \leq \tau(z_1) \sum_{a_1 \leq A_1} \tau(a_1) \ll \tau(z_1)A_1 \log A_1,$$

which can be absorbed in the terms on the right hand side in the conclusion of Lemma 7. In the remaining sum

$$S_0 = \sum_{\substack{a_1 \leq A_1 \\ (a_1;d_3z_2)=1}} \sum_{\substack{f \mid a_1z_1 \\ (f;d_3)=1}} \frac{A_2\mu(f)}{d_3f},$$

we exchange the summations and then find that

$$S_0 = \frac{A_2}{d_3} \sum_{\substack{f \leq A_1z_1 \\ (f;d_3)=1}} \frac{\mu(f)}{f} \sum_{\substack{a_1 \leq A_1 \\ (a_1;d_3z_2)=1 \\ a_1z_1 \equiv 0 \bmod f}} 1.$$

In the current situation, the inner sum is empty unless $(f;d_3z_2) = 1$. Since $a_1z_1 \equiv 0 \bmod f$ is the same as $\frac{f}{(f;z_1)} \mid a_1$, we may write $f_1 = f/(f;z_1)$ to rewrite the previous sum as

$$\begin{aligned} S_0 &= \frac{A_2}{d_3} \sum_{\substack{f \leq A_1z_1 \\ (f;d_3z_2)=1}} \frac{\mu(f)}{f} \sum_{\substack{a \leq A_1/f_1 \\ (a;d_3z_2)=1}} 1 \\ &= \frac{A_2}{d_3} \sum_{\substack{f \leq A_1z_1 \\ (f;d_3z_2)=1}} \frac{\mu(f)}{f}\left(\frac{\varphi(d_3z_2)}{d_3z_2f_1}A_1 + O(\tau(d_3z_2))\right). \end{aligned}$$

Here, we have applied (19), a formula that we prove momentarily. The error term $O(\tau(d_3 z_2))$ sums up to a total contribution not exceeding

$$\frac{A_2}{d_3}\tau(d_3)\tau(z_2)\sum_{f\le A_1 z_1}\frac{1}{f} \ll \tau(z_2)A_2\log A_1 z_1,$$

which is one of the terms on the right-hand side in the conclusion of Lemma 7. The leading term makes a total contribution of

$$S_1 = A_1 A_2 \frac{\varphi(d_3 z_2)}{d_3^2 z_2}\sum_{\substack{f\le A_1 z_1\\ (f;d_3 z_2)=1}}\frac{\mu(f)(f;z_1)}{f^2}. \tag{27}$$

In order to estimate the expression S_1 from below, we first complete the sum over f to a series, and estimate the error. In fact, we have

$$\sum_{\substack{f=1\\ (f;d_3 z_2)=1}}^{\infty}\frac{\mu(f)(f;z_1)}{f^2} = \prod_{p\nmid d_3 z_2}\left(1-\frac{(p;z_1)}{p^2}\right) \gg \frac{\varphi(z_1)}{z_1}.$$

Moreover,

$$\sum_{\substack{f>A_1 z_1\\ (f;d_3 z_2)=1}}\frac{\mu(f)^2(f;z_1)}{f^2} \le z_1\sum_{f>A_1 z_1}\frac{1}{f^2} \ll \frac{1}{A_1},$$

so that (27) yields

$$S_1 \gg A_1 A_2 \frac{\varphi(d_3 z_2)\varphi(z_1)}{d_3^2 z_1 z_2} + O(A_2).$$

On collecting together, one readily confirms the claim in Lemma 7.

The proof of Lemma 8 is an exercise in multiplicative number theory. For (19), we merely note that

$$\sum_{\substack{d\le D\\ (c;d)=1}} 1 = \sum_{f\mid c}\mu(f)\left(\frac{D}{f}-O(1)\right) = \frac{\varphi(c)}{c}D + O(\tau(c)).$$

The proof of (20) is equally lowbrow, but a bit more involved. A brief sketch will suffice. By Möbius inversion, there is a multiplicative function h with

$$\frac{\varphi(d)^4}{d^4} = \sum_{r\mid d} h(r). \tag{28}$$

One finds that $h(r)=0$ unless r is square-free. When p is a prime, one has

$$h(p) = \left(\frac{p-1}{p}\right)^4 - 1 = -\frac{4}{p} + O\left(\frac{1}{p^2}\right). \tag{29}$$

We routinely deduce that

$$\sum_{d\le D} |h(d)| \ll \prod_{p\le D}\left(1+\frac{4}{p}\right) \ll (\log D)^4, \tag{30}$$

and that for any $c\in\mathbb{N}$ one has

$$\sum_{\substack{d=1\\(c;d)=1}}^{\infty} \frac{h(d)}{d} = \prod_{p\nmid c}\left(1+\frac{h(p)}{p}\right), \tag{31}$$

the sum being absolutely convergent. Let $H(c)$ be the real number described in (31). Then by (29), one finds that there exists a positive number C such that $C\le H(c)\le 1$ holds for all $c\in\mathbb{N}$. By (28) and (19),

$$\sum_{\substack{d\le D\\(c;d)=1}} \frac{\varphi(d)^4}{d^4} = \sum_{r\le D} h(r) \sum_{\substack{d\le D/r\\(c;dr)=1}} 1 = \sum_{\substack{r\le D\\(r;c)=1}} h(r)\left(\frac{\varphi(c)D}{cr} + O(\tau(c))\right).$$

By (30), the error term $O(\tau(c))$ sums to $O(\tau(c)(\log D)^4)$. In the main term, we complete the sum over r to (31), and control the resulting error by (30) and partial summation. It follows that

$$\sum_{\substack{d\le D\\(c;d)=1}} \frac{\varphi(d)^4}{d^4} = \frac{\varphi(c)}{c}H(c)D + O(\tau(c)(\log D)^4).$$

By partial summation, one finds that there exists a certain real number $H_0(c)$ such that

$$\sum_{\substack{d\le D\\(c;d)=1}} \frac{\varphi(d)^4}{d^5} = \frac{\varphi(c)}{c}H(c)\log D + H_0(c) + O\left(\tau(c)\frac{(\log D)^4}{D}\right)$$

holds uniformly in c. This implies (20).

7 An Alternative Argument

The referee has kindly pointed out the following variant of the proof of the upper bound, and we express our gratitude for allowing us to include a sketch of the argument here. One splits the ranges for a_i, d_i into dyadic intervals $A_j < a_j \le 2A_j$, $D_j < d_j \le 2D_j$. For each such range, the estimate of Lemma 3 from [5] yields $\ll 1 + A_1A_2A_3/\max(A_iD_i)$ solutions a_1, a_2, a_3 of (5) for each choice of d_1, d_2, d_3, and therefore $\ll D_1D_2D_3 + \prod(A_iD_i)/\max(A_iD_i)$ in total. Alternatively, we can count the number of d's for each choice of a's giving $\ll A_1A_2A_3 + \prod(A_iD_i)/\max(A_iD_i)$ in total. Combining the two estimates, one gets a bound

$$\ll \min(A_1A_2A_3, D_1D_2D_3) + \frac{\prod(A_iD_i)}{\max(A_iD_i)} \ll \left(\prod(A_iD_i)\right)^{2/3}.$$

One then has to sum over the z_i for $z_1 \ll \min(X/A_1, Y/D_2D_3)$, etc. Put $E_i = D_1D_2D_3/D_i$ so that the number of triples z_1, z_2, z_3 is $\ll \prod \min(X/A_i, Y/E_i)$. We conclude that for each 6-tuple of dyadic ranges A_i, D_i the contribution is

$$\ll \prod_{i=1}^{3} A_i^{2/3} E_i^{1/3} \min\left(\frac{X}{A_i}, \frac{Y}{E_i}\right).$$

Notice now that as (d_1, d_2, d_3) runs over $\mathbb{N}^3$ the values $(d_2+d_3, d_1+d_3, d_1+d_2)$ take each value in $\mathbb{N}^3$ at most once. Hence, we can replace a summation in which the D_i run over powers of 2, by a sum in which the E_i run over powers of 2. Our estimate then factorizes giving

$$\ll \left(\sum_{A,E} A^{2/3} E^{1/3} \min\left(\frac{X}{A}, \frac{Y}{E}\right)\right)^3,$$

where A, E run over powers of 2. When $X/A \le Y/E$ the summand is $XA^{-1/3}E^{1/3}$. We sum first for $E \le AY/X$ to get $\ll XA^{-1/3}.(AY/X)^{1/3} = X^{2/3}Y^{1/3}$. Then summing for $A \ll X$ gives $\ll X^{2/3}Y^{1/3}\log X$. The case $Y/E \le X/A$ similarly gives $\ll X^{2/3}Y^{1/3}\log Y$, and these lead at once to $V(X,Y) \ll X^2Y(\log XY)^3$.

Acknowledgements First author in part supported by a Volkswagen Lichtenberg Fellowship and a Starting Grant of the European Research Council.

References

1. V.V. Batyrev, Y.I. Manin. Sur le nombre des points rationnels de hauteur borné des variétés algébriques. *Math. Ann.* 286 (1990), 27–43.
2. V. Blomer and J. Brüdern. On a certain senary cubic form. Forthcoming.
3. J. Franke, Y.I. Manin, Y. Tschinkel. Rational points of bounded height on Fano varieties. *Invent. Math.* 95 (1989), 421–435.
4. D.R. Heath-Brown. Diophantine approximation with square-free numbers. *Math. Z.* 187 (1984), 335–344.
5. D.R. Heath-Brown. The density of rational points on Cayley's cubic surface. Proceedings of the Session in Analytic Number Theory and Diophantine Equations, 33 pp., *Bonner Math. Schriften*, 360, Univ. Bonn, Bonn, 2003.
6. M. Robbiani. On the number of rational points of bounded height on smooth bilinear hypersurfaces in biprojective space. *J. London Math. Soc.* (2) 63 (2001), 33–51.
7. C.V. Spencer. The Manin conjecture for $x_0y_0 + \cdots + x_sy_s = 0$. *J. Number Theory* 129 (2009), 1505–1521.

Affine Gindikin–Karpelevich Formula via Uhlenbeck Spaces

Alexander Braverman, Michael Finkelberg, and David Kazhdan

Abstract We prove a version of the Gindikin–Karpelevich formula for untwisted affine Kac–Moody groups over a local field of positive characteristic. The proof is geometric and it is based on the results of [Braverman, Finkelberg, and Gaitsgory, Uhlenbeck spaces via affine Lie algebras, Progr. Math., 244, 17–135, 2006] about intersection cohomology of certain Uhlenbeck-type moduli spaces (in fact, our proof is conditioned upon the assumption that the results of [Braverman, Finkelberg, and Gaitsgory, Uhlenbeck spaces via affine Lie algebras, Progr. Math., 244, 17–135, 2006] are valid in positive characteristic; we believe that generalizing [Braverman, Finkelberg, and Gaitsgory, Uhlenbeck spaces via affine Lie algebras, Progr. Math., 244, 17–135, 2006] to the case of positive characteristic should be essentially straightforward but we have not checked the details). In particular, we give a geometric explanation of certain combinatorial differences between finite-dimensional and affine case (observed earlier by Macdonald and Cherednik), which here manifest themselves by the fact that the affine Gindikin–Karpelevich formula has an additional term compared to the finite-dimensional case. Very roughly speaking, that additional term is related to the fact that the loop group of an affine Kac-Moody group (which should be thought of as some kind of "double loop group") does not behave well from algebro-geometric point of view; however, it

A. Braverman (✉)
Department of Mathematics, Brown University, 151 Thayer St., Providence, RI 02912, USA
e-mail: braval@math.brown.edu

M. Finkelberg
Department of Mathematics, IMU, IITP and State University Higher School of Economics, 20 Myasnitskaya st, Moscow 101000, Russia
e-mail: fnklberg@gmail.com

D. Kazhdan
Einstein Institute of Mathematics, Edmond J. Safra Campus, Givat Ram The Hebrew University of Jerusalem, Jerusalem 91904, Israel
e-mail: kazhdan@math.huji.ac.il

V. Blomer and P. Mihăilescu (eds.), *Contributions in Analytic and Algebraic Number Theory*, Springer Proceedings in Mathematics 9, DOI 10.1007/978-1-4614-1219-9_2,

has a better behaved version, which has something to do with algebraic surfaces. A uniform (i.e. valid for all local fields) and unconditional (but not geometric) proof of the affine Gindikin–Karpelevich formula is going to appear in [Braverman, Kazhdan, and Patnaik, The Iwahori-Hecke algebra for an affine Kac-Moody group (in preparation)].

Dedicated to S. Patterson on the occasion of his 60th birthday.

1 The Problem

1.1 Classical Gindikin–Karpelevich Formula

Let $\mathscr{K}$ be a non-archimedian local field with ring of integers $\mathscr{O}$ and let G be a split semi-simple group over $\mathscr{O}$. The classical Gindikin–Karpelevich formula describes explicitly how a certain intertwining operator acts on the spherical vector in a principal series representation of $G(\mathscr{K})$.[1] In more explicit terms, it can be formulated as follows.

Let us choose a Borel subgroup B of G and an opposite Borel subgroup B_-; let U,U_- be their unipotent radicals. In addition, let Λ denote the coroot lattice of G, $R_+\subset\Lambda$ – the set of positive coroots, Λ_+ – the subsemigroup of Λ generated by R_+. Thus any $\gamma\in\Lambda_+$ can be written as $\sum a_i\alpha_i$, where α_i are the simple roots. We shall denote by $|\gamma|$ the sum of all the a_i.

Set now $\mathrm{Gr}_G=G(\mathscr{K})/G(\mathscr{O})$. Then it is known that $\mathscr{U}(\mathscr{K})$-orbits on Gr are in one-to-one correspondence with elements of Λ (this correspondence will be reviewed in Sect. 2); for any $\mu\in\Lambda$, we shall denote by S^μ the corresponding orbit. The same thing is true for $U_-(\mathscr{K})$-orbits. For each $\gamma\in\Lambda$, we shall denote by T^γ the corresponding orbit. It is well known that $T^\gamma\cap S^\mu$ is non-empty iff $\mu-\gamma\in\Lambda_+$ and in that case the above intersection is finite. The Gindikin–Karpelevich formula allows one to compute the number of points in $T^{-\gamma}\cap S^0$ for $\gamma\in\Lambda_+$ (it is easy to see that the above intersection is naturally isomorphic to $T^{-\gamma+\mu}\cap S^\mu$ for any $\mu\in\Lambda$). The answer is most easily stated in terms of the corresponding generating function:

Theorem 1. *(Gindikin–Karpelevich formula)*

$$\sum_{\gamma\in\Lambda_+}\#(T^{-\gamma}\cap S^0)q^{-|\gamma|}e^{-\gamma}=\prod_{\alpha\in R_+}\frac{1-q^{-1}e^{-\alpha}}{1-e^{-\alpha}}.$$

---

[1]More precisely, the Gindikin-Karpelevich formula answers the analogous question for real groups; its analog for p-adic groups (usually also referred to as Gindikin-Karpelevich formula) is proved e.g., in Chap. 4 of [6].

1.2 Formulation of the Problem in the General Case

Let now G be a split symmetrizable Kac–Moody group functor in the sense of [8] and let $\mathfrak{g}$ be the corresponding Lie algebra. We also let $\widehat{G}$ denote the corresponding "formal" version of G (cf. page 198 in [8]). The notations $\Lambda, \Lambda_+, R_+, \mathrm{Gr}_G, S^\mu, T^\gamma$ make sense for $\widehat{G}$ without any changes (cf. Sect. 2 for more detail).

Conjecture 1. For any $\gamma \in \Lambda_+$, the intersection $T^{-\gamma} \cap S^0$ is finite.

This conjecture will be proved in [2] when G is of affine type. In this paper, we are going to prove the following result:

Theorem 2. *Assume that $\mathcal{K} = k((t))$ where k is finite. Then Conjecture 1 holds.*

So now (at least when $\mathcal{K}$ is as above) we can ask the following

Question: Compute the generating function[2]

$$I_{\mathfrak{g}}(q) = \sum_{\gamma \in \Lambda_+} \#(T^{-\gamma} \cap S^0)\, q^{-|\gamma|} e^{-\gamma}.$$

One possible motivation for the above question is as follows: when G is finite-dimensional, Langlands [6] has observed that the usual Gindikin–Karpelevich formula (more precisely, some generalization of it) is responsible for the fact that the constant term of Eisenstein series induced from a parabolic subgroup of G is related to some automorphic L-function. Thus, we expect that generalizing the Gindikin–Karpelevich formula to general Kac-Moody group will eventially become useful for studying Eisenstein series for those groups. This will be pursued in further publications.

We do not know the answer for general G. In the case when G is finite-dimensional, the answer is given by Theorem 1. In this paper we are going to reprove that formula by geometric means and give a generalization to the case when G is untwisted affine.

1.3 The Affine Case

Let us now assume that $\mathfrak{g} = \mathfrak{g}'_{\mathrm{aff}}$, where $\mathfrak{g}'$ is a simple finite-dimensional Lie algebra. The Dynkin diagram of $\mathfrak{g}$ has a canonical ("affine") vertex and we let $\mathfrak{p}$ be the corresponding maximal parabolic subalgebra of $\mathfrak{g}$. Let $\mathfrak{g}^\vee$ denote the Langlands dual algebra and let $\mathfrak{p}^\vee$ be the corresponding dual parabolic. We denote by $\mathfrak{n}(\mathfrak{p}^\vee)$ its (pro)nilpotent radical.

Let (e,h,f) be a principal $sl(2)$-triple in $(\mathfrak{g}')^\vee$. Since the Levi subalgebra of $\mathfrak{p}^\vee$ is $\mathbb{C} \oplus \mathfrak{g}' \oplus \mathbb{C}$ (where the first multiple is central in $\mathfrak{g}^\vee$ and the second is responsible

[2] The reason that we use the notation $I_{\mathfrak{g}}$ rather than I_G is that it is clear that this generating function depends only on $\mathfrak{g}$ and not on G.

for the "loop rotation"), this triple acts on $\mathfrak{n}(\mathfrak{p}^\vee)$ and we let $\mathscr{W} = (\mathfrak{n}(\mathfrak{p}^\vee))^f$ (the centralizer of f in $\mathfrak{n}(\mathfrak{p}^\vee)$). We are going to regard $\mathscr{W}$ as a complex (with zero differential) and with grading coming from the action of h (thus, $\mathscr{W}$ is negatively graded). In addition, $\mathscr{W}$ is endowed with an action of $\mathbb{G}_m$, coming from the loop rotation in $\mathfrak{g}^\vee$. In the case when $\mathfrak{g}'$ is simply laced we have $(\mathfrak{g}')^\vee \simeq \mathfrak{g}'$ and $\mathfrak{n}(\mathfrak{p}^\vee) = t\cdot\mathfrak{g}'[t]$ (i.e., $\mathfrak{g}'$-valued polynomials, which vanish at 0). Hence, $\mathscr{W} = t\cdot(\mathfrak{g}')^f[t]$ and the above $\mathbb{G}_m$-action just acts by rotating t. Let $d_1,\dots,d_r$ be the exponents of $\mathfrak{g}'$ (here $r = \operatorname{rank}(\mathfrak{g}')$). Then $(\mathfrak{g}')^f$ has a basis $(x_1,\dots,x_r)$, where each x_i is placed in the degree $-2d_i$. We let Fr act on $\mathscr{W}$ by requiring that it acts by $q^{i/2}$ on elements of degree i. Also for any $n\in\mathbb{Z}$, let $\mathscr{W}(n)$ be the same graded vector space but with Frobenius action multiplied by q^{-n}.

Consider now $\operatorname{Sym}^*(\mathscr{W})$. We can again consider it as a complex concentrated in degrees ≤ 0 endowed with an action of Fr and $\mathbb{G}_m$. For each $n\in\mathbb{Z}$, we let $\operatorname{Sym}^*(\mathscr{W})_n$ be the part of $\operatorname{Sym}^*(\mathscr{W})$ on which $\mathbb{G}_m$ acts by the character $z\mapsto z^n$. This is a finite-dimensional complex with zero differential, concentrated in degrees ≤ 0 and endowed with an action of Fr.

We are now ready to formulate the main result. Let δ denote the minimal positive imaginary coroot of $\mathfrak{g}$. Set

$$\Delta_{\mathscr{W}}(z) = \sum_{n=0}^{\infty} \operatorname{Tr}(\mathrm{Fr}, \operatorname{Sym}^*(\mathscr{W})_n) z^n.$$

In particular, when $\mathfrak{g}'$ is simply laced we have

$$\Delta(z) = \prod_{i=1}^{r}\prod_{j=0}^{\infty}(1-q^{-d_i}z^j)^{-1}.$$

Theorem 3. *(Affine Gindikin–Karpelevich formula)*
Assume that the results of [1] are valid over k *and let* $\mathscr{K} = \mathsf{k}((t))$. *Then*

$$I_{\mathfrak{g}}(q) = \frac{\Delta_{\mathscr{W}}(e^{-\delta})}{\Delta_{\mathscr{W}(1)}(e^{-\delta})}\prod_{\alpha\in R_+}\left(\frac{1-q^{-1}e^{-\alpha}}{1-e^{-\alpha}}\right)^{m_\alpha}.$$

Here, m_α *denotes the multiplicity of the coroot* α.

Remark. Although formally the paper [1] is written under the assumption that char $\mathsf{k} = 0$, we believe that adapting all the constructions of [1] to the case char $\mathsf{k} = p$ should be more or less straightforward. We plan to discuss it in a separate publication.

Let us make two remarks about the above formula: first, we see that it is very similar to the finite-dimensional case (of course in that case $m_\alpha = 1$ for any α) with the exception of a "correction term" (which is equal to $\frac{\Delta_{\mathscr{W}}(e^{-\delta})}{\Delta_{\mathscr{W}(1)}(e^{-\delta})}$). Roughly speaking, this correction term has to do with imaginary coroots of $\mathfrak{g}$. The second remark is that the same correction term appeared in the work of Macdonald [7] from purely combinatorial point of view (cf. also [3] for a more detailed study). The main purpose of this note is to explain how the term $\frac{\Delta_{\mathscr{W}}(e^{-\delta})}{\Delta_{\mathscr{W}(1)}(e^{-\delta})}$ appears naturally from geometric point of view (very roughly speaking it is related to the fact that affine Kac–Moody groups over a local field of positive characteristic can be studied using various moduli spaces of bundles on an algebraic surface). The relation between the present work and the constructions of [3] and [7] will be discussed in [2].

2 Interpretation via Maps from $\mathbb{P}^1$ to $\mathscr{B}$

2.1 Generalities on Kac–Moody Groups

In what follows all schemes will be considered over a field k which at some point will be assumed to be finite. Our main reference for Kac–Moody groups is [8]. Assume that we are given a symmetrizable Kac–Moody root data and we denote by G (resp. $\widehat{G}$) the corresponding minimal (resp. formal) Kac–Moody group functor (cf. [8], page 198); we have the natural embedding $G \hookrightarrow \widehat{G}$. We also let W denote the corresponding Weyl group and we let $\ell : W \to \mathbb{Z}_{\geq 0}$ be the corresponding length function.

The group G is endowed with closed subgroup functors $U \subset B, U_- \subset B_-$ such that the quotients B/U and B_-/U_- are naturally isomorphic to the Cartan group H of G; also H is isomorphic to the intersection $B \cap B_-$. Moreover, both U_- and B_- are still closed as subgroup functors of $\widehat{G}$. On the other hand, B and U are not closed in $\widehat{G}$ and we denote by $\widehat{B}$ and $\widehat{U}$ their closures.

The quotient G/B has a natural structure of an ind-scheme which is ind-proper; the same is true for the quotient $\widehat{G}/\widehat{B}$ and the natural map $G/B \to \widehat{G}/\widehat{B}$ is an isomorphism. This quotient is often called *the thin flag variety of* G. Similarly, one can consider the quotient $\mathscr{B} = \widehat{G}/B_-$; it is called *the thick flag variety of* G or *Kashiwara flag scheme*. As is suggested by the latter name, $\mathscr{B}$ has a natural scheme structure. The orbits of B on $\mathscr{B}$ are in one-to-one correspondence with the elements of the Weyl group W; for each $w \in W$, we denote by $\mathscr{B}_w$ the corresponding orbit. The codimension of $\mathscr{B}_w$ is $\ell(w)$; in particular, $\mathscr{B}_e$ is open. There is a unique H-invariant point $y_0 \in \mathscr{B}_e$. The complement to $\mathscr{B}_e$ is a divisor in $\mathscr{B}$ whose components are in one-to-one correspondence with the simple roots of G.

In what follows Λ will denote the coroot lattice of G, $R_+ \subset \Lambda$ – the set of positive coroots, Λ_+ – the subsemigroup of Λ generated by R_+. Thus, $\gamma \in \Lambda_+$ can be written as $\sum a_i \alpha_i$ where α_i are the simple coroots. We shall denote by $|\gamma|$ the sum of all the a_i.

In what follows we shall assume that G is "simply connected," which means that Λ is equal to the full cocharacter lattice of H.

2.2 Some Further Notations

For any variety X and any $\gamma \in \Lambda_+$ we shall denote by $\mathrm{Sym}^\gamma X$ the variety parametrizing all unordered collections $(x_1,\gamma_1),\dots,(x_n,\gamma_n)$, where $x_j \in X, \gamma_j \in \Lambda_+$ such that $\sum \gamma_j = \gamma$.

Assume that k is finite and let $\mathscr{S}$ be a complex of ℓ-adic sheaves on a variety X over k. We set

$$\chi_{\mathsf{k}}(\mathscr{S}) = \sum_{i\in\mathbb{Z}} (-1)^i \mathrm{Tr}(\mathrm{Fr}, H^i(\overline{X},\mathscr{S})),$$

where $\overline{X} = X \underset{\mathrm{Spec}\,\mathsf{k}}{\times} \mathrm{Spec}\,\overline{\mathsf{k}}$.

We shall denote by $(\overline{\mathbb{Q}}_l)_X$ the constant sheaf with fiber $\overline{\mathbb{Q}}_l$. According to the Grothendieck–Lefschetz fixed point formula, we have

$$\chi_{\mathsf{k}}((\overline{\mathbb{Q}}_l)_X) = \#X(\mathsf{k}).$$

2.3 Semi-Infinite Orbits

As in the introduction, we set $\mathscr{K} = \mathsf{k}((t))$, $\mathscr{O} = \mathsf{k}[[t]]$. We let $\mathrm{Gr} = \widehat{G}(\mathscr{K})/\widehat{G}(\mathscr{O})$, which we are just going to consider as a set with no structure. Each $\lambda \in \Lambda$ is a homomorphism $\mathbb{G}_m \to H$; in particular, it defines a homomorphism $\mathscr{K}^* \to H(\mathscr{K})$. We shall denote the image of t under the latter homomorphism by t^λ. Abusing the notation, we shall denote its image in Gr by the same symbol. Set

$$S^\lambda = \widehat{U}(\mathscr{K})\cdot t^\lambda \subset \mathrm{Gr}; \quad T^\lambda = U_-(\mathscr{K})\cdot t^\lambda \subset \mathrm{Gr}.$$

Lemma 1. Gr *is equal to the disjoint union of all the* S^λ.

Proof. This follows from the *Iwasawa decomposition* for G of [5]; we include a different proof for completeness. Since $\Lambda \simeq \widehat{U}(\mathscr{K})\backslash\widehat{B}(\mathscr{K})/\widehat{B}(\mathscr{O})$, the statement of the lemma is equivalent to the assertion that the natural map $\widehat{B}(\mathscr{K})/\widehat{B}(\mathscr{O}) \to \widehat{G}(\mathscr{K})/\widehat{G}(\mathscr{O})$ is an isomorphism; in other words, we need to show that $\widehat{B}(\mathscr{K})$ acts transitively on Gr. But this is equivalent to saying that $\widehat{G}(\mathscr{O})$ acts transitively on $\widehat{G}(\mathscr{K})/\widehat{B}(\mathscr{K})$, which means that the natural map $\widehat{G}(\mathscr{O})/\widehat{B}(\mathscr{O}) \to \widehat{G}(\mathscr{K})/\widehat{B}(\mathscr{K})$ is an isomorphism. However, the left-hand side is $(\widehat{G}/\widehat{B})(\mathscr{O})$ and the right-hand side is $(\widehat{G}/\widehat{B})(\mathscr{K})$ and the assertion follows from the fact that the ind-scheme $\widehat{G}/\widehat{B}$ satisfies the valuative criterion of properness.

The statement of the lemma is definitely false if we use T^μ's instead of S^λ's since the scheme $\widehat{G}/B_-$ does not satisfy the valuative ctiterion of properness. Let us say that an element $g(t) \in \widehat{G}(\mathscr{K})$ is good if its projection to $\mathscr{B}(\mathscr{K}) = B_-(\mathscr{K})\backslash\widehat{G}(\mathscr{K})$

comes from a point of $\mathscr{B}(\mathscr{O})$. Since $\mathscr{B}(\mathscr{O}) = B_-(\mathscr{O})\backslash\widehat{G}(\mathscr{O})$, it follows that the set of good elements of $\widehat{G}(\mathscr{K})$ is just equal to $B_-(\mathscr{K}) \cdot G(\mathscr{O})$, which immediately proves the following result:

Lemma 2. *The preimage of* $\bigcup_{\gamma\in\Lambda} T^\gamma$ *in* $\widehat{G}(\mathscr{K})$ *is equal to the set of good elements of* $\widehat{G}(\mathscr{K})$.

2.4 Spaces of Maps

Recall that the Picard group of $\mathscr{B}$ can be naturally identified with $\Lambda^\vee$ (the dual lattice to Λ). Thus for any map $f : \mathbb{P}^1 \to \mathscr{B}$, we can talk about the degree of f as an element $\gamma \in \Lambda$. The space of such maps is non-empty iff $\gamma \in \Lambda_+$. We say that a map $f : \mathbb{P}^1 \to \mathscr{B}$ is *based* if $f(\infty) = y_0$. Let $\mathscr{M}^\gamma$ be the space of based maps $f : \mathbb{P}^1 \to \mathscr{B}$ of degree γ. It is shown in the Appendix to [1] that this is a smooth scheme of finite type over k of dimension $2|\gamma|$. We have a natural ("factorization") map $\pi^\gamma : \mathscr{M}^\gamma \to \mathrm{Sym}^\gamma \mathbb{A}^1$, which is related to how the image of a map $\mathbb{P}^1 \to \mathscr{B}$ intersects the complement to $\mathscr{B}_e$. In particular, if we set

$$\mathscr{F}^\gamma = (\pi^\gamma)^{-1}(\gamma \cdot 0),$$

then $\mathscr{F}^\gamma$ consists of all the based maps $f : \mathbb{P}^1 \to \mathscr{B}$ of degree γ such that $f(x) \in \mathscr{B}_e$ for any $x \neq 0$.

Theorem 4. *There is a natural identification* $\mathscr{F}^\gamma(\mathsf{k}) \simeq T^{-\gamma} \cap S^0$.

Since $\mathscr{F}^\gamma$ is a scheme of finite type over k, it follows that $\mathscr{F}^\gamma(\mathsf{k})$ is finite and thus Theorem 4 implies Theorem 2.

The proof of Theorem 4 is essentially a repetition of a similar proof in the finite-dimensional case, which we include here for completeness.

Proof. First of all, let us construct an embedding of the union of all the $\mathscr{F}^\gamma(\mathsf{k})$ into $S^0 = \widehat{U}(\mathscr{K})/\widehat{U}(\mathscr{O})$. Indeed, an element of $\bigcup_{\gamma\in\Lambda_+} \mathscr{F}^\gamma$ is uniquely determined by its restriction to $\mathbb{G}_m \subset \mathbb{P}^1$; this restriction is a map $f : \mathbb{G}_m \to \mathscr{B}_e$ such that $\lim_{x\to\infty} f(x) = y_0$. We may identify $\mathscr{B}_e$ with $\widehat{U}$ (by acting on y_0). Thus, we get

$$\bigcup_{\gamma\in\Lambda_+} \mathscr{F}^\gamma \subset \{u : \mathbb{P}^1\backslash\{0\} \to \widehat{U} \mid u(\infty) = e\}. \tag{1}$$

We have a natural map from the set of k-points of the right-hand side of (1) to $\widehat{U}(\mathscr{K})$; this map sends every u as above to its restriction to the formal punctured neighbourhood of 0. We claim that after projecting $\widehat{U}(\mathscr{K})$ to $S^0 = \widehat{U}(\mathscr{K})/\widehat{U}(\mathscr{O})$, this map becomes an isomorphism. Recall that $\widehat{U}$ is a group-scheme, which can be written as a projective limit of finite-dimensional unipotent group-schemes U_i;

moreover, each U_i has a filtration by normal subgroups with successive quotients isomorphic to $\mathbb{G}_a$. Hence, it is enough to prove that the above map is an isomorphism when $U = \mathbb{G}_a$. In this case, we just need to check that any element of the quotient $\mathsf{k}((t))/\mathsf{k}[[t]]$ has unique lift to a polynomial $u(t) \in \mathsf{k}[t,t^{-1}]$ such that $u(\infty) = 0$, which is obvious.

Now Lemma 2 implies that a map $u(t)$ as above extends to a map $\mathbb{P}^1 \to \mathscr{B}$ if and only if the corresponding element of S^0 lies in the intersection with some $T^{-\gamma}$.

It remains to show that $\mathscr{F}^\gamma(\mathsf{k})$ is exactly equal to $S^0 \cap T^{-\gamma}$ as a subset of S^0. Let $\Lambda^\vee$ be the weight lattice of G and let $\Lambda^\vee_+$ denote the set of dominant weights of G. For each $\lambda^\vee \in \Lambda^\vee_+$, we can consider the Weyl module $L(\lambda^\vee)$, defined over $\mathbb{Z}$; in particular, $L(\lambda^\vee)(\mathscr{K})$ and $L(\lambda^\vee)(\mathscr{O})$ make sense. By the definition $L(\lambda^\vee)$ is the module of global sections of a line bundle $\mathscr{L}(\lambda^\vee)$ on $\mathscr{B}$. Moreover, we have a weight decomposition

$$L(\lambda^\vee) = \bigoplus_{\mu^\vee \in \Lambda^\vee} L(\lambda^\vee)_{\mu^\vee},$$

where each $L(\lambda^\vee)_{\mu^\vee}$ is a finitely generated free $\mathbb{Z}$-module and $L(\lambda^\vee)_{\lambda^\vee} := l_{\lambda^\vee}$ has rank one. Geometrically, $l_{\lambda^\vee}$ is the fiber of $\mathscr{L}(\lambda^\vee)$ at y_0 and the corresponding projection map from $L(\lambda^\vee) = \Gamma(\mathscr{B}, \mathscr{L}(\lambda^\vee))$ to $l_{\lambda^\vee}$ is the restriction to y_0.

Let $\eta_{\lambda^\vee}$ denote the projection of $L(\lambda^\vee)$ to $l_{\lambda^\vee}$. This map is U_--equivariant (where U_- acts trivially on $l_{\lambda^\vee}$).

Lemma 3. *The projection of a good element $g \in G(\mathscr{K})$ lies in T^ν (for some $\nu \in \Lambda$) if and only if for any $\lambda^\vee \in \Lambda^\vee$ we have:*

$$\eta_{\lambda^\vee}(g(L(\lambda^\vee)(\mathscr{O}))) \subset t^{\langle \nu,\lambda^\vee\rangle} l_{\lambda^\vee}(\mathscr{O}); \qquad \eta_{\lambda^\vee}(g(L(\lambda^\vee)(\mathscr{O}))) \not\subset t^{\langle \nu,\lambda^\vee\rangle - 1} l_{\lambda^\vee}(\mathscr{O}). \tag{2}$$

Proof. First of all, we claim that if the projection of g lies in T^ν then the above condition is satisfied. Indeed, it is clearly satisfied by t^ν; moreover, (2) is clearly invariant under left multiplication by $U_-(\mathscr{K})$ and under right multiplication by $G(\mathscr{O})$. Hence any $g \in U_-(\mathscr{K}) \cdot t^\nu \cdot G(\mathscr{O})$ satisfies (2).

On the other hand, assume that a good element $g \in G(\mathscr{K})$ satisfies (2). Since g lies in $U_-(\mathscr{K}) \cdot t^{\nu'} \cdot G(\mathscr{O})$ for some ν', it follows that g satisfies (2) when ν is replaced by ν'. However, it is clear that this is possible only if $\nu = \nu'$.

It is clear that in (2) one can replace $g(L(\lambda^\vee)(\mathscr{O}))$ with $g(L(\lambda^\vee)(\mathsf{k}))$ (since the latter generates the former as an $\mathscr{O}$-module).

Let now f be an element of $\mathscr{F}^\gamma$. Then $f^*\mathscr{L}(\lambda^\vee)$ is isomorphic to the line bundle $\mathscr{L}(\langle \gamma, \lambda^\vee \rangle)$ on $\mathbb{P}^1$. On the other hand, the bundle $\mathscr{L}(\lambda^\vee)$ is trivialized on $\mathscr{B}_e$ by means of the action of U; more precisely, the restriction of $\mathscr{L}(\lambda^\vee)$ is canonically identified with the trivial bundle with fiber $l_{\lambda^\vee}$. Let now $s \in L(\lambda^\vee)(\mathsf{k})$; we are going to think of it as a section of $L(\lambda^\vee)$ on $\mathscr{B}$. In particular, it gives rise to a function $\tilde{s} : \mathscr{B}_e \to l_{\lambda^\vee}$. Let also $u(t)$ be the element of $U(\mathscr{K})$, corresponding to f. Then $\eta_{\lambda^\vee}(u(t)(s))$ can be described as follows: we consider the composition $\tilde{s} \circ f$ and restrict it to the formal neighbourhood of $0 \in \mathbb{P}^1$ (we get an element of $l_{\lambda^\vee}(\mathscr{K})$).

On the other hand, since $f \in \mathscr{F}^\gamma$, it follows that $f^*\mathscr{L}(\lambda^\vee)$ is trivialized away from 0 and any section of it can be thought of as a function $\mathbb{P}^1\backslash\{0\}$ with pole of order $\leq \langle\gamma,\lambda^\vee\rangle$ at 0. Hence, $\widetilde{s}\circ f$ has pole of order $\leq \langle\gamma,\lambda^\vee\rangle$ at 0.

To finish the proof it is enough to show that for some s the function $\widetilde{s}\circ f$ has pole of order exactly $\langle\gamma,\lambda^\vee\rangle$ at 0 (indeed if $f \in T^{-\gamma'}$ for some $\gamma' \in \Lambda$, then by (2) $\widetilde{s}\circ f$ has pole of order $\leq \langle\gamma',\lambda^\vee\rangle$ at 0 and for some s, it has pole of order exactly $\langle\gamma',\lambda^\vee\rangle$, which implies that $\gamma=\gamma'$). To prove this, let us note that since $\mathscr{L}(\lambda^\vee)$ is generated by global sections, the line bundle $f^*\mathscr{L}(\lambda^\vee)$ is generated by sections of the form f^*s, where s is a global section of $\mathscr{L}(\lambda^\vee)$. This implies that for any $\overline{s} \in \Gamma(\mathbb{P}^1, f^*\mathscr{L}(\lambda^\vee))$ there exists a section $s \in \Gamma(\mathscr{B},\mathscr{L}(\lambda^\vee))$ such that the ratio $s/\overline{s}$ is a rational function on $\mathbb{P}^1$, which is invertible at 0. Taking $\overline{s}$ such that its pole with respect to the above trivialization of $f^*\mathscr{L}(\lambda^\vee)$ is exactly equal to $\langle\gamma',\lambda^\vee\rangle$ and taking s as above, we see that the pole of f^*s with respect to the above trivialization of $f^*\mathscr{L}(\lambda^\vee)$ is exactly equal to $\langle\gamma',\lambda^\vee\rangle$.

3 Proof of Theorem 1 via Quasi-Maps

3.1 Quasi-Maps

We shall denote by $\mathscr{QM}^\gamma$ the space of based *quasi-maps* $\mathbb{P}^1 \to \mathscr{B}$. According to [4], we have the stratification

$$\mathscr{QM}^\gamma = \bigcup_{\gamma'\leq\gamma} \mathscr{M}^{\gamma'} \times \mathrm{Sym}^{\gamma-\gamma'}\mathbb{A}^1.$$

The factorization morphism π_0^γ extends to the similar morphism $\overline{\pi}^\gamma : \mathscr{QM}^\gamma \to \mathrm{Sym}^\gamma$ and we set $\overline{\mathscr{F}}^\gamma = (\overline{\pi}^\gamma)^{-1}(0)$. Thus, we have

$$\overline{\mathscr{F}}^\gamma = \bigcup_{\gamma'\leq\gamma} \mathscr{F}^{\gamma'}. \tag{3}$$

There is a natural section $i^\gamma : \mathrm{Sym}^\gamma\mathbb{A}^1 \to \mathscr{QM}^\gamma$. According to [4], we have

Theorem 5. *1. The restriction of* $\mathrm{IC}_{\mathscr{QM}^\gamma}$ *to* $\mathscr{F}^{\gamma'}$ *is isomorphic to* $(\overline{\mathbb{Q}}_l)_{\mathscr{F}^{\gamma'}}[2](1)^{\otimes|\gamma'|} \otimes \mathrm{Sym}^*(\mathfrak{n}_+^\vee[2](1))_{\gamma-\gamma'}$.

2. There exists a $\mathbb{G}_m$*-action on* $\mathscr{QM}^\gamma$*, which contracts it to the image of* i^γ*. In particular, it contracts* $\overline{\mathscr{F}}^\gamma$ *to one point (corresponding to* $\gamma'=0$ *in (3)).*

3. Let s_γ *denote the embedding of* $\gamma\cdot 0$ *into* $\mathrm{Sym}^\gamma\mathbb{A}^1$*. Then*

$$s_\gamma^* i_\gamma^! \mathrm{IC}_{\mathscr{QM}^\gamma} = \mathrm{Sym}^*(\mathfrak{n}_+)_\gamma$$

(here the right hand is a vector space concentrated in cohomological degree 0 and with trivial action of Fr*).*

The assertion (2) implies that $\pi_!^\gamma \mathrm{IC}_{\mathscr{QM}^\gamma} = i_\gamma^! \mathrm{IC}_{\mathscr{QM}^\gamma}$ and hence

$$H_c^*(\overline{\mathscr{F}}, \mathrm{IC}_{\mathscr{QM}^\gamma}|_{\overline{\mathscr{F}}^\gamma}) = s_\gamma^* \pi_!^\gamma \mathrm{IC}_{\mathscr{QM}^\gamma} = s_\gamma^* i_\gamma^! \mathrm{IC}_{\mathscr{QM}^\gamma} = \mathrm{Sym}^*(\mathfrak{n}_+)_\gamma.$$

Thus, setting, $\mathscr{S}^\gamma = \mathrm{IC}_{\mathscr{QM}^\gamma}|_{\overline{\mathscr{F}}^\gamma}$ we get

$$\sum_{\gamma \in \Lambda_+} \chi_{\mathsf{k}}(\mathscr{S}^\gamma) e^{-\gamma} = \prod_{\alpha \in R_+} \frac{1}{1 - e^{-\alpha}}. \tag{4}$$

On the other hand, according to (1) we have

$$\chi_{\mathsf{k}}(\mathscr{S}^\gamma) = \sum_{\gamma' \leq \gamma} (\#\mathscr{F}^{\gamma'}) q^{-|\gamma'|} \mathrm{Tr}(\mathrm{Fr}, \mathrm{Sym}^*(\mathfrak{n}_+^\vee[2](1))_{\gamma - \gamma'}),$$

which implies that

$$\sum_{\gamma \in \Lambda_+} \chi_{\mathsf{k}}(\mathscr{S}^\gamma) e^{-\gamma} = \frac{\sum_{\gamma \in \Lambda_+} \#\mathscr{F}^\gamma(\mathsf{k}) q^{-|\gamma|} e^{-\gamma}}{\prod_{\alpha \in R_+} 1 - q^{-1} e^{-\alpha}} = \frac{I_{\mathfrak{g}}(q)}{\prod_{\alpha \in R_+} 1 - q^{-1} e^{-\alpha}}. \tag{5}$$

Hence,

$$I_{\mathfrak{g}}(q) = \prod_{\alpha \in R_+} \frac{1 - q^{-1} e^{-\alpha}}{1 - e^{-\alpha}}.$$

4 Proof of Theorem 3

4.1 Flag Uhlenbeck Spaces

We now assume that $G = (G')_{\mathrm{aff}}$ where G' is some semi-simple simply connected group. We want to follow the pattern of Sect. 3. Let $\gamma \in \Lambda_+$. As is discussed in [1], the corresponding space of quasi-maps behaves badly when G is replaced by G_{aff}. However, in this case one can use the corresponding *flag Uhlenbeck space* $\mathscr{U}^\gamma$. In fact, as was mentioned in the Introduction, in [1] only the case of k of characteristic 0 is considered. In what follows we are going to assume that the results of *loc. cit.* are valid also in positive characteristic.

The flag Uhlenbeck space $\mathscr{U}^\gamma$ has properties similar to the space of quasi-maps $\mathscr{QM}^\gamma$ considered in the previous section. Namely, we have:

a. $\mathscr{U}^\gamma$ is an affine variety of dimension $2|\gamma|$, which contains $\mathscr{M}^\gamma$ as a dense open subset.
b. There is a factorization map $\pi^\gamma : \mathscr{U}^\gamma \to \mathrm{Sym}^\gamma \mathbb{A}^1$; it has a section $i_\gamma : \mathrm{Sym}^\gamma \mathbb{A}^1 \to \mathscr{U}^\gamma$.
c. $\mathscr{U}^\gamma$ is endowed with a $\mathbb{G}_m$-action, which contracts $\mathscr{U}^\gamma$ to the image of i_γ.

These properties are identical to the corresponding properties of $\mathscr{QM}^\gamma$ from the previous section. The next (stratification) property, however, is different (and it is in fact responsible for the additional term in Theorem 3). Namely, let δ denote the minimal positive imaginary coroot of G'_{aff}. Then we have

d. There exists a stratification

$$\mathscr{U}^\gamma = \bigcup_{\gamma' \in \Lambda_+, n \in \mathbb{Z}, \gamma - \gamma' - n\delta \in \Lambda_+} (\mathscr{M}^{\gamma - \gamma' - n\delta} \times \mathrm{Sym}^{\gamma'} \mathbb{A}^1) \times \mathrm{Sym}^n(\mathbb{G}_m \times \mathbb{A}^1). \tag{6}$$

In particular, if we now set $\overline{\mathscr{F}}^\gamma = (\pi^\gamma)^{-1}(\gamma \cdot 0)$, we get

$$\overline{\mathscr{F}}^\gamma = \left(\bigcup_{\gamma' \in \Lambda_+, n \in \mathbb{Z}, \gamma - \gamma' - n\delta \in \Lambda_+} \mathscr{F}^{\gamma'} \right) \times \mathrm{Sym}^n(\mathbb{G}_m). \tag{7}$$

4.2 *Description of the IC-Sheaf*

In [1], we describe the IC-sheaf of $\mathscr{U}^\gamma$. To formulate the answer, we need to introduce some notation. Let $\mathscr{P}(n)$ denote the set of partitions of n. In other words, any $P \in \mathscr{P}(n)$ is an unordered sequence $n_1, \dots, n_k \in \mathbb{Z}_{>0}$ such that $\sum n_i = n$. We set $|P| = k$. For a variety X and any $P \in \mathscr{P}(n)$, we denote by $\mathrm{Sym}^P(X)$ the locally closed subset of $\mathrm{Sym}^n(X)$ consisting of all formal sums $\sum n_i x_i$ where $x_i \in X$ and $x_i \neq x_j$ for $i \neq j$. The dimension of $\mathrm{Sym}^P(X)$ is $|P| \cdot \dim X$. Let also

$$\mathrm{Sym}^*(W[2](1))_P = \bigotimes_{i=1}^{k} \mathrm{Sym}^*(\mathscr{W}[2](1))_{n_i}.$$

Theorem 6. *The restriction of* $\mathrm{IC}_{\mathscr{U}^\gamma}$ *to* $\mathscr{M}^{\gamma-\gamma'-n\delta} \times \mathrm{Sym}^{\gamma'}(\mathbb{A}^1) \times \mathrm{Sym}^P(\mathbb{G}_m \times \mathbb{A}^1)$ *is isomorphic to constant sheaf on that scheme tensored with*

$$\mathrm{Sym}^*(\mathfrak{n}_+)_{\gamma'} \otimes \mathrm{Sym}^*(\mathscr{W})_P[2|\gamma - \gamma' - n\delta|](|\gamma - \gamma' - n\delta|).$$

Corollary 1. *The restriction of* $\mathrm{IC}_{\mathscr{U}^\gamma}$ *to* $\mathscr{F}^{\gamma-\gamma'-n\delta} \times \mathrm{Sym}^P(\mathbb{G}_m)$ *is isomorphic to the constant sheaf tensored with*

$$\mathrm{Sym}^*(\mathfrak{n}_+^\vee)_{\gamma'} \otimes \mathrm{Sym}^*(\mathscr{W})_P[2|\gamma - \gamma' - n\delta|](|\gamma - \gamma' - n\delta|).$$

Let now $\mathscr{S}^\gamma$ denote the restriction of $\mathrm{IC}_{\mathscr{U}^\gamma}$ to $\overline{\mathscr{F}}^\gamma$. Then as in (4) we get

$$\sum_{\gamma \in \Lambda_+} \chi_{\mathsf{k}}(\mathscr{S}^\gamma) e^{-\gamma} = \prod_{\alpha \in R_+} \frac{1}{(1 - e^{-\alpha})^{m_\alpha}}. \tag{8}$$

On the other hand, arguing as in (5) we get that

$$\sum_{\gamma\in\Lambda_+}\chi_{\mathsf{k}}(\mathscr{S}^\gamma)e^{-\gamma}=A(q)\frac{I_{\mathfrak{g}}(q)}{\prod\limits_{\alpha\in R_+}(1-q^{-1}e^{-\alpha})^{m_\alpha}},\tag{9}$$

where

$$A(q)=\sum_{n=0}^{\infty}\sum_{P\in\mathscr{P}(n)}\mathrm{Tr}(\mathrm{Fr},H_c^*(\mathrm{Sym}^P(\mathbb{G}_m),\overline{\mathbb{Q}}_l)\otimes\mathrm{Sym}^*(\mathscr{W}[2](1))_P)e^{-n\delta}.$$

This implies that

$$I_{\mathfrak{g}}(q)=A(q)\prod_{\alpha\in R_+}\left(\frac{1-q^{-1}e^{-\alpha}}{1-e^{-\alpha}}\right)^{m_\alpha}.$$

It remains to compute $A(q)$. However, it is clear that

$$A(q)=\sum_{n=0}^{\infty}\mathrm{Tr}(\mathrm{Fr},\mathrm{Sym}^n(H_c^*(\mathbb{G}_m))\otimes\mathscr{W}[2](1))e^{-n\delta}=\frac{\Delta_{\mathscr{W}}(e^{-\delta})}{\Delta_{\mathscr{W}(1)}(e^{-\delta})}.\tag{10}$$

This is true since $H_c^i(\mathbb{G}_m)=0$ unless $i=1,2$, and we have

$$H_c^1(\mathbb{G}_m)=\overline{\mathbb{Q}}_l,\quad H_c^2(\mathbb{G}_m)=\overline{\mathbb{Q}}_l(-1),$$

and thus if we ignore the cohomological $\mathbb{Z}$-grading, but only remember the corresponding $\mathbb{Z}_2$-grading, then we just have

$$\mathrm{Sym}^*(H_c^*(\mathbb{G}_m)\otimes\mathscr{W}[2](1))=\mathrm{Sym}^*(\mathscr{W})\otimes\Lambda^*(\mathscr{W}(1)),$$

whose character is exactly the right-hand side of (10).

Acknowledgements We thank I. Cherednik, P. Etingof and M. Patnaik for very helpful discussions. A. B. was partially supported by the NSF grant DMS-0901274. M. F. was partially supported by the RFBR grant 09-01-00242 and the Science Foundation of the SU-HSE awards No.T3-62.0 and 10-09-0015. D. K. was partially supported by the BSF grant 037.8389.

References

1. A. Braverman, M. Finkelberg and D. Gaitsgory, *Uhlenbeck spaces via affine Lie algebras*, The unity of mathematics, 17–135, Progr. Math., **244**, Birkhäuser Boston, Boston, MA, 2006.
2. A. Braverman, D. Kazhdan and M. Patnaik, *The Iwahori-Hecke algebra for an affine Kac-Moody group*, in preparation.
3. I. Cherednik and X. Ma, *A new take on spherical, Whittaker and Bessel functions*, arXiv:0904.4324.

4. B. Feigin, M. Finkelberg, A. Kuznetsov, I. Mirkovic, *Semi-infinite flags. II. Local and global intersection cohomology of quasimaps' spaces. Differential topology, infinite-dimensional Lie algebras, and applications*, 113–148, Amer. Math. Soc. Transl. Ser. 2, 194, Amer. Math. Soc., Providence, RI, 1999.
5. S. Gaussent and G. Rousseau, *Kac-Moody groups, hovels and Littelmann paths*, Annales de l'Institut Fourier, **58** no. 7 (2008), p. 2605–2657
6. R. P. Langlands, *Euler products*, A James K. Whittemore Lecture in Mathematics given at Yale University, 1967. Yale Mathematical Monographs, **1**. Yale University Press, New Haven, Conn.-London, 1971.
7. I. G. Macdonald, *A formal identity for affine root systems. Lie groups and symmetric spaces*, 195-211, Amer. Math. Soc. Transl. Ser. 2, **210**, Amer. Math. Soc., Providence, RI, 2003.
8. J. Tits, *Groups and group functors attached to Kac-Moody data*, Workshop Bonn 1984 (Bonn, 1984), 193–223, Lecture Notes in Math., **1111**, Springer, Berlin, 1985

Daniel's Twists of Hooley's Delta Function

Jörg Brüdern

Abstract Following ideas of Daniel, a function analogous to Hooley's Delta function is constructed for multiplicative functions with values in the unit disc. When the multiplicative function is of oscillatory nature, moments of the new Delta function are smaller than those for Hooley's original. Similar ideas apply to incomplete convolutions if the multiplicative function satisfies a more rigid condition that is best expressed in terms of its generating Dirichlet series. The most prominent example where the theory applies is the Möbius function, thus providing some new insights into its value distribution.

Prologue

In a highly original memoir, Hooley [4] undertook a thorough study of a family of functions that encode important information about the distribution of divisors. Nowadays known as Hooley's Delta function, the simplest of these is defined by

$$\Delta(n) = \max_u \#\{d \mid n : u < \log d \le u+1\}. \tag{1}$$

Amongst other things, Hooley established the upper bound

$$\sum_{n \le x} \Delta(n) \ll x(\log x)^{4/\pi - 1} \tag{2}$$

and worked out several applications in diverse areas of number theory, including divisor sums and problems of Waring's type. His work was taken further by Hall and Tenenbaum, resulting in a very subtle estimate ([7] and [3], Theorem 70). In particular, the exponent $4/\pi - 1$ in (2) may be replaced by any positive number.

J. Brüdern (✉)
Mathematisches Institut, Universität Göttingen, Bunsenstrasse 3–5, 37073 Göttingen, Germany
e-mail: bruedern@uni-math.gwdg.de

V. Blomer and P. Mihăilescu (eds.), *Contributions in Analytic and Algebraic Number Theory*, Springer Proceedings in Mathematics 9, DOI 10.1007/978-1-4614-1219-9_3,

Tenenbaum [8, 9] gave further applications and generalisations, and the average order of $\Delta(n)$ also features in work of Vaughan [10, 11] on the asymptotic formula in Waring's problem.

During a seminar in Stuttgart, on November 24, 2000, Stephan Daniel proposed a far-reaching generalisation of Hooley's Delta function. He attached a multiplicative weight to the divisors d in (1) and showed how further savings can be made when the weight has mean value 0 in some suitable quantitative sense. In particular, the Möbius function and Dirichlet characters are admissible weights. With the non-principal character modulo 4 in the rôle of the weight, Daniel's twist of the Delta function is related to the number of representations by sums of two squares in the same way as Hooley's original is to the ordinary divisor function. Hence, in all cases where Hooley's Delta function can be used to evaluate a divisor sum, one may hope to do so with the divisor function replaced by $r(n)$, the number of representations of the natural number n as sums of two squares of integers. In this spirit, Daniel developed his work on the divisor problem for binary forms [1], and announced an asymptotic formula for the sum

$$\sum_{0<|F(u,v)|\leq x} r(|F(u,v)|) \tag{3}$$

in which F denotes an irreducible binary quartic form with integer coefficients, and where u, v run over $\mathbb{Z}$.

To my great dismay, several months after the event, Daniel left the academic world, and his work was never published. This was much regretted by workers in the field, especially because an asymptotic formula for integer solutions of the conditions

$$F(u,v) = a^2 + b^2 \leq x, \tag{4}$$

with F as in (3), is one of the missing pieces for a complete resolution of Manin's conjecture for Châtelet surfaces. In March 2010, I rediscovered a set of notes that I took during Daniel's seminar, and after extensive discussions with him. These contained precise statements of his results and fairly thorough sketches of the principal arguments. The first chapter of this paper gives an account of his work on mean value estimates for twisted Delta functions, as far as I have been able to reconstruct it. I am grateful to Stephan Daniel for his kind permission to prepare the material for publication. The main result, Theorem 1 below, is entirely due to Daniel, and he alone should receive the full credit. Also, the main ideas for the proofs in chapter 1 are his. However, I am of course fully responsible for any errors that I may have introduced or overlooked. The notes also contained a mean value estimate for a certain maximal function associated with the Möbius function, and this will be the theme of chapter 2. Daniel's bound is stated as (85) below. My attempts to demonstrate it led to a more general result concerning incomplete convolutions of multiplicative functions that is formulated as Theorem 2, and that contains (85) as a special case.

It is hoped that Daniel's contributions will prove useful in many applications, beyond the obvious one to thin averages of $r(n)$. While the forensic efforts to recover Daniel's asymptotic formula for the sum (3) were still in progress, I learned that de la Bretèche and Tenenbaum also succeeded to establish a mean square estimate for twists of the Delta function with a Dirichlet character, or the Möbius function. Their bounds are somewhat different to Daniel's, and do not cover a situation as general as Theorem 1 does, but for the application to the diophantine problem (4), their estimates are equally good. A detailed account of the diophantine applications will be contained in their forthcoming memoir, so that there is no need to reproduce that part of Daniel's work here.

1 Twisted Delta Functions

1.1 Introduction

Before embarking on the main theme, it is appropriate to introduce some notational conventions. Standard notation is used for common arithmetical functions. Thus, the number of divisors of the natural number n is $\mathrm{d}(n)$, the number of prime factors is $\omega(n)$, and $\mu(n)$ denotes the Möbius function. The greatest common factor of a and b is (a,b), and $[a,b]$ is their lowest common multiple. The letter p, without or with subscripts, always denotes prime numbers. If $f: X \to \mathbb{C}$ and $g: X \to \mathbb{R}$ are functions such that there is a positive number K with $|f| \le Kg$, then, following Landau and Vinogradov, this is abbreviated as $f = O(g)$, or $f \ll g$. The "implicit constant" in these symbols is an admissible value for K.

Finally, the "ε-convention" is applied in this paper: whenever ε occurs in a statement, it is asserted that the statement is true for any fixed positive number ε. Note that, with this convention activated, the bounds $A \ll x^{\varepsilon}$ and $B \ll x^{\varepsilon}$ imply that $AB \ll x^{\varepsilon}$, for example. Implicit constants may depend on ε.

Let $\mathscr{H}$ denote the set of multiplicative functions $h: \mathbb{N} \to \mathbb{C}$ with $|h(n)| \le 1$ for all $n \in \mathbb{N}$. Then define

$$\Delta_h(n;u,v) = \sum_{\substack{d|n \\ u<\log d \le u+v}} h(d), \tag{5}$$

$$\Delta_h(n) = \max_{u} \max_{0<v\le 1} |\Delta_h(n;u,v)|. \tag{6}$$

Note that when $h(n) = 1$ for all n, then $\Delta_h = \Delta$. The main theorem concerns a weighted mean of Δ_h when h oscillates. To describe the set of weights properly, first fix positive real numbers c, C, δ with $C \ge 1$ (the *parameter set* in the sequel). Then let $\mathscr{F} = \mathscr{F}(c,C,\delta)$ denote the set of multiplicative functions $\rho: \mathbb{N} \to [0,\infty)$ satisfying the following two conditions:

- For all primes p and all $k \geq 2$, one has

$$\rho(p) \leq C, \quad \rho(p^k) \leq Cp^{(1-\delta)k-1}, \tag{7}$$

- For $1 < \sigma \leq 2$, one has

$$\sum_p \frac{\rho(p)\log p}{p^\sigma} \leq \frac{c}{\sigma - 1} + C. \tag{8}$$

These conditions are of familiar type for those working with averages of multiplicative functions over thin sequences. Typical choices in applications are $\rho(n) = 1$ for all n, or, when $f \in \mathbb{Z}[x]$ is irreducible, the number $\rho_f(n)$ of incongruent solutions to $f(a) \equiv 0 \bmod n$. In both cases, one may take $c = 1$ in (8).

Now let $h \in \mathscr{H}$, and let $\mathscr{F}_h$ be the set of all $\rho \in \mathscr{F}$ that satisfy the inequality

$$\Big|\sum_{n\leq x} \mu(n)^2 \rho(n) h(n)\Big| \leq Cx(\log x)^{-c} \tag{9}$$

for all $x \geq 2$. Note that if $\mathscr{F}_h$ contains functions that mimic the characteristic function of a dense set, then (9) forces h to oscillate on that set.

For positive numbers c and t, let

$$\alpha = \max\Big(c, c + \frac{1}{2}(c-1)t, 2^t c - \frac{3}{2}t\Big), \quad \beta = \Big(\frac{1}{2}ct\Big)^{1/2} \tag{10}$$

and when $z > 30$, define the functions

$$\mathscr{L}(z) = \exp((\log z)^{1/2} \log\log z), \quad \mathscr{L}^*(z) = \exp((\log z)^{1/2}(\log\log z)^{1/2}).$$

For $0 \leq z \leq 30$, put $\mathscr{L}(z) = \mathscr{L}^*(z) = 1$.

Theorem 1 (Daniel). *Fix a parameter set and a real number t with $1 \leq t \leq 2$. Then there is a number D such that for any $h \in \mathscr{H}$ and $\rho \in \mathscr{F}_h$, one has*

$$\sum_{n\leq x} \rho(n)\Delta_h(n)^t \ll x(\log x)^{\alpha-1}\mathscr{L}(\log x)^\beta \mathscr{L}^*(\log x)^D.$$

The implicit constant depends only on the parameter set and on t.

Note that $c = 1$ implies $\alpha = 1$. Hence, according to an earlier comment, if $h \in \mathscr{H}$ is such that $\rho \equiv 1$ or ρ_f for some irreducible integer polynomial f are in $\mathscr{F}_h$, then for these ρ one has

$$\sum_{n\leq x} \rho(n)\Delta_h(n)^t \ll x\mathscr{L}(\log x)^{\sqrt{t/2}+o(1)} \ll x(\log x)^\varepsilon. \tag{11}$$

Similar results for $\Delta(n)$ are necessarily inflated by powers of $\log x$. It is an immediate corollary of Lemma 2.2 of Tenenbaum [9] that for $t \geq 1$ one has

$$\sum_{n\leq x} \rho_f(n)\Delta(n)^t \ll x(\log x)^{2^t-t-1+\varepsilon}. \tag{12}$$

Thus, if one would use the trivial inequality $\Delta_h(n) \leq \Delta(n)$ in (11), then a factor of approximate size $\log x$ is lost in the important special case $t = 2$. Moreover, the obvious inequality

$$\mathrm{d}(n) \leq \Delta(n)\log 3n \tag{13}$$

and the classical estimate

$$\sum_{n\leq x} \rho_f(n)\mathrm{d}(n)^t \asymp x(\log x)^{2^t-1} \tag{14}$$

show that the exponent of $\log x$ in (12) is the best possible. Thus, Daniel's bound (11) with $t = 2$ is a genuine improvement over what can be achieved with mean values of $\Delta(n)$.

It is perhaps of interest that the factor $\mu(n)^2$ may not be omitted from (9). In fact, with $\rho \equiv 1$ and $h(n) = (-1)^{n+1}$, one has

$$\sum_{n\leq x} \rho(n)h(n) \ll x(\log x)^{-c}, \tag{15}$$

but also $\Delta_h(n) = \Delta(n)$ for all odd n, so that (13) and a suitable variant of (14) yield

$$\sum_{n\leq x} \Delta_h(n)^2 \geq \sum_{m\leq x/2} \Delta(2m-1)^2 \gg x\log x.$$

This bound is in conflict with (11), which is a special case of Theorem 1. Consequently, the conclusion of Theorem 1 is no longer valid if the condition (9) is replaced with the seemingly more natural (15).

1.2 A Simplicistic Lemma

In the later stages of the advance toward Theorem 1, reference will be made to the following inequality between certain Dirichlet series. Its proof is routine.

Lemma 1. *Let $f : \mathbb{N} \to [0,\infty)$ be an arithmetical function with $f(n) = 0$ whenever n is not square-free. Suppose that there exists a number $A \geq 1$ such that the inequality $f(pn) \leq Af(n)$ holds for all primes p and for all natural numbers n. Let $k \in \mathbb{N}$. Then there exists a number B depending only on A and k such that for all $1 < \sigma \leq 2$, one has*

$$\sum_{n=1}^{\infty} \frac{f(n)}{n^\sigma}(\log\log 9n)^k \leq B\Big(\log\frac{3}{\sigma-1}\Big)^{2k} \sum_{n=1}^{\infty} \frac{f(n)}{n^\sigma}. \tag{16}$$

Proof. By hypotheses, $f(n) \le f(1)A^{\omega(n)}$. Hence, both series in (16) converge for $\sigma > 1$. For square-free n, one has

$$\log n = \sum_{p|n} \log p, \tag{17}$$

and therefore,

$$\log 9n = \log 9 + \sum_{p|n} \log p \le \log 1890 + \sum_{\substack{p|n \\ p\ge 11}} \log p.$$

For real numbers u, v with $u \ge 2$, $v \ge 2$ one has $\log(u+v) \le \log u + \log v$, as one readily confirms. Recursive application yields

$$\log\log 9n \le 3 + \sum_{\substack{p|n \\ p\ge 11}} \log\log p,$$

and consequently,

$$(\log\log 9n)^k \le 2^k\Big(3^k + \Big(\sum_{\substack{p|n \\ p\ge 11}} \log\log p\Big)^k\Big).$$

It follows that the sum on the left-hand side of (16) does not exceed

$$6^k F(\sigma) + 2^k \sum_{\substack{p_1,\dots,p_k \\ p_j\ge 11}} \sum_{\substack{n=1 \\ [p_1,\dots,p_k]|n}}^{\infty} \frac{f(n)}{n^\sigma} \prod_{j=1}^{k} \log\log p_j,$$

where

$$F(\sigma) = \sum_{n=1}^{\infty} \frac{f(n)}{n^\sigma}.$$

The hypotheses concerning f show that

$$\sum_{\substack{n=1 \\ [p_1,\dots,p_k]|n}}^{\infty} \frac{f(n)}{n^\sigma} \le A^k [p_1,\dots,p_k]^{-\sigma} F(\sigma),$$

whence the expression

$$(6^k + (2A)^k U(\sigma))F(\sigma)$$

with

$$U(\sigma) = \sum_{\substack{p_1,\dots,p_k \\ p_j \geq 11}} \frac{(\log\log p_1)\dots(\log\log p_k)}{[p_1,\dots,p_k]^\sigma}$$

is an upper bound for the left-hand side of (16). The lemma will therefore follow from the inequality

$$U(\sigma) \ll \left(\log \frac{3}{\sigma - 1}\right)^{2k} \tag{18}$$

that is now derived.

Let

$$V_j(\sigma) = \sum_{p \geq 11} \frac{(\log\log p)^j}{p^\sigma}.$$

Since $[p_1,\dots,p_k]$ is the product of the distinct primes among $p_1,\dots,p_k$, one finds that

$$U(\sigma) \ll \sum_{l=1}^{k} \sum_{\substack{j_1,\dots,j_l \geq 1 \\ j_1+\dots+j_l = k}} V_{j_1}(\sigma)\dots V_{j_l}(\sigma). \tag{19}$$

Here, the implicit constant depends only on k. By partial summation,

$$V_j(\sigma) = \sigma \int_{11}^{\infty} \sum_{11 \leq p \leq x} (\log\log p)^j \frac{\mathrm{d}x}{x^{1+\sigma}}.$$

To obtain an upper bound, recall that $\sigma \leq 2$ and replace $\log\log p$ by $\log\log x$. Then by Chebyshev's estimate,

$$V_j(\sigma) \leq 8 \int_{11}^{\infty} \frac{(\log\log x)^j}{x^\sigma \log x} \mathrm{d}x = 8 \int_{(\sigma-1)\log 11}^{\infty} \frac{(\log y/(\sigma-1))^j}{y\mathrm{e}^y} \mathrm{d}y; \tag{20}$$

for the last identity, the substitution $y = (\sigma - 1)\log x$ was used. To estimate this integral, first observe that the function $(\log y/(\sigma-1))^j \mathrm{e}^{-y/2}$ is decreasing on the interval $y \geq 2j^2$, as one readily verifies by differentiation. Therefore, one now has

$$\int_{2j^2}^{\infty} \frac{(\log y/(\sigma-1))^j}{y\mathrm{e}^y} \mathrm{d}y \leq \left(\log \frac{2j^2}{\sigma-1}\right)^j \mathrm{e}^{-j^2} \int_0^\infty \mathrm{e}^{-y/2} \mathrm{d}y.$$

This is crude, but suffices. On the interval $(\sigma-1)\log 11 \leq y \leq 2j^2$, the function $(\log y/(\sigma-1))^j$ is increasing, so that

$$\int_{(\sigma-1)\log 11}^{2j^2} \frac{(\log y/(\sigma-1))^j}{y\mathrm{e}^y} \mathrm{d}y \leq \left(\log \frac{2j^2}{\sigma-1}\right)^j \int_{(\sigma-1)\log 11}^{\infty} \mathrm{e}^{-y} \frac{\mathrm{d}y}{y}.$$

The integral that remained on the right-hand side here is $O(\log(3/(\sigma-1)))$, by straightforward estimates. On collecting together, it follows that whenever $1 \leq j \leq k$, one has

$$V_j(\sigma) \ll \Big(\log\frac{3}{\sigma-1}\Big)^{j+1},$$

with an implicit constant depending only on k. Now (18) follows from (19).

1.3 Mean Values of the Weights

In this preparatory section, certain mean values for functions $h \in \mathscr{H}$ and $\rho \in \mathscr{F}$ are examined. The main result is Lemma 4, which plays a pivotal rôle in the transition from the oscillatory condition (9) to the estimate in Theorem 1.

Lemma 2. *Let t be a non-negative real number, and let $\rho \in \mathscr{F}$. Then uniformly for $1 < \sigma \leq 2$ one has*

$$\sum_{n=1}^{\infty} \mu(n)^2\rho(n)\mathrm{d}(n)^t n^{-\sigma} \ll (\sigma-1)^{-2^t c}.$$

The implicit constant depends only on t and C.

Proof. The Dirichlet series on the left-hand side of the proposed inequality equals

$$\prod_p(1+\rho(p)2^t p^{-\sigma}) \leq \prod_p(1+\rho(p)p^{-\sigma})^{2^t}.$$

By (7), one finds that

$$\log\prod_p(1+\rho(p)p^{-\sigma}) \leq \sum_p \rho(p)p^{-\sigma} \leq \int_\sigma^2 \sum_p \frac{\rho(p)\log p}{p^s}\,\mathrm{d}s + \sum_p \frac{C}{p^2}.$$

Now, by rough estimates and (8),

$$\log\prod_p(1+\rho(p)p^{-\sigma}) \leq \int_\sigma^2 \frac{c}{s-1}\,\mathrm{d}s + 2C = -c\log(\sigma-1)+2C,$$

and the lemma follows immediately.

Lemma 3. *Fix a parameter set. Then there exists a sequence of positive real numbers $C(k)$ such that for any $h \in \mathscr{H}$ and $\rho \in \mathscr{F}_h$, any $k \in \mathbb{N}$ and $x \geq 1$, one has*

$$\Big|\sum_{\substack{n\leq x\\(n,k)=1}} \mu(n)^2\rho(n)h(n)\Big| \leq C(k)x(\log 3x)^{-c}. \tag{21}$$

The proof is by induction on k. The case $k = 1$ follows from (9). Now let $k > 1$, and write $\Xi_k(x)$ for the sum on the left-hand side of (21). Then a routine rearrangement gives

$$\Xi_k(x) = \sum_{l|k} \mu(l) \sum_{m \le x/l} \mu(lm)^2 \rho(lm) h(lm) = \sum_{l|k} \mu(l)\rho(l)h(l)\Xi_l(x/l).$$

By induction hypothesis, the individual contribution of terms with $l < k$, $l \le x$ does not exceed

$$\rho(l)C(l)\frac{x}{l}\left(\log\frac{3x}{l}\right)^{-c},$$

and when $l > x$, then $\Xi_l(x/l) = 0$. Since the bound

$$\frac{x}{l}\left(\log\frac{3x}{l}\right)^{-c} \ll x(\log 3x)^{-c}$$

holds uniformly in l with the implicit constant depending only on c, it follows that the inequality

$$|\Xi_k(x)| \le \rho(k)|\Xi_k(x/k)| + D(k)x(\log 3x)^{-c}$$

is valid with some constant $D(k)$ depending at most on k and the parameter set. This inequality may be iterated υ times, until one reaches $x/k^{\upsilon} < 1$, and one then finds

$$|\Xi_k(x)| \le D'(k) \sum_{\upsilon : k^{\upsilon} \le x} \frac{xk^{-\upsilon}}{(\log 3xk^{-\upsilon})^c} \le C(k)x(\log x)^{-c}$$

when $D'(k)$ and $C(k)$ are suitably large. This confirms the claim in the lemma.

The previous lemma will enter the further proceedings only through the next lemma, which is a more precise statement. Lemma 3 will be used in the proof as a presieving device. Before the result can be formulated precisely, two multiplicative functions are to be introduced. Let

$$\upsilon(k) = \prod_{p|k}(1 - p^{-1/4})^{-1}. \tag{22}$$

Moreover, when σ is a real number, $h \in \mathscr{H}$ and $\rho \in \mathscr{F}_h$, let

$$\theta_\sigma(k) = \theta_\sigma(k;\rho) = \prod_{p|k}(1 + \rho(p)p^{-\sigma})^{-1}. \tag{23}$$

Lemma 4. *Let $h \in \mathscr{H}$ and $\rho \in \mathscr{F}_h$. Then whenever $k \in \mathbb{N}$ and $1 \le \sigma \le 2$, one has*

$$\sum_{\substack{n \le x \\ (n,k)-1}} \mu(n)^2 \rho(n) h(n) \theta_\sigma(n) \ll \upsilon(k)x(\log 3x)^{-c}. \tag{24}$$

The implicit constant depends only on the parameter set.

Proof. Throughout this proof, let K denote the product of all primes not exceeding C^2. Moreover, again only within this proof, let a denote a number with the property that $p \mid a$ implies $p \le C^2$, and let b denote a number with $(b,K) = 1$. This convention also applies when subscripts are present. Note that with this convention activated, any natural number n has a unique factorisation $n = ab$.

Now let $\Upsilon_k(x)$ denote the sum on the left of (24), and factor n as above. Then $(a,b) = 1$, and one finds that

$$\Upsilon_k(x) = \sum_{\substack{a \le x \\ (a,k)=1}} \mu(a)^2 \rho(a) h(a) \theta_\sigma(a) \Upsilon_k^*(x/a), \tag{25}$$

where

$$\Upsilon_k^*(y) = \sum_{\substack{b \le y \\ (b,k)=1}} \mu(b)^2 \rho(b) h(b) \theta_\sigma(b).$$

For $p > C^2$, one has $\rho(p)p^{-1} \le C^{-1} < 1$. On writing $(1+\rho(p)p^{-\sigma})^{-1}$ as a geometric series, it follows that whenever $\sigma \ge 1$, one can rewrite (23) as

$$\theta_\sigma(b) = \prod_{p \mid b} \sum_{l=0}^{\infty} \Big(-\frac{\rho(p)}{p^\sigma} \Big)^l = \sum_{\substack{m=1 \\ p\mid m \Rightarrow p\mid b}}^{\infty} \frac{\psi(m)}{m^\sigma}$$

with

$$\psi(m) = \prod_{p^l \| m} (-\rho(p))^l.$$

This may be injected into the formula for $\Upsilon_k^*(y)$. Exchanging the order of summation then produces

$$\Upsilon_k^*(y) = \sum_{m=1}^{\infty} \frac{\psi(m)}{m^\sigma} \sum_{\substack{b \le y \\ (b,k)=1 \\ p\mid m \Rightarrow p\mid b}} \mu(b)^2 \rho(b) h(b). \tag{26}$$

Note that the inner sum here is empty unless $(m,kK) = 1$. For a natural number n, let

$$n^* = \prod_{p \mid n} p$$

be the square-free kernel. Then (26) becomes

$$\begin{aligned} \Upsilon_k^*(y) &= \sum_{\substack{m=1 \\ (m,kK)=1}}^{\infty} \frac{\psi(m)}{m^\sigma} \sum_{\substack{b \le y \\ (b,k)=1 \\ m^* \mid b}} \mu(b)^2 \rho(b) h(b) \\ &= \sum_{\substack{m=1 \\ (m,kK)=1}}^{\infty} \frac{\psi(m)}{m^\sigma} \mu(m^*)^2 \rho(m^*) h(m^*) \Upsilon_{km^*}^{\dagger}(y/m^*), \end{aligned} \tag{27}$$

where

$$\Upsilon_d^\dagger(z) = \sum_{\substack{b\le z\\(b,d)=1}} \mu(b)^2\rho(b)h(b). \tag{28}$$

In preparation for the removal of the coprimality condition, consider the Dirichlet series

$$G_d(s) = \sum_{\substack{b=1\\(b,d)=1}}^{\infty} \mu(b)^2\rho(b)h(b)b^{-s}$$

that converges absolutely in $\operatorname{Re} s > 1$. Its Euler product is

$$G_d(s) = \prod_{\substack{p>C^2\\p\nmid d}} (1+\rho(p)h(p)p^{-s}) = G_1(s)\prod_{\substack{p>C^2\\p\mid d}} (1+\rho(p)h(p)p^{-s})^{-1},$$

and one has

$$\prod_{\substack{p>C^2\\p\mid d}} (1+\rho(p)h(p)p^{-s})^{-1} = \sum_{\substack{b=1\\b^*\mid d}}^{\infty} \frac{\nu(b)}{b^s} \tag{29}$$

with

$$\nu(n) = \prod_{p^l\|n} (-\rho(p)h(p))^l.$$

It follows that $G_d(s)$ is the product of $G_1(s)$ and the Dirichlet series described in (29). On comparing cofficients, one obtains a convolution formula that transforms the sum in (28) into

$$\Upsilon_d^\dagger(z) = \sum_{\substack{b_1b_2\le z\\b_2^*\mid d}} \mu(b_1)^2\rho(b_1)h(b_1)\nu(b_2).$$

More explicitly, this can be rewritten as

$$\Upsilon_d^\dagger(z) = \sum_{\substack{b\le z\\b^*\mid d}} \nu(b) \sum_{\substack{n\le z/b\\(n,K)=1}} \mu(n)^2\rho(n)h(n).$$

Here, the inner sum is of the type considered in Lemma 3, and the choice of K depends only on C. Hence, by Lemma 3,

$$\Upsilon_d^\dagger(z) \ll \sum_{\substack{b<z\\b^*\mid d}} |\nu(b)|\frac{z}{b}\left(\log\frac{3z}{b}\right)^{-c}, \tag{30}$$

where the implicit constant depends at most on the parameter set. Now note that

$$b^{-1/4}\Big(\log\frac{3z}{b}\Big)^{-c} \ll (\log 3z)^{-c} \tag{31}$$

holds uniformly in b, with the implicit constant depending only on c. Moreover, since $h\in\mathscr{H}$, a comparison of the definitions of v and ψ yields the inequality $v(b)\le |\psi(b)|$, and consequently, (30) implies that

$$\Upsilon_d^{\dagger}(z) \ll \frac{z}{(\log 3z)^c}\sum_{b^*|d}\frac{|\psi(b)|}{b^{3/4}} = \frac{z}{(\log 3z)^c}\prod_{\substack{p|d\\ p>C^2}}\sum_{l=0}^{\infty}\frac{\rho(p)^l}{p^{3l/4}}.$$

However, for $p>C^2$, one has $\rho(p)p^{-1/2}\le 1$, and therefore,

$$\sum_{l=0}^{\infty}\frac{\rho(p)^l}{p^{3l/4}} \le \sum_{l=0}^{\infty}p^{-l/4} = \upsilon(p). \tag{32}$$

It follows that

$$\Upsilon_d^{\dagger}(z) \ll \upsilon(d)z(\log 3z)^{-c}.$$

Note that this bound crucially depends on Lemma 3, and hence on the oscillatory hypothesis (9).

The estimation of $\Upsilon_k^*(y)$ now proceeds through (27), which combines with the final bound for $\Upsilon_d^{\dagger}(z)$ to

$$\Upsilon_k^*(y) \ll y \sum_{\substack{m^*\le y\\ (m,kK)=1}}\frac{|\psi(m)|}{m^{\sigma}m^*}\rho(m^*)|h(m^*)|\upsilon(km^*)\Big(\log\frac{3y}{m^*}\Big)^{-c}.$$

Here, recall the currently active convention about the letter b, that is now used as a substitute for m. Then since $|h(m^*)|\le 1$ and $\sigma\ge 1$ by hypotheses, the previous bound simplifies to

$$\begin{aligned}\Upsilon_k^*(y) &\ll \upsilon(k)y\sum_{\substack{b^*\le y\\ (b,k)=1}}\frac{|\psi(b)|\rho(b^*)\upsilon(b^*)}{bb^*}\Big(\log\frac{3y}{b^*}\Big)^{-c}\\ &\ll \frac{\upsilon(k)y}{(\log 3y)^c}\sum_{\substack{b^*\le y\\ (b,k)=1}}\frac{|\psi(b)|\rho(b^*)\upsilon(b^*)}{b^{3/4}b^*}.\end{aligned} \tag{33}$$

For the final inequality, note that again (31) was used in conjunction with $b^* \mid b$. The sum over b on the right of (33) is bounded above by the corresponding Euler product

$$\prod_{C^2<p\le y}\Big(1+\frac{\rho(p)\upsilon(p)}{p}\sum_{l=1}^{\infty}\frac{\rho(p)^l}{p^{3l/4}}\Big),$$

whence by (32)

$$\Upsilon_k^*(y) \ll \frac{\upsilon(k)y}{(\log 3y)^c}\prod_p\Big(1+\frac{\rho(p)^2\upsilon(p)^2}{p^{7/4}}\Big).$$

The inequality $\rho(p)\upsilon(p) \le C\upsilon(2)$ shows that the product converges and that it is bounded in terms of C, so it may be absorbed into the implicit constant. The consequential estimate for $\Upsilon_k^*(y)$ may then be imported into (25) to infer that

$$\Upsilon_k(x) \ll \sum_{\substack{a\le x\\(a,k)=1}} \mu(a)^2\rho(a)\theta_\sigma(a)\upsilon(k)\,\frac{x}{a}\Big(\log\frac{3x}{a}\Big)^{-c}.$$

However, square-free values of a are divisors of K. Therefore, $\mu(a)^2\rho(a)$ is bounded in terms of C. Moreover, one also has $|\theta_\sigma(a)| \le 1$, so that it now follows that $\Upsilon_k(x) \ll \upsilon(k)x(\log 3x)^{-c}$, as required.

1.4 Classical Propinquity Estimates

The quantity

$$\Delta(n;u,v) = \#\{d \mid n : u < \log d \le u+v\} \tag{34}$$

is a special case of (5). The next lemma exploits the expectation that the maximal function

$$\Delta^{(v)}(n) = \max_u \Delta(n;u,v) \tag{35}$$

is essentially bounded when v is small in terms of n.

Lemma 5. *Let $\rho \in \mathscr{F}(c,C,\delta)$ and $0 < v \le (\log x)^{-4(C^2+1)}$. Then one has*

$$\sum_{n\le x}\frac{\mu(n)^2\rho(n)\Delta^{(v)}(n)^2}{n} \ll (\log x)^c.$$

The implicit constant depends only on the parameter set.

Proof. The propinquity function

$$\mathrm{d}^{(\nu)}(n) = \#\{(d_1,d_2) : d_1 \mid n,\, d_2 \mid n,\, d_1 < d_2 \le \mathrm{e}^{\nu} d_1\}$$

serves well as an upper bound for $\Delta^{(\nu)}(n)$. In fact, one obviously has

$$\Delta^{(\nu)}(n)^2 \le \Delta^{(\nu)}(n) + 2\mathrm{d}^{(\nu)}(n),$$

whence

$$\Delta^{(\nu)}(n)^2 \le 1 + 4\mathrm{d}^{(\nu)}(n).$$

Hence, the sum in question does not exceed

$$\sum_{n\le x} \frac{\mu(n)^2\rho(n)}{n} + 4\sum_{n\le x} \frac{\mu(n)^2\rho(n)\mathrm{d}^{(\nu)}(n)}{n},$$

and for the first summand here, Rankin's trick and Lemma 2 yield the acceptable bound

$$\sum_{n\le x} \frac{\mu(n)^2\rho(n)}{n} \le \mathrm{e} \sum_{n=1}^{\infty} \frac{\mu(n)^2\rho(n)}{n^{1+1/\log x}} \ll (\log x)^c.$$

For the second sum, Cauchy's inequality produces

$$\sum_{n\le x} \frac{\mu(n)^2\rho(n)\mathrm{d}^{(\nu)}(n)}{n} \le \Big(\sum_{n\le x} \frac{\mu(n)^2\rho(n)^2\mathrm{d}^{(\nu)}(n)}{n}\Big)^{\frac{1}{2}} \Big(\sum_{n\le x} \frac{\mathrm{d}^{(\nu)}(n)}{n}\Big)^{\frac{1}{2}}.$$

The estimation of the first factor on the right is straightforward. By (7), one has $\mu(n)^2\rho(n)^2 \le C^{2\omega(n)}$, and the inequality $\mathrm{d}^{(\nu)}(n) \le \mathrm{d}(n)^2$ is obvious. This yields

$$\sum_{n\le x} \frac{\mu(n)^2\rho(n)^2\mathrm{d}^{(\nu)}(n)}{n} \le \prod_{p\le x}\Big(1+\frac{4C^2}{p}\Big) \ll (\log x)^{4C^2}.$$

For the second factor, the definition of $\mathrm{d}^{(\nu)}(n)$ may be opened and the order of summation be reversed. This gives

$$\sum_{n\le x} \frac{\mathrm{d}^{(\nu)}(n)}{n} = \sum_{\substack{d_1<d_2\le \mathrm{e}^{\nu}d_1 \\ d_2\le x}} \sum_{\substack{n\le x \\ [d_1,d_2]\mid n}} \frac{1}{n} \ll (\log x) \sum_{\substack{d_1<d_2\le \mathrm{e}^{\nu}d_1 \\ d_2\le x}} \frac{1}{[d_1,d_2]}.$$

The remaining sum can be transformed by $d = (d_1, d_2)$, $d_j = df_j$, and is then seen not to exceed

$$\sum_{d \le x} \sum_{\substack{f_1 < f_2 \le e^v f_1 \\ f_2 \le x/d}} \frac{1}{d f_1 f_2} \ll (\log x) \sum_{\substack{f_1 < f_2 \le e^v f_1 \\ f_1 \le x}} \frac{1}{f_1^2}$$

$$\ll (\log x) \sum_{f_1 \le x} \frac{[e^v f_1 - f_1]}{f_1^2} \ll (\log x)^2 (e^v - 1).$$

For $0 \le v \le 1$, one has $e^v - 1 \ll v$. The estimates established so far combine to

$$\sum_{n \le x} \frac{\mu(n)^2 \rho(n) \Delta^{(v)}(n)^2}{n} \ll (\log x)^{2C^2+2} v^{1/2} + (\log x)^c.$$

This proves Lemma 5.

An estimate for weighted moments of the classical Delta function is the theme of the next lemma. The result is very similar to Lemma 2.2 of Tenenbaum [9], but uniformity with respect to the weights needs attention.

Lemma 6. *Fix a parameter set and a real number $t \ge 1$. Then there is a number B such that for $\rho \in \mathscr{F}$ and $1 < \sigma \le 2$ one has*

$$\sum_{n=1}^{\infty} \frac{\mu(n)^2 \rho(n) \Delta(n)^t}{n^\sigma} \ll (\sigma - 1)^{-\max(2^t c - t, c)} \mathscr{L}^*((\sigma - 1)^{-1})^B.$$

The implicit constant and B depend only on the parameter set and on t.

Proof. The method of moments and differential inequalities is central to the argument; see Chap. 7 of [3] for an account. For easy reference, notation is in line with this source as far as is possible. For any natural number, consider the moment

$$M_q(n) = \int_{-\infty}^{\infty} \Delta(n; u, 1)^q \, du, \tag{36}$$

where $\Delta(n; u, v)$ is as in (34). By Theorem 72 of Hall and Tenenbaum [3], one has

$$2^{1-q} \Delta(n)^q \le M_q(n) \le d(n)^q, \tag{37}$$

whence in view of (7), the Dirichlet series

$$L(\sigma) = \sum_{n=1}^{\infty} \frac{\mu(n)^2 \rho(n) M_q(n)^{t/q}}{n^\sigma} \tag{38}$$

converges for $\sigma > 1$, and the inequality

$$\sum_{n=1}^{\infty} \frac{\mu(n)^2 \rho(n) \Delta(n)^t}{n^\sigma} \le 2^t L(\sigma) \tag{39}$$

holds. For later use, it is worth remarking that (7) and the second inequality in (37) imply that

$$L(2) \le \prod_p (1 + 2^t C p^{-2}), \tag{40}$$

where the important point is that the right-hand side is independent of q.

The Dirichlet series $L(\sigma)$ has been examined in detail by Tenenbaum [9] in the special case where $\rho = \rho_f$ and f is an irreducible polynomial with integer coefficients. His method can be followed closely to complete the proof of Lemma 6. The reader is referred to [9] for more details of the argument to follow.

The argument leading from [9, (2.9)] to [9, (2.10)] is still valid in the more general context, and an inspection of the proceedings there reveals that Tenenbaum's constant C_0 depends only on an upper bound for $\rho(p)$, and hence only on C, in the notation of this paper. One then substitutes the inequality (2.11) of [9] by (8) of this paper. Arguing as on p. 220 of [9], one confirms that for all $\sigma > 1$ one has

$$-L'(\sigma) \le 2^{t/q} L(\sigma) \Big(\frac{c}{\sigma - 1} + C \Big) + C_0^{t/q} 4^t \Big(\frac{c}{\sigma - 1} + C \Big)^{1-t/q} J(\sigma),$$

where

$$J(\sigma) = \sum_{n=1}^{\infty} \mu(n)^2 M_q(n)^{t(q-2)/q(q-1)} \rho(n) \mathrm{d}(n)^{t/(q-1)} n^{-\sigma}.$$

Hölder's inequality is now applied to estimate $J(\sigma)$ in terms of $L(\sigma)$ and a sum that has occurred in Lemma 2. With the estimate from Lemma 2 in the rôle of [9, (2.12)], the argument of [9] then yields

$$\begin{aligned} -L'(\sigma) \le{} & 2^{t/q} L(\sigma) \Big(\frac{c}{\sigma - 1} + C \Big) \\ & + K C_0^{t/q} 4^t \Big(\frac{c}{\sigma - 1} + C \Big)^{1-t/q} (\sigma - 1)^{-2^t c/(q-1)} L(\sigma)^{(q-2)/(q-1)}, \end{aligned}$$

in which K denotes a constant dependent at most on the parameter set. As in the argument leading to [9, (2.13)], this may then be used to derive the preliminary differential inequality

$$-L'(\sigma) \le c 2^{t/q} L(\sigma)(\sigma - 1)^{-1} + C_1 L(\sigma)^{(q-2)/(q-1)} (\sigma - 1)^{-1+(t/q)-2^t c/(q-1)},$$

where C_1 depends only on the parameter set and on t. For a technical reason, a slightly weaker inequality is easier to handle. For any $\gamma \geq 2^t c - t + t/q$, the inequality

$$-L'(\sigma) \leq c2^{t/q}L(\sigma)(\sigma-1)^{-1} + C_1 L(\sigma)^{(q-2)/(q-1)}(\sigma-1)^{-1-\gamma/(q-1)} \qquad (41)$$

still holds whenever $1 < \sigma \leq 2$, and one may replace C_1 by any larger number if necessary. It will be convenient to choose

$$\gamma = \max(c2^t - t + t/q, c + (1 + ct\log 2)/q),$$

because one then has

$$\gamma - c2^{t/q} \geq c\Big(1 + \frac{1+t\log 2}{q} - 2^{t/q}\Big) \geq \frac{c}{2q}, \qquad (42)$$

provided only that q is sufficiently large in terms of t. The function

$$X(\sigma) = \Big(\frac{C_1}{\gamma - c2^{t/q}}\Big)^{q-1}(\sigma-1)^{-\gamma}$$

satisfies (41) with equality when X replaces L. The preceding inequality proves the denominator in the definition of X positive, at least when q is large. Also, when $q > t$, one has $\gamma - c2^{t/q} \leq \max(c2^t, 2c+1)$. Now choose $C_1 \geq 2\max(c2^t, 2c+1)$ so large that (41) holds; according to an earlier comment, this is possible. Note that C_1 still depends only on the parameter set and on t. The inequality $C_1/(\gamma - c2^{t/q}) \geq 2$ has been enforced, and so, for large q, one has $X(2) \geq 2^{q-1}$. By (40), it follows that $X(2) \geq L(2)$ holds for large values of q. By Lemma 70.2 of Hall and Tenenbaum [3] and (42), this shows that the inequality

$$L(\sigma) \leq X(\sigma) \leq (2C_1 q/c)^q (\sigma-1)^{-\gamma}$$

holds throughout the range $1 < \sigma \leq 2$. With $C_2 = 2C_1/c$, the previous inequality can then be rewritten as

$$L(\sigma) \leq (\sigma-1)^{-\max(c2^t - t, c)} \exp(q\log C_2 q + C_3 q^{-1}\log(\sigma-1)^{-1}), \qquad (43)$$

where again C_3 only depends on the parameter set and on t. If one were allowed to choose

$$q = \left[\left(\frac{\log(\sigma-1)^{-1}}{\log\log(\sigma-1)^{-1}}\right)^{1/2}\right]$$

in (43), then in view of (39), the proof of Lemma 6 would be complete. In fact, when σ decreases to 1, then q goes to infinity, and the lower bound for q for which (43) holds depends only on t and the parameter set. Thus, the above choice is definitely admissible for $\sigma - 1$ sufficiently small. For larger values of σ, one chooses the smallest q for which (43) holds. Now $\sigma - 1$ is bounded below, and again, the conclusion of Lemma 6 follows from (43) and (39).

1.5 The Moment Method

Preparatory work now completed, the remaining sections of this chapter are devoted to the principal steps in the proof of Theorem 1. The strongest upper bounds for the average order of Hooley's Delta function known hitherto have been obtained through the method of moments and a differential inequalities technique, both developed by Hall and Tenenbaum (see Chap. 7 of [3] and the comments therein). Their strategy also underpinned the proof of Lemma 6 in the previous section. In the classical case, the first inequality in (37) is central to the passage from moments $M_q(n)$ to the Delta function. In the presence of a twist $h \in \mathscr{H}$, this step is less obvious, and one also has to keep track of the parameter v in (5) and (6). It is therefore appropriate to define the maximal function

$$\Delta_h^{(v)}(n) = \max_u |\Delta_h(n;u,v)|, \tag{44}$$

and, whenever q is a natural number, the moment

$$M_{h;q}^{(v)}(n) = \int_{-\infty}^{\infty} |\Delta_h(n;u,v)|^q \, du. \tag{45}$$

The next lemma is an attempt to provide a useful substitute for (37).

Lemma 7. *Let n be a natural number and $h \in \mathscr{H}$. Then for any positive real number v, one has*

$$\Delta_h^{(v)}(n) \le \Delta^{(v)}(n).$$

Moreover, when q and K are natural numbers, then

$$\Delta_h(n) \le 3K\Delta^{(2^{-K})}(n) + 2K^2 \max_{0\le k\le K} 2^{k/q} M_{h;q}^{(2^{-k})}(n)^{1/q}.$$

Proof. The first inequality follows immediately from $|h(m)| \le 1$ for all m. For the other, more important inequality, rather more care is required. For given data $u \in \mathbb{R}$, $0 < v \le 1$ and $K \in \mathbb{N}$, first choose a subset $\mathscr{K} \subset \{1, \dots, K\}$ such that

$$0 \le v - \sum_{k\in\mathscr{K}} 2^{-k} < 2^{-K}.$$

Then by the triangle inequality and (35), again using $|h| \le 1$, one obtains the bound

$$|\Delta_h(n;u,v)| \le \sum_{k\in\mathscr{K}} |\Delta_h(n;u_k,2^{-k})| + \Delta^{(2^{-K})}(n)$$

in which

$$u_k = u + \sum_{\substack{l\in\mathscr{K}\\ l<k}} 2^{-l}.$$

In particular, it follows that

$$\Delta_h(n) \le \sum_{k=1}^{K} \Delta_h^{(2^{-k})}(n) + \Delta^{(2^{-K})}(n). \tag{46}$$

This will be complemented with another inequality that again is a consequence of the triangle inequality. The starting point is (5). For any $v > 0$ and real numbers u, w with $u < w \le u+v$, one observes that

$$\Delta_h(n;u,v) = \Delta_h(n;u,w-u) + \Delta_h(n;w,v) - \Delta_h(n;u+v,w-u).$$

Now apply the triangle inequality and integrate over admissible values of w. This brings in the integrals

$$I_h(n;u,v) = \int_0^v |\Delta_h(n;u,\upsilon)|\,\mathrm{d}\upsilon, \tag{47}$$

and the resulting inequality reads

$$v|\Delta_h(n;u,v)| \le I_h(n;u,v) + \int_u^{u+v} |\Delta_h(n;w,v)|\,\mathrm{d}w + I_h(n;u+v,v). \tag{48}$$

A similar argument applies to the integral (47) and produces an iterative inequality. In fact, when $\frac{1}{2}v < \upsilon \le v$, one begins with the identity

$$\Delta_h(n;u,v) = \Delta_h\Big(n;u,\upsilon - \frac{1}{2}v\Big) + \Delta_h\Big(n;u+\upsilon-\frac{1}{2}v,\frac{1}{2}v\Big).$$

On isolating the contribution from the interval $0 \le \upsilon \le \frac{1}{2}v$ in (47), integration of the previous identity over $\frac{1}{2}v \le \upsilon \le v$ demonstrates that

$$I_h(n;u,v) \le 2I_h\Big(n;u,\frac{1}{2}v\Big) + \int_u^{u+\frac{1}{2}v} \Big|\Delta_h\Big(n;w,\frac{1}{2}v\Big)\Big|\,\mathrm{d}w.$$

One may replace v by $\frac{1}{2}v$ here, and substitute the result for the first term on the right-hand side of the original inequality. Iterating this process J times yields

$$I_h(n;u,v) \le 2^J I_h(n;u,2^{-J}v) + \sum_{j=1}^{J} 2^{j-1} \int_u^{u+2^{-j}v} |\Delta_h(n;w,2^{-j}v)|\,\mathrm{d}w.$$

An inspection of (47) and (44) reveals the alternative inequality $I_h(n;u,v) \le v\Delta^{(v)}(n)$. Hence, on combining these estimates with (48), one derives the bound

$$v|\Delta_h(n;u,v)| \le 2v\Delta^{(2^{-J}v)}(n) + \sum_{j=0}^{J} 2^j \Big(\int_u^{u+2^{-j}v} + \int_{u+v}^{u+v+2^{-j}v} \Big) |\Delta_h(n,w,2^{-j}v)|\,\mathrm{d}w.$$

Hölder's inequality now produces

$$v|\Delta_h(n;u,v)| \le 2v\Delta^{(2^{-J}v)}(n) + \sum_{j=0}^{J} 2^j (2^{1-j}v)^{1-1/q} M_{h;q}^{(2^{-j}v)}(n)^{1/q},$$

which simplifies to

$$|\Delta_h(n;u,v)| \le 2\Delta^{(2^{-J}v)}(n) + 2\sum_{j=0}^{J} \Big(\frac{2^j}{v}\Big)^{1/q} M_{h;q}^{(2^{-j}v)}(n)^{1/q}.$$

Here, the right-hand side is independent of u, and therefore an upper bound for $\Delta_h^{(v)}(n)$. Now choose $v = 2^{-k}$, $J = K - k$ and insert the result into (46) to deduce that

$$\Delta_h(n) \le (2K+1)\Delta^{(2^{-K})}(n) + 2\sum_{k=1}^{K}\sum_{j=0}^{K-k} 2^{(j+k)/q} M_{h;q}^{(2^{-j-k})}(n)^{1/q}.$$

The lemma is now immediate.

A convoluted moment is now to be estimated. For $0 \le j \le q$ and primes p, this is defined by

$$N_{h;q}^{(v)}(n;j,p) = \int_{-\infty}^{\infty} |\Delta_h(n;u,v)|^{q-j} |\Delta_h(n,u-\log p,v)|^j\,\mathrm{d}u. \tag{49}$$

One may use Hölder's inequality to separate the two factors in the integrand. By (45), this yields the simple bound

$$N_{h;q}^{(v)}(n;j,p) \le M_{h;q}^{(v)}(n) \tag{50}$$

that will be used later only for $j = 1$ and $j = q-1$. In the special case $h = 1$ and $v = 1$, superior upper bounds are part of Theorem 73 of Hall and Tenenbaum [3]. The following lemma serves as an appropriate replacement.

Lemma 8. *Uniformly in $q \geq 5$, $2 \leq j \leq q-2$, $0 < v \leq 1$ and $h \in \mathscr{H}$, one has*

$$\sum_p \frac{\log p}{p} N_{h;q}^{(v)}(n;j,p)$$
$$\ll \left(M_{h;2}^{(v)}(n) + \Delta(n)^2(\log\log 9n)^4\right)\left(v d(n)^2\right)^{2/(q-2)} M_{h;q}^{(v)}(n)^{(q-4)/(q-2)}.$$

Proof. By (49) and Hölder's inequality, it is immediately clear that it suffices to establish Lemma 8 in the two cases $j = 2$ and $j = q-2$. First consider $j = 2$. Let

$$\mathrm{E}_h(n;u,v) = \sum_p \frac{\log p}{p} |\Delta_h(n;u-\log p, v)|^2.$$

Note that by (5) the sum over p is over a finite range, and that

$$\sum_p \frac{\log p}{p} N_{h;q}^{(v)}(n;2,p) = \int_{-\infty}^{\infty} \mathrm{E}_h(n;u,v)|\Delta_h(n;u,v)|^{q-2}\,\mathrm{d}u. \tag{51}$$

Moreover, by (5) again,

$$\mathrm{E}_h(n;u,v) = \sum_p \frac{\log p}{p} \sum_{\substack{d_1|n, d_2|n \\ \mathrm{e}^u < pd_j \leq \mathrm{e}^{u+v}}} h(d_1)\overline{h(d_2)}.$$

For a pair d_1, d_2 it will be convenient to write

$$d^+ = \max(d_1,d_2), \quad d^- = \min(d_1,d_2). \tag{52}$$

Then

$$\mathrm{E}_h(n;u,v) = \sum_{d_1|n, d_2|n} h(d_1)\overline{h(d_2)} \sum_{\mathrm{e}^u/d^- < p \leq \mathrm{e}^{u+v}/d^+} \frac{\log p}{p}. \tag{53}$$

For an efficient evaluation of the inner sum over p, first note that the prime number theorem coupled with partial summation shows that there are certain real numbers E, κ with $\kappa > 0$ and

$$\sum_{p \leq x} \frac{\log p}{p} = \log x + E + O\big(\exp(-\kappa\sqrt{\log x})\big), \tag{54}$$

throughout the range $x \geq 1$. Now note that the sum over p in (53) is non-empty only in cases where $d^+/d^- \leq \mathrm{e}^v$, and if this is so, then by (54), this sum in (53) equals

$$\log \frac{\mathrm{e}^v d^-}{d^+} + O\left(\exp\left(-\kappa\left(\log\frac{\mathrm{e}^u}{d^-}\right)^{1/2}\right)\right).$$

This is useful only when d^- is not too large. Therefore, let

$$\gamma(n) = \exp\left(- 3\left(\frac{\log\log 9n}{\kappa}\right)^2\right), \tag{55}$$

and let $\mathrm{E}'_h(n;u,v)$ be the sum (53) with the additional constraint $d_1 \le \mathrm{e}^u\gamma(n)$. Write $\mathrm{E}''_h(n;u,v)$ for the sum with the complementary condition $d_1 > \mathrm{e}^u\gamma(n)$ so that

$$\mathrm{E}_h(n;u,v) = \mathrm{E}'_h(n;u,v) + \mathrm{E}''_h(n;u,v).$$

When $d_1 \le \mathrm{e}^u\gamma(n)$, it follows that $\mathrm{e}^u/d^- \ge \gamma(n)^{-1}$, and consequently,

$$\mathrm{E}'_h(n;u,v) = \sum_{\substack{d_1|n, d_2|n\\ d_1 \le \mathrm{e}^u\gamma(n)\\ d^+/d^- \le \mathrm{e}^v}} h(d_1)\overline{h(d_2)}\left(\log\frac{\mathrm{e}^v d^-}{d^+} + O\left(\frac{1}{(\log 9n)^3}\right)\right).$$

Since h takes values in the complex unit disc only, the error term sums to at most

$$\ll (\log 9n)^{-3} \sum_{d^-|n} \sum_{\substack{d^+|n\\ d^- < d^+ \le d^- \mathrm{e}^v}} 1 \ll \mathrm{d}(n)\Delta^{(v)}(n)(\log 9n)^{-3},$$

and (13) now yields

$$\mathrm{E}'_h(n;u,v) = \sum_{\substack{d_1|n, d_2|n\\ d_1 \le \mathrm{e}^u\gamma(n)\\ d^+/d^- \le \mathrm{e}^v}} h(d_1)\overline{h(d_2)} \log\frac{\mathrm{e}^v d^-}{d^+} + O(\Delta^{(v)}(n)\Delta(n)).$$

The leading term on the right-hand side here is readily seen to equal the integral

$$\int_{-\infty}^{\infty}\Bigg(\sum_{\substack{d|n\\ d \le \mathrm{e}^u\gamma(n)\\ \mathrm{e}^w < d \le \mathrm{e}^{w+v}}} h(d)\Bigg)\Bigg(\sum_{\substack{f|n\\ \mathrm{e}^w < f \le \mathrm{e}^{w+v}}} \overline{h(f)}\Bigg)\,\mathrm{d}w.$$

By Cauchy's inequality, the modulus of this expression does not exceed

$$\Bigg(\int_{-\infty}^{\infty}\Bigg|\sum_{\substack{d|n\\ d \le \mathrm{e}^u\gamma(n)\\ \mathrm{e}^w < d \le \mathrm{e}^{w+v}}} h(d)\Bigg|^2 \mathrm{d}w\Bigg)^{1/2} M_{h;2}^{(v)}(n)^{1/2} = J^{1/2} M_{h;2}^{(v)}(n)^{1/2},$$

say. An inspection of the summation conditions in the integrand of J reveals that the artificially introduced constraint $d \le \mathrm{e}^u\gamma(n)$ may be omitted in the initial range $\mathrm{e}^{w+v} \le \mathrm{e}^u\gamma(n)$, and forces the integrand to vanish when $\mathrm{e}^w > \mathrm{e}^u\gamma(n)$. Hence, by (45),

$$J \leq M_{h;2}^{(v)}(n) + \int_{e^{-v}\gamma(n) \leq e^{w-u} \leq \gamma(n)} \Big| \sum_{\substack{d|n \\ e^w < d \leq e^u \gamma(n)}} h(d) \Big|^2 \mathrm{d}w$$

$$\leq M_{h;2}^{(v)}(n) + v\Delta^{(v)}(n)^2.$$

This implies the inequality

$$\mathrm{E}'_h(n;u,v) \leq M_{h;2}^{(v)}(n) + v^{1/2} M_{h;2}^{(v)}(n)^{1/2} \Delta^{(v)}(n) + O(\Delta^{(v)}(n)\Delta(n))$$

that simplifies to

$$\mathrm{E}'_h(n;u,v) \ll M_{h;2}^{(v)}(n) + \Delta^{(v)}(n)\Delta(n). \tag{56}$$

Here, the implicit constant was inherited only from the use of the prime number theorem, and is therefore an absolute one.

The estimation of $\mathrm{E}''_h(n;u,v)$ is straightforward. Since the sum over p in (53) is empty unless $2d^+ \leq e^{u+v}$, and since one has $d^- \geq d_1 \geq e^u \gamma(n)$ in the current context, the trivial bound $|h(d_1)h(d_2)| \leq 1$ already gives

$$\mathrm{E}''_h(n;u,v) \leq \sum_{\substack{d_1|n, d_2|n \\ e^u\gamma(n) < d_1 \leq e^{u+v} \\ e^{-v} \leq d_1/d_2 \leq e^v}} \sum_{p \leq e^v/\gamma(n)} \frac{\log p}{p}.$$

The sum over p contributes $O\big((\log\log 9n)^2\big)$. For any fixed d_1, the sum over d_2 does not exceed $\Delta^{(2v)}(n) \leq 2\Delta^{(v)}(n)$. Then split the range $e^u\gamma(n) < d_1 \leq e^{u+v}$ into $O\big((\log\log 9n)^2\big)$ intervals of length at most 1 for $\log d_1$ to finally confirm that

$$\mathrm{E}''_h(n;u,v) \ll \Delta(n)\Delta^{(v)}(n)(\log\log 9n)^4.$$

This combines with (56) to

$$\mathrm{E}_h(n;u,v) \ll M_{h;2}^{(v)}(n) + \Delta(n)\Delta^{(v)}(n)(\log\log 9n)^4, \tag{57}$$

and (51) yields

$$\sum_p \frac{\log p}{p} N_{h;q}^{(v)}(n;2,p) \ll \big(M_{h;2}^{(v)}(n) + \Delta(n)\Delta^{(v)}(n)(\log\log 9n)^4\big) M_{h;q-2}^{(v)}(n). \tag{58}$$

By (45) and Hölder's inequality,

$$M_{h;q-2}^{(v)}(n) \leq M_{h;2}^{(v)}(n)^{2/(q-2)} M_{h;q}^{(v)}(n)^{(q-4)/(q-2)}, \tag{59}$$

and in the notation introduced in (52), one has

$$M_{h;2}^{(v)}(n) = \sum_{\substack{d_1|n, d_2|n \\ d^+/d^- \le e^v}} h(d_1)\overline{h(d_2)} \log \frac{e^v d^-}{d^+}. \tag{60}$$

For each of the terms to be summed here, one has $|h(d_1)h(d_2)| \log \frac{e^v d^-}{d^+} \le v$ so that in fact

$$M_{h;2}^{(v)}(n) \le 2v\Delta^{(v)}(n)\mathrm{d}(n) \le 2v\mathrm{d}(n)^2. \tag{61}$$

One may now combine (58), (59) and (61) to confirm the case $j = 2$ of Lemma 8.

It remains to consider $j = q-2$. The substitution $u' = u - \log p$ in (49) leads to an expression similar to (51), but with $u - \log p$ in the definition of E now replaced by $u + \log p$. Little change is necessary in the following argument to establish the appropriate analogue of (57), and after that point, the case $j = q-2$ of Lemma 8 follows *mutatis mutandis*.

The crucial step is right ahead. A differential inequality will be derived for a Dirichlet series analogous to (38), but with M_q replaced by $M_{h;q}^{(v)}$. It will be important to have at hand the special case $q = 2$ that is directly accessible.

Lemma 9. *Fix a parameter set. Let $h \in \mathscr{H}$ and $\rho \in \mathscr{F}_h$. Then, for $1 < \sigma \le 2$ and $0 < v \le 1$, one has*

$$\sum_{n=1}^{\infty} \mu(n)^2 \rho(n) M_{h;2}^{(v)}(n) n^{-\sigma} \ll v(\sigma-1)^{-2c} \log \frac{3}{\sigma-1}.$$

The implicit constant depends only on the parameter set.

Proof. Let $Z(\sigma)$ denote the Dirichlet series that is to be estimated. By (7), one has $\mu(n)^2\rho(n) \ll C^{\omega(n)}$, and by (45), (36) and (37), one has $M_{h;2}^{(v)}(n) \le M_2(n) \le \mathrm{d}(n)^2$, so that $Z(\sigma)$ is absolutely convergent for $\sigma > 1$. Hence, by (60),

$$Z(\sigma) = \sum_{\substack{d_1,d_2=1 \\ d^+/d^- \le e^v}}^{\infty} h(d_1)\overline{h(d_2)} \log \frac{e^v d^-}{d^+} \sum_{\substack{n=1 \\ [d_1,d_2]|n}}^{\infty} \frac{\mu(n)^2\rho(n)}{n^{\sigma}}.$$

Now substitute $n = m[d_1,d_2]$ in the inner sum, and write the remaining sum over m as a product. This yields

$$Z(\sigma) = P(\sigma) \sum_{\substack{d_1,d_2=1 \\ d^+/d^- \le e^v}}^{\infty} h(d_1)\overline{h(d_2)} \frac{\rho([d_1,d_2])\mu(d_1)^2\mu(d_2)^2}{[d_1,d_2]^{\sigma}} \theta_{\sigma}(d_1d_2;\rho) \log \frac{e^v d^-}{d^+},$$

where $\theta_{\sigma}(k;\rho)$ is defined by (23), and

$$P(\sigma) = \prod_p (1 + \rho(p)p^{-\sigma}). \tag{62}$$

For later use, note that Lemma 2 delivers the bound

$$\mathrm{P}(\sigma) = \sum_{n=1}^{\infty} \frac{\mu(n)^2 \rho(n)}{n^{\sigma}} \ll (\sigma - 1)^{-c}. \tag{63}$$

Returning to $\mathrm{Z}(\sigma)$, the sum over d_1, d_2 is now arranged according to the value of $d = (d_1, d_2)$. Then

$$\mathrm{Z}(\sigma) = \mathrm{P}(\sigma) \sum_{d=1}^{\infty} \frac{\mu(d)^2 \rho(d) |h(d)|^2}{d^{\sigma}} \theta_{\sigma}(d) \mathrm{Z}_d(\sigma) \tag{64}$$

in which $\mathrm{Z}_d(\sigma)$ is defined by

$$\mathrm{Z}_d(\sigma) = \sum_{\substack{d_1, d_2 = 1 \\ d^+/d^- \leq \mathrm{e}^{\nu} \\ (d_1, d_2) = (d_1 d_2, d) = 1}}^{\infty} \frac{\lambda(d_1) \overline{\lambda(d_2)}}{d_1^{\sigma} d_2^{\sigma}} \log \frac{\mathrm{e}^{\nu} d^-}{d^+},$$

and where in the interest of brevity, from now on the shorthand

$$\lambda(d) = \lambda_{\sigma}(d) = \mu(d)^2 \rho(d) h(d) \theta_{\sigma}(d) \tag{65}$$

is used. The diagonal term $d_1 = d_2 = 1$ contributes ν. The remaining terms come in conjugate pairs so that

$$\mathrm{Z}_d(\sigma) = \nu + 2\mathrm{Re} \sum_{\substack{d_1 = 1 \\ (d_1, d) = 1}}^{\infty} \frac{\lambda(d_1)}{d_1^{\sigma}} \sum_{\substack{d_1 < d_2 \leq d_1 \mathrm{e}^{\nu} \\ (d_2, d d_1) = 1}} \frac{\overline{\lambda(d_2)}}{d_2^{\sigma}} \log \frac{\mathrm{e}^{\nu} d_1}{d_2}.$$

Here, the sum over d_2 is empty unless $[d_1(\mathrm{e}^{\nu} - 1)] \geq 1$ holds, and the previous identity therefore becomes

$$\mathrm{Z}_d(\sigma) = \nu + 2\mathrm{Re} \sum_{\substack{d_1 \geq (\mathrm{e}^{\nu} - 1)^{-1} \\ (d_1, d) = 1}} \frac{\lambda(d_1)}{d_1^{\sigma}} \int_{d_1}^{\mathrm{e}^{\nu} d_1} \sum_{\substack{d_1 < d_2 \leq w \\ (d_2, d d_1) = 1}} \frac{\overline{\lambda(d_2)}}{d_2^{\sigma}} \frac{\mathrm{d}w}{w}.$$

To obtain an estimate for the sum over d_2, one first removes $d_2^{-\sigma}$ by partial summation and then applies Lemma 4. Uniformly for $d_1 \leq w \leq 3d_1$, this yields the bound

$$\ll \upsilon(d d_1) \left(\frac{w^{1-\sigma}}{(\log 3w)^c} + \sigma \int_{d_1}^{w} \frac{y}{(\log 3y)^c} \frac{\mathrm{d}y}{y^{1+\sigma}} \right) \ll \upsilon(d d_1) \frac{d_1^{1-\sigma}}{(\log 3d_1)^c}$$

for the relevant sum that may now be integrated against $\mathrm{d}w/w$. The implicit constant in this estimate stems from Lemma 4, and depends therefore only on the parameter set, this remaining true for the consequential formula

$$\mathrm{Z}_d(\sigma) = v + O\left(v\upsilon(d) \sum_{d_1 \geq (\mathrm{e}^v - 1)^{-1}} \frac{\upsilon(d_1)|\lambda(d_1)|}{d_1^\sigma(\log 3d_1)^c} \right).$$

Now consider the Dirichlet series

$$\mathrm{H}(\sigma) = \sum_{n=1}^{\infty} \frac{\mu(n)^2\upsilon(n)\rho(n)}{n^\sigma}, \quad \mathrm{H}^*(\sigma) = \sum_{n=1}^{\infty} \frac{\mu(n)^2\upsilon(n)\rho(n)}{n^\sigma(\log 3n)^c}. \tag{66}$$

The trivial bound $|h(d_1)\theta_\sigma(d_1)| \leq 1$ for $1 < \sigma \leq 2$ first shows that $|\lambda(d_1)| \leq \mu(d_1)^2\rho(d_1)$, and then that

$$\mathrm{Z}_d(\sigma) = v + O(v\upsilon(d)\mathrm{H}^*(\sigma)), \tag{67}$$

in which once again the implicit constant depends only on the parameter set. It remains to estimate $\mathrm{H}^*(\sigma)$. By (7) and (22), it is clear that $\mathrm{H}(\sigma)$, and hence also $\mathrm{H}^*(\sigma)$, converges for $\sigma > 1$. Also, for $\sigma > 1$, (22) shows that $\upsilon(p) \leq 1 + 8p^{-1/4}$. By (7), it now follows that

$$\mathrm{H}(\sigma) = \prod_p \left(1 + \frac{\upsilon(p)\rho(p)}{p^\sigma}\right) \leq \prod_p \left(1 + \frac{\rho(p)}{p^\sigma}\right)\left(1 + \frac{8C}{p^{5/4}}\right) \leq \mathrm{P}(\sigma)\prod_p \left(1 + \frac{8C}{p^{5/4}}\right).$$

The second factor on the right is a convergent product depending only on C. Hence, by (63),

$$\mathrm{H}(\sigma) \ll (\sigma - 1)^{-c}.$$

Here, we choose $\sigma = 1 + (\log 3x)^{-1}$ and apply Rankin's trick to deduce that

$$\sum_{n \leq x} \frac{\mu(n)^2\upsilon(n)\rho(n)}{n} \ll (\log 3x)^c.$$

By partial summation, this implies

$$\sum_{n \leq x} \frac{\mu(n)^2\upsilon(n)\rho(n)}{n(\log 3n)^c} \ll 1 + \int_1^x \frac{\mathrm{d}y}{y\log 3y} \ll \log\log 9x,$$

and another partial summation yields

$$\mathrm{H}^*(\sigma) = (\sigma - 1)\int_1^\infty x^{-\sigma} \sum_{n \leq x} \frac{\mu(n)^2\upsilon(n)\rho(n)}{n(\log 3n)^c}\,\mathrm{d}x.$$

Now estimate the integrand and then integrate by parts to deduce that

$$\mathrm{H}^*(\sigma) \ll (\sigma - 1)\int_1^\infty x^{-\sigma}(\log\log 9x)\,\mathrm{d}x \ll 1 + \int_1^\infty \frac{\mathrm{d}x}{x^\sigma \log 9x}.$$

The final integral here is of the same type as the one considered in (20), and the argument given there yields

$$\mathrm{H}^*(\sigma) \ll \log\frac{3}{\sigma - 1}.$$

By (67), it follows that

$$\mathrm{Z}_d(\sigma) \ll v\upsilon(d)\log\frac{3}{\sigma - 1}.$$

This may be inserted into (64) and combined with (63). The estimates obtained so far then produce

$$\mathrm{Z}(\sigma) \ll v\mathrm{P}(\sigma)\mathrm{H}(\sigma)\log\frac{3}{\sigma - 1} \ll v(\sigma - 1)^{-2c}\log\frac{3}{\sigma - 1},$$

as required to complete the proof of the lemma.

The reader is invited to analyse the proof of Lemma 9 in cases where cancellations from oscillatory h are not available. The most natural situation is $h \equiv \rho \equiv 1$, the average order of Hooley's Delta function. Then the best one can hope for, for the sum considered in Lemma 9, is $O((\sigma - 1)^{-3})$. For comparison, Lemma 9 provides a superior bound almost as good as $(\sigma - 1)^{-2}$ in all cases, where $c = 1$. Thus, the oscillatory properties of h are not only coded into the sum considered in Lemma 9, but are also extractable in a simple manner.

The result is now fed into an estimation of a Dirichlet series that plays the same rôle in the proof of Theorem 1 as the function $L(\sigma)$ did in Lemma 8. The relevant series is given by

$$L_h(\sigma) = L_h^{(v)}(\sigma;\rho,q,t) = \sum_{n=1}^\infty \mu(n)^2\rho(n)M_{h;q}^{(v)}(n)^{t/q}n^{-\sigma}, \tag{68}$$

where $\sigma > 1$ and the parameters range over $0 < v \le 1$, $t \ge 1$, $q \in \mathbb{N}$, $h \in \mathscr{H}$ and $\rho \in \mathscr{F}_h$. For notational simplicity, most parameters are often suppressed in the sequel. Yet, it is important to estimate this sum not also uniformly for $1 < \sigma \le 2$, but also with respect to v and q.

By (45) and (36), one has $M_{h;q}^{(v)}(n) \le M_q(n)$ so that the series $L(\sigma)$ defined in (38) is an upper bound for $L_h(\sigma)$. In particular, it follows that $L_h(\sigma)$ converges for $\sigma > 1$, and a short calculation based on (17) reveals that

$$-L'(\sigma) = \sum_{n=1}^\infty \frac{\mu(n)^2\rho(n)}{n^\sigma}\sum_{p\nmid n}\frac{\rho(p)\log p}{p^\sigma}M_{h;q}^{(v)}(pn)^{t/q}. \tag{69}$$

The next goal is an estimate for the inner sum here. By (5), when $p \nmid n$, one has

$$|\Delta_h(pn;u,v)| \le |\Delta_h(n;u,v)| + |\Delta_h(n;u-\log p,v)|.$$

One takes the q-th power, expands by the binomial theorem and then integrates over u to confirm that

$$\begin{aligned} M_{h;q}^{(v)}(pn) &\le 2M_{h;q}^{(v)}(n) + \sum_{j=1}^{q-1}\binom{q}{j} N_{h;q}^{(v)}(n;j,p) \\ &\le 2(q+1)M_{h;q}^{(v)}(n) + \sum_{j=2}^{q-2}\binom{q}{j} N_{h;q}^{(v)}(n;j,p), \end{aligned} \tag{70}$$

where for the last inequality, the summands $j=1$ and $j=q-1$ were estimated by (50). For the remaining j, one multiplies by $\rho(p)(\log p)p^{-\sigma}$ and notes that $\rho(p) \le C$ and $1 < \sigma \le 2$ in the current context. Hence, by Lemma 8,

$$\begin{aligned} \sum_{p\nmid n} \frac{\rho(p)\log p}{p^\sigma} M_{h;q}^{(v)}(pn) &\le 2(q+1)M_{h;q}^{(v)}(n) \sum_{p\nmid n} \frac{\rho(p)\log p}{p^\sigma} \\ &\quad + C'2^q\Gamma_h^{(v)}(n)\big(v\mathrm{d}(n)^2\big)^{2/(q-2)} M_{h;q}^{(v)}(n)^{(q-4)/(q-2)}, \end{aligned} \tag{71}$$

where the shorthand

$$\Gamma_h^{(v)}(n) = M_{h;2}^{(v)}(n) + \Delta(n)^2(\log\log 9n)^4 \tag{72}$$

has been used in the interest of brevity, and where C' denotes the product of C with the constant implicit in Lemma 8.

Now restrict to the range $1 \le t \le 2$ and suppose that $q \ge 5$ so that $q > t$. An estimate for the inner sum in (69) is readily available. One first uses Hölder's inequality to bring in the bound in (71), then applies the inequality

$$(\xi + \eta)^\upsilon \le \xi^\upsilon + \eta^\upsilon \tag{73}$$

that is valid for $0 \le \upsilon \le 1$ and non-negative real numbers ξ, η. This procedure yields

$$\begin{aligned} \sum_{p\nmid n} \frac{\rho(p)\log p}{p^\sigma} M_{h;q}^{(v)}(pn)^{t/q} &\le \big((2q+2)M_{h;q}^{(v)}(n)\big)^{t/q} \sum_{p\nmid n} \frac{\rho(p)\log p}{p^\sigma} \\ &+ \big(C'2^q\Gamma_h^{(v)}(n)\big)^{t/q}\big(vd(n)^2\big)^{2t/q(q-2)} M_{h;q}^{(v)}(n)^{t(q-4)/q(q-2)} \Big(\sum_p \frac{\rho(p)\log p}{p^\sigma}\Big)^{1-(t/q)}. \end{aligned}$$

By (9) and crude estimates, the above does not exceed

$$\Big((2q+2)M_{h;q}^{(v)}(n)\Big)^{t/q}\left(\frac{c}{\sigma-1}+C\right)$$
$$+C''\Gamma_h^{(v)}(n)^{t/q}\big(v\mathrm{d}(n)^2\big)^{2t/q(q-2)}M_{h;q}^{(v)}(n)^{t(q-4)/q(q-2)}(\sigma-1)^{t/q-1},$$

where C'' is a positive number depending at most on the parameter set and on t. When inserted into (69), this yields

$$-L_h'(\sigma)\le(2q+2)^{t/q}L_h(\sigma)\Big(\frac{c}{\sigma-1}+C\Big)+C''v^{2t/q(q-2)}(\sigma-1)^{t/q-1}J_h(\sigma), \quad (74)$$

where

$$J_h(\sigma)=\sum_{n=1}^{\infty}\frac{\mu(n)^2\rho(n)}{n^\sigma}\Gamma_h^{(v)}(n)^{t/q}d(n)^{4t/q(q-2)}M_{h;q}^{(v)}(n)^{t(q-4)/q(q-2)}.$$

By Hölder's inequality and (68), one finds that

$$J_h(\sigma)\le\left(\sum_{n=1}^{\infty}\frac{\mu(n)^2\rho(n)\mathrm{d}(n)^t}{n^\sigma}\right)^{4/q(q-2)}K(\sigma)^{2/q}L_h(\sigma)^{(q-4)/(q-2)} \quad (75)$$

with

$$K(\sigma)=\sum_{n=1}^{\infty}\frac{\mu(n)^2\rho(n)\Gamma_h^{(v)}(n)^{t/2}}{n^\sigma}.$$

This procedure brings back $L_h(\sigma)$, and at the same time separates $\mathrm{d}(n)$ and $\Gamma_h^{(v)}(n)$. An estimate for $K(\sigma)$ is available by (72) and (73) (take $\upsilon=\frac{1}{2}t$), namely

$$K(\sigma)\le\sum_{n=1}^{\infty}\frac{\mu(n)^2\rho(n)M_{h;q}^{(v)}(n)^{t/2}}{n^\sigma}+\sum_{n=1}^{\infty}\frac{\mu(n)^2\rho(n)\Delta(n)^t(\log\log 9n)^{2t}}{n^\sigma}.$$

For the first summand on the right-hand side, one may apply Hölder's inequality to interpolate between the estimates provided by Lemmata 2 and 9. This yields the upper bound

$$\ll\Big(v(\sigma-1)^{-2c}\log\frac{3}{\sigma-1}\Big)^{t/2}\big((\sigma-1)^{-c}\big)^{1-t/2}\ll(\sigma-1)^{-c(1+t/2)}\log\frac{3}{\sigma-1}$$

for the term under consideration. For the second summand, note that $\Delta(pn)\le 2\Delta(n)$ holds, as a consequence of Hooley [4], his formula (5). Therefore, Lemma 1 is applicable with $A=2^t$ and $k=4$. This procedure combined with Lemma 6 yields the upper bound

$$\ll \left(\log \frac{3}{\sigma-1}\right)^8 \sum_{n=1}^{\infty} \frac{\mu(n)^2\rho(n)\Delta(n)^t}{n^\sigma}$$

$$\ll \left(\log \frac{3}{\sigma-1}\right)^8 (\sigma-1)^{-\max(2^t c-t,c)} \mathscr{L}^*((\sigma-1)^{-1})^B.$$

Collecting together, this shows that

$$K(\sigma) \ll (\sigma-1)^{-\max(2^t c-t,c(1+t/2))} \mathscr{L}^*((\sigma-1)^{-1})^{2B}.$$

Note that this is somewhat crude: the potentially small factor $v^{t/2}$ was neglected, and B was increased to $2B$ to absorb powers of $\log 3/(\sigma-1)$.

The Dirichlet series that remained explicit in (75) was estimated in Lemma 2. Therefore, by (74) and the preceding estimate for $K(\sigma)$, it now follows that

$$-L_h'(\sigma) \leq (2q+2)^{t/q} L_h(\sigma)\Big(\frac{c}{\sigma-1}+C\Big)$$
$$+ C'' v^{2t/q(q-2)} \mathscr{L}^*((\sigma-1)^{-1})^{2B/(q-2)} (\sigma-1)^{-1-2\eta/(q-2)} L_h(\sigma)^{(q-4)/(q-2)}, \quad (76)$$

where

$$\eta = \frac{1}{q}\Big(2^{t+1}c + (q-2)\max\Big(2^t c - \frac{3}{2}t, c + \frac{1}{2}(c-1)t\Big)\Big).$$

A crude upper bound for $L_h(\sigma)$ is enough to remove the term involving C from (76). In fact, by (45) and trivial estimates followed by an application of (73), one first observes that

$$M_{h;q}^{(v)}(n) \leq M_{h,2}^{(v)}(n)\mathrm{d}(n)^{q-2} \leq 2v\mathrm{d}(n)^q,$$

and then infers from Lemma 2 the bound

$$L_h(\sigma) \leq (2v)^{t/q} \sum_{n=1}^{\infty} \frac{\mu(n)^2\rho(n)\mathrm{d}(n)^t}{n^\sigma} \ll v^{t/q}(\sigma-1)^{-2^t c}. \quad (77)$$

Since $t \leq 2$, one has $(2q+2)^{t/q} \ll 1$, and it then readily follows that

$$(2q+2)^{t/q} L_h(\sigma) \ll v^{2t/q(q-2)} (\sigma-1)^{-1-2\eta/(q-2)} L_h(\sigma)^{(q-4)/(q-2)}.$$

Consequently, there is a number C''' such that

$$-L_h'(\sigma) \leq (2q+2)^{t/q} L_h(\sigma)\frac{c}{\sigma-1}$$
$$+ C''' v^{2t/q(q-2)} \mathscr{L}^*((\sigma-1)^{-1})^{4B/(q-2)} (\sigma-1)^{-1-2\eta/(q-2)} L_h(\sigma)^{(q-4)/(q-2)}.$$

The further proceedings are similar to those within the proof of Lemma 6. One may replace η with any larger number, and such is given by

$$\eta' = \frac{8c}{q} \max\left(2^t c - \frac{3}{2}t, c + \frac{1}{2}(c-1)t, c(2q+2)^{t/q}\right), \tag{78}$$

so that the differential inequality

$$-L_h'(\sigma) \leq c(2q+2)^{t/q} L_h(\sigma)(\sigma-1)^{-1}$$
$$+ C''' v^{2t/q(q-2)} \mathscr{L}^*((\sigma-1)^{-1})^{4B/(q-2)}(\sigma-1)^{-1-2\eta'/(q-2)} L_h(\sigma)^{(q-4)/(q-2)}$$

holds throughout the range $1 < \sigma \leq 2$. The factor $\mathscr{L}^*((\sigma-1)^{-1})^{4B/(q-2)}$ disturbs the argument used in the proof of Lemma 6, but monotonicity tames its influence. Fix a number σ_0 with $1 < \sigma_0 < 2$. Then for $\sigma_0 \leq \sigma \leq 2$, one has

$$-L_h'(\sigma) \leq c(2q+2)^{t/q} L_h(\sigma)(\sigma-1)^{-1}$$
$$+ C''' v^{2t/q(q-2)} \mathscr{L}^*((\sigma_0-1)^{-1})^{4B/(q-2)}(\sigma-1)^{-1-2\eta'/(q-2)} L_h(\sigma)^{(q-4)/(q-2)}. \tag{79}$$

The function

$$X_h(\sigma) = X_h(\sigma;\sigma_0) = v^{t/q} \mathscr{L}^*((\sigma_0-1)^{-1})^{2B} \left(\frac{C'''}{\eta' - c(2q+2)^{t/q}}\right)^{(q-2)/2} (\sigma-1)^{-\eta'}$$

satisfies (79) with equality, as one readily checks. Here, it is important to note that (78) guarantees $\eta' - c(2q+2)^{t/q} > 0$. Moreover, again by (78),

$$X_h(2) = v^{t/q} \mathscr{L}^*((\sigma_0-1)^{-1})^{2B} \left(\frac{C'''}{\eta' - c(2q+2)^{t/q}}\right)^{(q-2)/2} \geq v^{t/q} \left(\frac{C'''}{8c}\right)^{(q-2)/2},$$

at least when q is large and $\sigma_0 - 1$ is sufficiently small. By (77), one has $L_h(2) \ll v^{t/q}$, so that for large q one concludes that $X_h(2) \geq L_h(2)$. By Lemma 70.2 of Hall and Tenenbaum [3] and (79), it follows that $L_h(\sigma) \leq X_h(\sigma,\sigma_0)$ holds for all $\sigma_0 \leq \sigma \leq 2$. In particular, one may take $\sigma = \sigma_0$ to deduce that the inequality

$$L_h(\sigma) \leq v^{t/q} \mathscr{L}^*((\sigma-1)^{-1})^{2B} \left(\frac{C'''q}{8c}\right)^{q/2} (\sigma-1)^{-\eta'} \tag{80}$$

is valid for $1 < \sigma \leq 2$. Now use the expansion $(2q+2)^{t/q} = 1 + \frac{t}{q}\log(2q+2) + \cdots$ to compare (78) with (10). This yields

$$\eta' \leq \alpha + \frac{ct}{q}\log(2q+2) + O\left(\frac{1}{q}\right)$$

in which the implicit constant depends only on c as long as t is constrained to $1 \leq t \leq 2$. Inserting this into (80), one now concludes as follows.

Lemma 10. *Let $h \in \mathscr{H}$ and $\rho \in \mathscr{F}_h$. There is a number q_0 depending only on the parameter set such that whenever $q \geq q_0$, $1 \leq t \leq 2$, $0 < v \leq 1$ and $1 < \sigma \leq 28/27$, one has*

$$L_h^{(v)}(\sigma;\rho,q,t) \leq v^{t/q}(\sigma-1)^{-\alpha}\exp\Psi,$$

where α is given by (10),

$$\Psi = \frac{q}{2}\log\frac{C'''q}{8c} + \left(\frac{ct}{q}\log(2q+2) + O\left(\frac{1}{q}\right)\right)\log(\sigma-1)^{-1}$$
$$+2B(\log(\sigma-1)^{-1})^{1/2}(\log\log(\sigma-1)^{-1})^{1/2},$$

and B, C''' denote suitable numbers depending only on the parameter set. The implicit constant depends only on c.

The observant reader will have noticed that the condition $\sigma \leq 28/27$ has been introduced only to ensure that $\log\log(\sigma-1)^{-1} > 1$.

By an argument very similar to the one in Lemma 6, the bound in Lemma 10 transforms into the following intermediate version of Theorem 1.

Lemma 11. *Let $1 \leq t \leq 2$. Let $h \in \mathscr{H}$ and $\rho \in \mathscr{F}_h$. Then*

$$\sum_{n\leq x}\frac{\mu(n)^2\rho(n)\Delta_h(n)^t}{n} \ll (\log x)^{\alpha}\mathscr{L}(\log x)^{\beta}\mathscr{L}^*(\log x)^D$$

in which the positive number D and the implicit constant depend only on the parameter set.

Proof. Let q and K denote natural numbers. Then by Lemma 7, the inequality

$$\Delta(n)^t \leq 18K^t\Delta^{(2^{-K})}(n)^t + 8K^{2t}\max_{0\leq k\leq K} 2^{kt/q}M_{h;q}^{(2^{-k})}(n)^{t/q}$$

holds for any $1 \leq t \leq 2$. By Rankin's trick,

$$\sum_{n\leq x}\frac{\mu(n)^2\rho(n)\Delta_h(n)^t}{n} \leq 18K^t\Sigma_1 + 24K^{2t}\Sigma_2 \tag{81}$$

with

$$\Sigma_1 = \sum_{n\leq x}\frac{\mu(n)^2\rho(n)\Delta^{(2^{-K})}(n)^t}{n},$$

$$\Sigma_2 = \sum_{k=0}^{K}2^{kt/q}L_h^{(2^{-k})}\left(1+\frac{1}{\log x};\rho,q,t\right).$$

It is convenient to choose

$$K = [4(C^2+1)\log\log x] + 1, \tag{82}$$

because then $2^{-K} \ll (\log x)^{-4(C^2+1)}$, and Lemma 5 then provides the estimate $\Sigma_1 \ll (\log x)^c$ which in view of (81), (82) and (10) is acceptable.

It remains to estimate Σ_2. By Lemma 10,

$$2^{kt/q} L_h^{(2^{-k})}\Big(1+\frac{1}{\log x};\rho,q,t\Big) \le (\log x)^{\alpha} \exp \Phi,$$

where

$$\Phi = \frac{q}{2}\log\frac{C'''q}{8c} + \left(\frac{ct}{q}\log(2q+2) + O\Big(\frac{1}{q}\Big)\right)\log x$$
$$+ 2B(\log\log x)^{1/2}(\log\log\log x)^{1/2} + O(1)$$

which is independent of k. One chooses

$$q = [(2ct\log\log x)^{1/2}] + 1$$

to confirm that

$$\Phi \le \left(\frac{1}{2}ct\log\log x\right)^{1/2} \log\log\log x + \frac{1}{2}D(\log\log x)^{1/2}(\log\log\log x)^{1/2}.$$

Here, D is a number depending only the parameter set and on t; recall that this was the case with B as well. This gives

$$\Sigma_2 \ll K(\log x)^{\alpha}\exp\Phi \ll K(\log x)^{\alpha}\mathscr{L}(\log x)^{\beta}\mathscr{L}^*(\log x)^{D/2},$$

and Lemma 11 follows from (81) and (82).

1.6 The Proof of Theorem 1

The scene is ready for the proof of Theorem 1, which will be deduced from Lemma 11 by an argument of Hooley [4], Chap. II.2.

Let m, n be coprime numbers. Then by (5),

$$|\Delta_h(mn;u,v)| \le \sum_{d|n} |\Delta_h(m;u-\log d,v)|.$$

It follows that $|\Delta_h(mn;u,v)| \le \mathrm{d}(n)\Delta_h(m)$, and hence that

$$\Delta_h(mn) \le \mathrm{d}(n)\Delta_h(m). \tag{83}$$

This should be compared with the inequality (5) in Hooley [4], in which coprimality is not required.

Consider the sum

$$S(x)=\sum_{n\le x}\mu(n)^2\rho(n)\Delta_h(n)^t\log n,$$

and apply (17) to infer that

$$S(x)=\sum_{\substack{pm\le x\\ p\nmid m}}\mu(m)^2\rho(pm)\Delta_h(pm)^t\log p.$$

By (7) and (83),

$$S(x)\le 2^tC\sum_{m\le x}\mu(m)^2\rho(m)\Delta_h(m)^t\sum_{p\le x/m}\log p.$$

By Chebyshev's estimate and Lemma 11,

$$S(x)\ll x(\log x)^{\alpha}\mathscr{L}(\log x)^{\beta}\mathscr{L}^*(\log x)^{D}.$$

The unwanted factor $\log n$ in the definition of $S(x)$ may be removed by partial summation, and one finds that

$$\sum_{n\le x}\mu(n)^2\rho(n)\Delta_h(n)^t\ll x(\log x)^{\alpha-1}\mathscr{L}(\log x)^{\beta}\mathscr{L}^*(\log x)^{D}. \tag{84}$$

Finally, the restriction to square-free numbers in (84) can be removed by a standard process. For any integer n, there is a unique factorisation $n=ml$ with $\mu(m)^2=1$, square-full l and $(m,l)=1$. Hence, by (83) and (84),

$$\begin{aligned}\sum_{n\le x}\rho(n)\Delta_h(n)^t &= \sum_{\substack{l\le x\\ p\mid l\Rightarrow p^2\mid l}}\rho(l)\sum_{\substack{m\le x/l\\ (m,l)=1}}\mu(m)^2\rho(m)\Delta_h(ml)^t\\ &\le x(\log x)^{\alpha-1}\mathscr{L}(\log x)^{\beta}\mathscr{L}^*(\log x)^{D}\sum_{\substack{l\le x\\ p\mid l\Rightarrow p^2\mid l}}\frac{\rho(l)\mathrm{d}(l)^t}{l}.\end{aligned}$$

By (7), the sum over l contributes at most

$$\prod_{p\le x}\left(1+\sum_{\upsilon=2}^{\infty}p^{-\upsilon}\rho(p^{\upsilon})(\upsilon+1)^t\right)\le\prod_{p}\left(1+C\sum_{\upsilon=2}^{\infty}p^{-\upsilon\delta-1}(\upsilon+1)^2\right).$$

The right-hand side here depends only on the parameter set. Theorem 1 now follows.

2 Incomplete Convolutions

2.1 The Main Result

The truncated convolution of Möbius' function, defined by

$$\mathrm{M}(n,u) = \sum_{\substack{d|n \\ \log d \le u}} \mu(d),$$

is a frequently recurring object in areas related to sieve theory. The size of the maximal function

$$\mathrm{M}(n) = \max_u |\mathrm{M}(n,u)|$$

is therefore of interest in not only a few applications. Following a pivotal contribution of Erdös and Kátai [2], research on $\mathrm{M}(n)$ focussed on bounds that would hold for almost all natural numbers n. In this spirit, Maier [5] showed that the inequality $\mathrm{M}(n) > (\log\log n)^{0.2875}$ fails for no more than $o(x)$ of the natural numbers n not exceeding x. Also, on combining the methods of [5] with recent results of Maier and Tenenbaum [6], one finds that $\mathrm{M}(n) < (\log\log n)^{\log 2+\varepsilon}$ holds for almost all n. In his Stuttgart seminar in 2000, Daniel mentioned that his method for twisted Delta functions can be modified to attack $\mathrm{M}(n)$ in mean, and announced the estimate

$$\sum_{n\le x} \mathrm{M}(n)^2 \ll x\mathscr{L}(\log x)^{1+\varepsilon}. \tag{85}$$

It is surprising that Daniel's remark, apparently, is the first attempt to control moments of $\mathrm{M}(n)$. Following Daniel's line of thought, it seems natural to discuss the underlying problem in the same general framework as in the first chapter of this memoir. Thus, for $h \in \mathscr{H}$, consider the incomplete convolution

$$\mathrm{B}_h(n,u) = \sum_{\substack{d|n \\ \log d \le u}} h(d) \tag{86}$$

and its cognate maximal function

$$\mathrm{B}_h(n) = \max_u |\mathrm{B}(n,u)|. \tag{87}$$

Note that $\mathrm{B}_\mu(n) = \mathrm{M}(n)$, and that the estimate (85) for the t-th moment of $\mathrm{M}(n)$ is the same as the one that Theorem 1 yields for $\Delta_\mu(n)$. However, one should not expect that means of $\mathrm{B}_h(n)$ are typically as small as the related mean of $\Delta_h(n)$. For an example, let χ denote the primitive Dirichlet character, modulo 4. Then Theorem 1 provides the estimate

$$\sum_{n\le x} \Delta_\chi(n)^2 \ll x(\log x)^\varepsilon.$$

Yet, whenever n is a product of primes congruent to 1, modulo 4, one has $\chi(d) = 1$ for all $d|n$, whence $\mathrm{B}_\chi(n) = \mathrm{d}(n)$, and some mundane analytic number theory reveals that

$$\sum_{n\le x} \mathrm{B}_\chi(n)^2 \ge \sum_{\substack{n\le x\\ p|n\Rightarrow p\equiv 1 \bmod 4}} \mathrm{d}(n)^2 \gg x\log x. \tag{88}$$

Consequently, the oscillatory condition (9) is not sufficient to imply an estimate of the desired type for $\mathrm{B}_h(n)$. For such an estimate to hold, a more rigid interplay between the multiplicative function h and the weight ρ is inevitable. To make this precise, let c, C, δ be a parameter set, and suppose that $h \in \mathscr{H}$ and $\rho \in \mathscr{F}_h$ conspire through the estimate

$$\sum_{n=1}^{\infty} \mu(n)^2\rho(n)h(n)n^{-\sigma} \ll \sqrt{\sigma-1} \tag{89}$$

that must be assumed to hold uniformly in the interval $1 < \sigma \le 2$. Perhaps less significantly, the oscillatory condition (9) will from now on be replaced by the stronger bound

$$\sum_{n\le x} \mu(n)^2\rho(n)h(n) \ll x(\log x)^{-1/\varepsilon}, \tag{90}$$

and in addition to (7), it will be convenient to suppose that

$$|h(p)|\rho(p) \le (1-\delta)p \tag{91}$$

holds for all primes p. In view of (7), this last condition is relevant only for small primes.

Let $\mathscr{F}_h^*(c,C,\delta)$ denote the set of all $\rho \in \mathscr{F}_h$ for which (89)–(91) hold.

Theorem 2. *Fix a parameter set and a real number t with $1 \le t \le 2$. Then there is a number D such that for any $h \in \mathscr{H}$ and $\rho \in \mathscr{F}_h^*$, one has*

$$\sum_{n\le x} \rho(n)\mathrm{B}_h(n)^t \ll x(\log x)^{\alpha-1}\mathscr{L}(\log x)^{\beta}\mathscr{L}^*(\log x)^{D}.$$

The implicit constant depends at most on t and the parameter set.

One may take $h = \mu$ and $\rho = 1$. Then the Dirichlet series on the left-hand side of (89) is $\zeta(\sigma)^{-1}$, where ζ is Riemann's zeta function. Hence, the critical condition (89) does indeed hold, and also the other hypotheses in Theorem 2 are satisfied with $\alpha = c = 1$ and $\delta = \frac{1}{2}$. Therefore, Theorem 2 contains Daniel's upper bound (85), as a very special case. On the other hand, by (88), the vanishing condition (89) cannot hold for the primitive character modulo 4. In fact, when a non-principal Dirichlet character χ is substituted for h in (89), and one takes $\rho = 1$, then the sum on the left-hand side of (89) is $L(s,\chi)/L(2s,\chi^2)$, which does not vanish at $s = 1$. Thus, the hypotheses of Theorem 2 are not satisfied for any character, as expected.

2.2 A Transistor for the Vanishing Condition

In this section, we consider a series of Dirichlet's type, for later use in an application of the moment method. The series will encode the principal implications of the vanishing condition (89). In the proceedings, Lemma 13 will play the same rôle as Lemma 4 did in the proof of Theorem 1. Since the oscillatory condition (9) is now available in the sharper form (90), Lemma 4 itself may be refined. Recall the definition of $\lambda(n) = \lambda_\sigma(n)$ in (65), which is a multiplicative function that vanishes unless n is square-free. Also, recall the definition of the function $\upsilon(n)$ in (22).

Lemma 12. *Fix a parameter set. Then uniformly for $h \in \mathscr{H}$, $\rho \in \mathscr{F}_h^*$, $k \in \mathbb{N}$, $1 \le \sigma \le 2$ and $x \ge 1$, one has*

$$\sum_{\substack{n \le x \\ (n,k)=1}} \lambda_\sigma(n) \ll \upsilon(k) x (\log 3x)^{-1/\varepsilon}.$$

Proof. This follows by rewriting the proofs of Lemmas 3 and 4, with the stronger hypothesis (90) in place of (9).

For primes p, a crude consequence of (7) is that $|\lambda(p)| \le 2C$, and consequently, for any natural number k, the series

$$\Lambda_k(\sigma) = \sum_{\substack{n=1 \\ (n,k)=1}}^{\infty} \lambda_\sigma(n) n^{-\sigma} \tag{92}$$

converges absolutely for $\sigma > 1$. Convergence for $\sigma = 1$ is implied by Lemma 12, but this is not needed later.

Lemma 13. *Fix a parameter set, and suppose that h and ρ satisfy the hypotheses of Theorem 2. Then uniformly in $\sigma \in (1,2]$ and $k \in \mathbb{N}$, one has*

$$\Lambda_k(\sigma) \ll \upsilon(k)\sqrt{\sigma - 1}.$$

Proof. As in the proof of Lemma 4, let K denote the product of all primes p with $p \le C^2$. The series defining $\Lambda_k(\sigma)$ expands as an Euler product. Brute force estimation of the Euler factors for the finitely many $p \le C^2$ then shows that

$$\Lambda_k(\sigma) \ll \left| \prod_{p \nmid kK} (1 + \lambda(p) p^{-\sigma}) \right|.$$

For primes p with $p > C^2$, one deduces from (7), (23) and (65) that

$$\begin{aligned} 1 + \lambda(p)p^{-\sigma} &= 1 + \rho(p)h(p)p^{-\sigma}(1 + \rho(p)p^{-\sigma})^{-1} \\ &= (1 + \rho(p)h(p)p^{-\sigma})(1 + f(p,\sigma)), \end{aligned} \tag{93}$$

where

$$f(p,\sigma) = \frac{\rho(p)^2 h(p) p^{-2\sigma}}{(1+\rho(p)h(p)p^{-\sigma})(1+\rho(p)p^{-\sigma})}.$$

Note that by (7), the denominator here is non-zero, and one finds that

$$|f(p,\sigma)| \leq C^2 p^{-2}\Big(1-\frac{1}{p}\Big)^{-1},$$

so that, throughout the interval $1<\sigma\leq 2$, one has the uniform estimate

$$\prod_{p\nmid kK} |1+f(p,\sigma)| \ll 1.$$

Consequently, by (93),

$$\Lambda_k(\sigma) \ll \Big|\prod_{p\nmid kK}(1+\rho(p)h(p)p^{-\sigma})\Big|.$$

The Euler factors here are those of the series in (89). The postulated bound for this series now yields

$$\Lambda_k(\sigma) \ll \sqrt{\sigma-1}\prod_{p|kK}|1+\rho(p)h(p)p^{-\sigma}|^{-1}.$$

For primes $p>C^2$ with $p|k$, recall that (7) yields $\rho(p)|h(p)|p^{-\sigma}\leq Cp^{-1}\leq p^{-1/2}$. Hence, by (22),

$$\Lambda_k(\sigma) \ll \upsilon(k)\sqrt{\sigma-1}\prod_{p\leq C^2}|1+\rho(p)h(p)p^{-\sigma}|^{-1}.$$

By (91), one has $|\rho(p)h(p)p^{-1}|\leq 1-\delta$. It follows that the product on the far right of the previous display is bounded above in terms of the parameter set, uniformly for $\sigma\geq 1$. This establishes the lemma.

2.3 Preparatory Work

This section prepares the ground for the application of a variant of the moment method. Appropriate analogues of Lemmas 7–9 will be developed.

Moments of $\mathrm{B}_h(n;u)$ have to be defined with some care. The sum (86) is empty for $u<0$, whence $\mathrm{B}_h(n;u)=0$ in this case. Moreover, when $u\geq \log n$, one has $\mathrm{B}_h(n;u)=g(n)$, where

$$g(n)=\sum_{d|n}h(d) \tag{94}$$

is the complete convolution associated with $h \in \mathscr{H}$. Thus, the maximum in (87) occurs for some $u \in [0, \log n]$. This suggests to study the moments

$$\mathsf{M}_{h;q}(n) = \int_0^{\log n} |\mathrm{B}_h(n;u)|^q \, \mathrm{d}u \tag{95}$$

by the techniques described in the first chapter. The following inequality bounds $\mathrm{B}_h(n)$ in terms of these moments. It will substitute Lemma 7.

Lemma 14. *Let q, n be natural numbers with $n \geq 3$. Then for any $h \in \mathscr{H}$, one has*

$$\mathrm{B}_h(n) \leq \mathsf{M}_{h;q}(n)^{1/q} + \Delta_h(n).$$

Proof. Let $u \in [0, \log n]$, and choose an interval $I \subset [0, \log n]$ of length 1, with $u \in I$. Then for any $u' \in I$, one has $|u - u'| \leq 1$, whence by (86) and (6), it follows that

$$|\mathrm{B}_h(n;u)| \leq |\mathrm{B}_h(n;u')| + \Delta_h(n).$$

This may be integrated over $u' \in I$ to conclude that

$$|\mathrm{B}_h(n;u)| \leq \int_I |\mathrm{B}_h(n;u')| \, \mathrm{d}u' + \Delta_h(n),$$

and Hölder's inequality yields

$$|\mathrm{B}_h(n;u)| \leq \Big(\int_I |\mathrm{B}_h(n;u')|^q \, \mathrm{d}u' \Big)^{1/q} + \Delta_h(n).$$

On the right-hand side, one may extend the integration to $[0, \log n]$. The lemma now follows from (95).

The next goal is an analogue of Lemma 9. The trivial bound $|\mathrm{B}_h(n;u)| \leq \mathrm{d}(n)$ and (95) show that

$$\mathsf{M}_{h;2}(n) \leq \mathrm{d}(n)^2 \log n. \tag{96}$$

Hence, the Dirichlet series

$$\mathsf{Z}(\sigma) = \sum_{n=1}^{\infty} \mu(n)^2 \rho(n) \mathsf{M}_{h;2}(n) n^{-\sigma} \tag{97}$$

converges for $\sigma > 1$.

Lemma 15. *Subject to the hypotheses of Theorem 2, in the range $1 < \sigma \leq 2$ one has*

$$\mathsf{Z}(\sigma) \ll (\sigma - 1)^{-2c}.$$

Proof. Let $\tau \geq 1$. Then recalling the definition of d^+ in (52), a straightforward integration of (86) yields

$$\int_0^{\log\tau} |\mathrm{B}_h(n;u)|^2 \,\mathrm{d}u = \sum_{\substack{d_1|n, d_2|n \\ d^+ \leq \tau}} h(d_1)\overline{h(d_2)} \log\frac{\tau}{d^+}. \tag{98}$$

The admissible choice $\tau = n$, via (94) and (95), produces the identity

$$\mathsf{M}_{h;2}(n) = \sum_{d_1|n, d_2|n} h(d_1)\overline{h(d_2)} \log\frac{n}{d^+} = (\log n)|g(n)|^2 - \sum_{d_1|n, d_2|n} h(d_1)\overline{h(d_2)} \log d^+. \tag{99}$$

This may be inserted into (97). One then finds that

$$\mathsf{Z}(\sigma) = \mathsf{Z}^{(1)}(\sigma) - \mathsf{Z}^{(2)}(\sigma),$$

where

$$\mathsf{Z}^{(1)}(\sigma) = \sum_{n=1}^{\infty} (\log n)\mu(n)^2 \rho(n)|g(n)|^2 n^{-\sigma},$$

$$\mathsf{Z}^{(2)}(\sigma) = \sum_{n=1}^{\infty} \sum_{d_1|n, d_2|n} (\log d^+) h(d_1)\overline{h(d_2)} \mu(n)^2 \rho(n) n^{-\sigma},$$

and (7) implies that both sums converge absolutely for $\sigma > 1$.

The sum $\mathsf{Z}^{(2)}(\sigma)$ will be analysed first. The treatment is initiated by reversing the order of summation. Then, writing $n = [d_1, d_2]m$ in the now inner sum, one infers that

$$\mathsf{Z}^{(2)}(\sigma) = \sum_{d_1, d_2 = 1}^{\infty} h(d_1)\overline{h(d_2)}(\log d^+) \sum_{m=1}^{\infty} \mu(m[d_1,d_2])^2 \rho(m[d_1,d_2]) m^{-\sigma}[d_1,d_2]^{-\sigma}.$$

Non-zero contributions to the sum over m require that $(m, d_1 d_2) = 1$, and that $[d_1, d_2]$ is square-free. The latter is equivalent to the constraint that d_1, d_2 be both square-free. Hence,

$$\mathsf{Z}^{(2)}(\sigma) = \sum_{d_1, d_2 = 1}^{\infty} \mu(d_1)^2 \mu(d_2)^2 \frac{\rho([d_1,d_2]) h(d_1)\overline{h(d_2)}}{[d_1,d_2]^{\sigma}} (\log d^+) \sum_{m=1}^{\infty} \mu(m)^2 \frac{\rho(m)}{m^{\sigma}}.$$

By (62) and (23), the sum over m equals

$$\prod_{p \nmid d_1 d_2} (1 + \rho(p) p^{-\sigma}) = \mathrm{P}(\sigma)\theta_{\sigma}(d_1 d_2; \rho),$$

whence $\mathsf{Z}^{(2)}(\sigma)$ may be rewritten as

$$\mathsf{P}(\sigma)\sum_{d_1,d_2=1}^{\infty}\mu(d_1)^2\mu(d_2)^2\rho([d_1,d_2])h(d_1)\overline{h(d_2)}\theta_\sigma(d_1d_2)(\log d^+)[d_1,d_2]^{-\sigma}.$$

Now write $d=(d_1,d_2)$ and $d_j=de_j$. Since d_1, d_2 are square-free, the numbers d, e_1, e_2 are square-free and coprime in pairs. Recalling (65), this shows that

$$\mathsf{Z}^{(2)}(\sigma)=\mathsf{P}(\sigma)\sum_{d=1}^{\infty}\sum_{\substack{e_1,e_2=1\\(e_1,e_2)=1\\(d,e_1e_2)=1}}^{\infty}\frac{\mu(d)^2\rho(d)|h(d)|^2\theta_\sigma(d^2)\lambda(e_1)\overline{\lambda(e_2)}}{(de_1e_2)^\sigma}\log de^+,$$

where $e^+=\max(e_1,e_2)$. By (23), $\theta_\sigma(d^2)=\theta_\sigma(d)$. Writing again d_j in place of e_j, the previous formula now recasts as

$$\mathsf{Z}^{(2)}(\sigma)=\mathsf{P}(\sigma)\sum_{d=1}^{\infty}\mu(d)^2\rho(d)|h(d)|^2\theta_\sigma(d)d^{-\sigma}\big((\log d)U_d(\sigma)+V_d(\sigma)\big),\tag{100}$$

in which

$$U_d(\sigma)=\sum_{\substack{d_1,d_2=1\\(d_1,d_2)=1\\(d,d_1d_2)=1}}^{\infty}\frac{\lambda(d_1)\overline{\lambda(d_2)}}{d_1^\sigma d_2^\sigma},\quad V_d(\sigma)=\sum_{\substack{d_1,d_2=1\\(d_1,d_2)=1\\(d,d_1d_2)=1}}^{\infty}\frac{\lambda(d_1)\overline{\lambda(d_2)}}{d_1^\sigma d_2^\sigma}\log d^+.$$

The condition that $(d_1,d_2)=1$ is removed by Möbius inversion to the effect that

$$U_d(\sigma)=\sum_{k=1}^{\infty}\mu(k)\sum_{\substack{d_1,d_2=1\\(d_1d_2,d)=1\\k|d_1,k|d_2}}^{\infty}\frac{\lambda(d_1)\overline{\lambda(d_2)}}{d_1^\sigma d_2^\sigma},$$

and then, on writing $d_j=kl_j$, the sums transforms further to

$$U_d(\sigma)=\sum_{\substack{k=1\\(k,d)=1}}^{\infty}\frac{\mu(k)}{k^{2\sigma}}\sum_{\substack{l_1,l_2=1\\(l_1l_2,d)=1}}^{\infty}\frac{\lambda(kl_1)\overline{\lambda(kl_2)}}{l_1^\sigma l_2^\sigma}=\sum_{\substack{k=1\\(k,d)=1}}^{\infty}\frac{\mu(k)}{k^{2\sigma}}\Big|\sum_{\substack{l=1\\(l,d)=1}}^{\infty}\frac{\lambda(kl)}{l^\sigma}\Big|^2.$$

However, $\lambda(kl)=0$ unless $(k,l)=1$, a condition that may now be added to the sum over l. But then, by (92),

$$U_d(\sigma)=\sum_{\substack{k=1\\(k,d)=1}}^{\infty}\mu(k)|\lambda(k)|^2k^{-2\sigma}|\Lambda_{kd}(\sigma)|^2.$$

By Lemma 13 and routine estimates, one finally deduces that for $\sigma > 1$ one has

$$U_d(\sigma) \ll (\sigma - 1)\upsilon(d)^2 \sum_{\substack{k=1\\(k,d)=1}}^{\infty} \upsilon(k)^2|\lambda(k)|^2 k^{-2} \ll (\sigma - 1)\upsilon(d)^2. \tag{101}$$

The treatment of the sum $V_d(\sigma)$ is similar, but the presence of the factor $\log d^+$ causes extra difficulties. The principal step is to separate the variables of summation. First note that diagonal terms, with $d_1 = d_2$, do not contribute to $V_d(\sigma)$ because, in this case, the condition that $(d_1,d_2) = 1$ implies $d_1 = d_2 = 1$, where $\log d^+ = 0$. Combining conjugate pairs then shows that

$$V_d(\sigma) = 2\mathrm{Re} \sum_{\substack{d_1,d_2=1\\(d_1d_2,d)=(d_1,d_2)=1\\d_1<d_2}}^{\infty} \frac{\lambda(d_1)\overline{\lambda(d_2)}}{d_1^{\sigma}d_2^{\sigma}} \log d_2.$$

As in the initial transformation of $U_d(\sigma)$, the coprimality condition $(d_1,d_2) = 1$ may be removed by Möbius inversion. Then again following the argument used for $U_d(\sigma)$ *mutatis mutandis*, the change of variables $d_j = kl_j$ produces the formula

$$\begin{aligned} V_d(\sigma) &= 2\mathrm{Re} \sum_{\substack{k=1\\(k,d)=1}}^{\infty} \mu(k)\frac{|\lambda(k)|^2}{k^{2\sigma}} \sum_{\substack{l_1<l_2\\(l_1l_2,kd)=1}} \frac{\lambda(l_1)\overline{\lambda(l_2)}}{l_1^{\sigma}l_2^{\sigma}} \log kl_2 \\ &= 2\mathrm{Re}\big(V_d^{(1)}(\sigma) + V_d^{(2)}(\sigma)\big), \end{aligned}$$

where

$$V_d^{(1)}(\sigma) = \sum_{\substack{k=1\\(k,d)=1}}^{\infty} \mu(k)(\log k)\frac{|\lambda(k)|^2}{k^{2\sigma}} \sum_{\substack{l_1<l_2\\(l_1l_2,kd)=1}} \frac{\lambda(l_1)\overline{\lambda(l_2)}}{l_1^{\sigma}l_2^{\sigma}},$$

$$V_d^{(2)}(\sigma) = \sum_{\substack{k=1\\(k,d)=1}}^{\infty} \mu(k)\frac{|\lambda(k)|^2}{k^{2\sigma}} \sum_{\substack{l_1<l_2\\(l_1l_2,kd)=1}} \frac{\lambda(l_1)\overline{\lambda(l_2)}}{l_1^{\sigma}l_2^{\sigma}} \log l_2.$$

By symmetry and (92), the first of the two sums reassembles to

$$\begin{aligned} 2\mathrm{Re}V_d^{(1)}(\sigma) &= \sum_{\substack{k=1\\(k,d)=1}}^{\infty} \mu(k)(\log k)\frac{|\lambda(k)|^2}{k^{2\sigma}} \sum_{\substack{l_1\neq l_2\\(l_1l_2,kd)=1}} \frac{\lambda(l_1)\overline{\lambda(l_2)}}{l_1^{\sigma}l_2^{\sigma}} \\ &= \sum_{\substack{k=1\\(k,d)=1}}^{\infty} \mu(k)(\log k)\frac{|\lambda(k)|^2}{k^{2\sigma}} \Bigg(|\Lambda_{kd}(\sigma)|^2 - \sum_{\substack{l=1\\(l,kd)=1}}^{\infty} \frac{|\lambda(l)|^2}{l^{2\sigma}}\Bigg). \end{aligned}$$

By Lemma 13 and straightforward estimates, it now follows that

$$2\mathrm{Re}\,V_d^{(1)}(\sigma) \ll \upsilon(d)^2.$$

In the sum defining $V_d^{(2)}(\sigma)$, it will be convenient to replace the condition $l_1 < l_2$ by $l_1 \le l_2$. This restores the diagonal terms $l_1 = l_2$ at the cost of an error

$$\sum_{\substack{k=1\\(k,d)=1}}^{\infty} \mu(k)\frac{|\lambda(k)|^2}{k^{2\sigma}} \sum_{\substack{l=1\\(l,kd)=1}}^{\infty} \frac{|\lambda(l)|^2}{l^{2\sigma}}\log l \ll 1,$$

as one readily checks. The summation over l_1 is performed first, starting from

$$V_d^{(2)}(\sigma) = \sum_{\substack{k=1\\(k,d)=1}}^{\infty} \mu(k)\frac{|\lambda(k)|^2}{k^{2\sigma}} \sum_{\substack{l_2=1\\(l_2,kd)=1}}^{\infty} \frac{\overline{\lambda(l_2)}}{l_2^{\sigma}}(\log l_2) \sum_{\substack{l_1\le l_2\\(l_1,kd)=1}} \frac{\lambda(l_1)}{l_1^{\sigma}} + O(1).$$

By partial summation,

$$V_d^{(2)}(\sigma) = V_d^{(3)}(\sigma) + \sigma V_d^{(4)}(\sigma) + O(1),$$

where

$$V_d^{(3)}(\sigma) = \sum_{\substack{k=1\\(k,d)=1}}^{\infty} \mu(k)\frac{|\lambda(k)|^2}{k^{2\sigma}} \sum_{\substack{l_2=1\\(l_2,kd)=1}}^{\infty} \frac{\overline{\lambda(l_2)}}{l_2^{2\sigma}}(\log l_2) \sum_{\substack{l_1\le l_2\\(l_1,kd)=1}} \lambda(l_1),$$

$$V_d^{(4)}(\sigma) = \sum_{\substack{k=1\\(k,d)=1}}^{\infty} \mu(k)\frac{|\lambda(k)|^2}{k^{2\sigma}} \sum_{\substack{l_2=1\\(l_2,kd)=1}}^{\infty} \frac{\overline{\lambda(l_2)}}{l_2^{\sigma}}(\log l_2) \int_1^{l_2} w^{-\sigma-1} \sum_{\substack{l_1\le w\\(l_1,kd)=1}} \lambda(l_1)\,\mathrm{d}w.$$

By Lemma 12, the innermost sum in $V_d^{(3)}$ contributes $O(\upsilon(kd)l_2(\log 3l_2)^{-1/\varepsilon})$, so that a crude upper bound is

$$\begin{aligned} V^{(3)}(\sigma) &\ll \upsilon(d) \sum_{\substack{k=1\\(k,d)=1}}^{\infty} \frac{|\lambda(k)|^2\upsilon(k)}{k^{2\sigma}} \sum_{\substack{l_2=1\\(l_2,kd)=1}}^{\infty} \frac{|\lambda(l_2)|}{l_2^{2\sigma-1}}(\log 3l_2)^{-1/\varepsilon} \\ &\ll \upsilon(d)\sum_{k=1}^{\infty} \frac{|\lambda(k)|^2\upsilon(k)}{k^2} \sum_{l_2=1}^{\infty} \frac{|\lambda(l_2)|}{l_2}(\log 3l_2)^{-1/\varepsilon}. \end{aligned}$$

The simple bound

$$\sum_{l\le x} \frac{|\lambda(l)|}{l} \le \prod_{p\le x}\left(1+\frac{|\lambda(p)|}{p}\right) \ll (\log x)^{2C}$$

combines with partial summation to show that the sum over l_2 in the previous display converges for $\varepsilon < (9C)^{-1}$, and one infers the bound

$$V_d^{(3)}(\sigma) \ll \upsilon(d).$$

A similar argument supplies a bound for $V^{(4)}(\sigma)$. Exchanging the integration with the sum over l_2 yields the formula

$$V_d^{(4)}(\sigma) = \sum_{\substack{k=1 \\ (k,d)=1}}^{\infty} \mu(k) \frac{|\lambda(k)|^2}{k^{2\sigma}} J_{kd}(\sigma) \tag{102}$$

with

$$J_q(\sigma) = \int_1^\infty w^{-\sigma-1} \Big(\sum_{\substack{l_1 \le w \\ (l_1,q)=1}} \lambda(l_1) \Big) \Big(\sum_{\substack{l_2 \ge w \\ (l_2,q)=1}} \frac{\overline{\lambda(l_2)}}{l_2^\sigma} \log l_2 \Big) \, dw.$$

Here, the summations over l_1 and l_2 are separated. By Lemma 12, the sum over l_1 is $O(\upsilon(q) w (\log 3w)^{-1/\varepsilon})$. Similarly, by partial summation and Lemma 12, a routine estimation shows that the sum over l_2 is $O(\upsilon(q))$, uniformly in $\sigma > 1$. Hence,

$$J_q(\sigma) \ll \upsilon(q)^2 \int_1^\infty w^{-\sigma} (\log 3w)^{-5} \, dw \ll \upsilon(q)^2,$$

and then, by (102) and straightforward estimates, one deduces that $V_d^{(4)}(\sigma) \ll \upsilon(d)^2$. On collecting together the bounds obtained for $V_d^{(l)}(\sigma)$, it follows that

$$V_d(\sigma) \ll \upsilon(d)^2.$$

This final bound may now be combined with (101) and (100) to complete the estimation of $\mathrm{Z}^{(2)}(\sigma)$. On recalling that $|h(d)|^2 \theta_\sigma(d) \le 1$, one first obtains the intermediate bound

$$\mathrm{Z}^{(2)}(\sigma) \ll \mathrm{P}(\sigma) \sum_{d=1}^{\infty} \mu(d)^2 \upsilon(d)^2 d^{-\sigma} \big((\log d)(\sigma - 1) + 1 \big). \tag{103}$$

The next steps are very similar to the argument used in the proof of Lemma 9. One considers the Dirichlet series

$$\mathrm{H}^\dagger(s) = \sum_{d=1}^{\infty} \mu(d)^2 \rho(d) \upsilon(d)^2 d^{-s}$$

as a function of the complex variable s. It has an Euler product in $\mathrm{Re}\, s > 1$, and differs from the product $\mathrm{H}(\sigma)$ defined in (66) only in that $\upsilon(d)$ is replaced by $\upsilon(d)^2$. Thus, a brief inspection of the argument used to bound $\mathrm{H}(\sigma)$ shows that an analogous estimation fully covers the current situation, and one finds that for real $\sigma \in (1,2]$ one has $\mathrm{H}^\dagger(\sigma) \ll (\sigma-1)^{-c}$. However, $\mathrm{H}^\dagger(s)$ is a Dirichlet series with real, non-negative coefficients, so that whenever $\sigma = \mathrm{Re}\, s > 1$ one has $|\mathrm{H}^\dagger(s)| \le \mathrm{H}^\dagger(\sigma)$. Consequently, observing that

$$-\frac{\mathrm{d}}{\mathrm{d}\sigma}\mathrm{H}^\dagger(\sigma) = \sum_{d=1}^{\infty} \mu(d)^2 \rho(d) \upsilon(d)^2 d^{-\sigma} \log d,$$

it follows from Cauchy's integral formula that

$$-\frac{\mathrm{d}}{\mathrm{d}\sigma}\mathrm{H}^\dagger(\sigma) = \frac{1}{4\pi^2} \int_{|w-\sigma|=\frac{1}{2}(\sigma-1)} \frac{\mathrm{H}^\dagger(w)}{(w-\sigma)^2}\,\mathrm{d}w \ll (\sigma-1)^{-1-c}.$$

With these bounds at hand, one deduces from (103) and (63) that

$$\mathsf{Z}^{(2)}(\sigma) \ll \mathrm{P}(\sigma)\big(\mathrm{H}^\dagger(\sigma) + (\sigma-1)(\mathrm{H}^\dagger)'(\sigma)\big) \ll (\sigma-1)^{-2c},$$

as required.

The estimation of $\mathsf{Z}^{(1)}(\sigma)$ is very similar. One considers the series

$$\mathsf{Z}^{(0)}(s) = \sum_{n=1}^{\infty} \mu(n)^2 \rho(n) |g(n)|^2 n^{-s},$$

and observes that

$$\mathsf{Z}^{(1)}(\sigma) = -\frac{\mathrm{d}}{\mathrm{d}\sigma}\mathsf{Z}^{(0)}(\sigma).$$

Since again $\mathsf{Z}^{(0)}(s)$ is defined by a Dirichlet series with real, non-zero coefficients, one may proceed as in the immediately preceding paragraph, and use Cauchy's formula to deduce the desired estimate

$$\mathsf{Z}^{(1)}(\sigma) \ll (\sigma-1)^{-2c}$$

from the bound

$$\mathsf{Z}^{(0)}(\sigma) \ll (\sigma-1)^{-1-2c}, \tag{104}$$

and the latter is not hard to prove. Indeed, by (94),

$$\mathsf{Z}^{(0)}(\sigma) = \sum_{n=1}^{\infty} \mu(n)^2 \rho(n) \sum_{d_1|n, d_2|n} h(d_1)\overline{h(d_2)} n^{-\sigma}.$$

The initial segment of the treatment of $\mathsf{Z}^{(2)}(\sigma)$ may now be copied, leading to an analogue of (100) that now reads

$$\mathsf{Z}^{(0)}(\sigma) = \mathrm{P}(\sigma) \sum_{d=1}^{\infty} \mu(d)^2 \rho(d) |h(d)|^2 \theta_\sigma(d) |U_d(\sigma)|^2.$$

One applies (101) and the bound $\theta_\sigma(d)|h(d)|^2 \le 1$ to deduce that

$$\mathsf{Z}^{(0)}(\sigma) = (\sigma-1)\mathrm{P}(\sigma)\mathrm{H}^\dagger(\sigma),$$

which implies the desired (104) by the same argument as in the last line of the preceeding paragraph. The proof of Lemma 15 is complete.

For the final technical lemma in this section, let $0 \le j \le q$ and p be a prime. Then consider the convoluted moment

$$\mathsf{N}_{h;q}(n;j,p) = \int_0^{\log n} |\mathrm{B}_h(n;u)|^{q-j} |\mathrm{B}_h(n,u-\log p)|^j \, du. \tag{105}$$

The analogue of (50) reads

$$\mathsf{N}_{h;q}(n;j,p) \le \mathsf{M}_{h;q}(n), \tag{106}$$

and it may be demonstrated by the same argument that was used to prove (50). The following lemma serves as a replacement for Lemma 8.

Lemma 16. *Uniformly in $q \ge 5$, $2 \le j \le q-2$, and $h \in \mathscr{H}$, one has*

$$\sum_p \frac{\log p}{p} \mathsf{N}_{h;q}(n;j,p)$$
$$\ll \left(\mathsf{M}_{h;2}(n) + \mathrm{B}_h(n)^2 (\log\log 9n)^2\right) \left(\mathrm{d}(n)^2 \log n\right)^{2/(q-2)} \mathsf{M}_{h;q}(n)^{(q-4)/(q-2)}.$$

Proof. As was the case with Lemma 8, it suffices to appeal to Hölder's inequality and establish Lemma 8 in the cases $j=2$ and $j=q-2$.

First consider $j=2$. For $p > \mathrm{e}^u$, one has $\mathrm{B}_h(n;u-\log p) = 0$. Hence, the sum over p in

$$\mathsf{E}_h(n;u) = \sum_p \frac{\log p}{p} |\mathrm{B}_h(n;u-\log p)|^2$$

is over a finite range, and one has

$$\mathsf{E}_h(n;u) = \sum_p \frac{\log p}{p} \Big| \sum_{\substack{d|n \\ \log pd \le u}} h(d) \Big|^2 = \sum_{d_1|n, d_2|n} h(d_1)\overline{h(d_2)} \sum_{p \le \mathrm{e}^u/d^+} \frac{\log p}{p}.$$

Now choose κ in accordance with (54), and let $\gamma = \gamma(n)$ be defined by (55). Consider the four mutually exclusive conditions

$$\text{(i): } d^+ \le \gamma \mathrm{e}^u \quad \text{(ii): } d_1 \le \gamma \mathrm{e}^u < d_2 \quad \text{(iii): } d_2 \le \gamma \mathrm{e}^u < d_1 \quad \text{(iv): } d_1, d_2 > \gamma \mathrm{e}^u,$$

and let $\mathsf{E}_h^{(i)}, \dots, \mathsf{E}_h^{(iv)}$ be the subsums of E_h with the corresponding condition on d_1, d_2 added to the sum over d_1, d_2. Then

$$\mathsf{E}_h(n;u) = \mathsf{E}_h^{(i)} + \cdots + \mathsf{E}_h^{(iv)}.$$

It may be helpful to observe that

$$\exp(-\kappa(\log\gamma^{-1})^{1/2}) \ll (\log 9n)^{-3},$$

so that, by (54),

$$\mathsf{E}_h^{(i)} = \sum_{\substack{d_1|n, d_2|n \\ d^+ \le \gamma e^u}} h(d_1)\overline{h(d_2)}\Big(\log\frac{e^u}{d^+} + E + O\big((\log 9n)^{-3}\big)\Big).$$

On comparing with (98), and using trivial estimates, this expression becomes

$$\mathsf{E}_h^{(i)} = \int_0^{\gamma e^u} |\mathrm{B}_h(n;u)|^2 \,\mathrm{d}u + (E + \log\gamma^{-1}) \sum_{\substack{d_1|n, d_2|n \\ d^+ \le \gamma e^u}} h(d_1)\overline{h(d_2)} + O\left(\frac{\mathrm{d}(n)^2}{(\log 9n)^3}\right),$$

and one infers the bound

$$\mathsf{E}_h^{(i)} \ll \mathsf{M}_{h;2}(n) + (\log\log 9n)^2 \mathrm{B}_h(n)^2 + \frac{\mathrm{d}(n)^2}{(\log 9n)^3}.$$

The treatment of $\mathsf{E}_h^{(ii)}$ is more direct. By definition,

$$\mathsf{E}_h^{(ii)} = \sum_{\substack{d_1|n \\ d_1 \le \gamma e^u}} h(d_1) \sum_{\substack{d_2|n \\ \gamma e^u < d_2 \le e^u}} \overline{h(d_2)} \sum_{p \le e^u/d_2} \frac{\log p}{p},$$

so that one obtains

$$\mathsf{E}_h^{(ii)} \le \mathrm{B}_h(n) \sum_{p \le \gamma^{-1}} \frac{\log p}{p} \Big| \sum_{\substack{d_2|n \\ \gamma e^u < d_2 \le e^u/p}} \overline{h(d_2)} \Big| \le 2\mathrm{B}_h(n)^2 \sum_{p \le \gamma^{-1}} \frac{\log p}{p},$$

which in turn implies

$$\mathsf{E}_h^{(ii)} \ll \mathrm{B}_h(n)^2 (\log\log 9n)^2.$$

By symmetry, the same bound holds for $\mathsf{E}_h^{(iii)}$. Also, a very similar procedure gives

$$\mathsf{E}_h^{(iv)} \le \sum_{p \le \gamma^{-1}} \frac{\log p}{p} \Big| \sum_{\substack{d_1|n, d_2|n \\ \gamma e^u < d_i \le e^u/p}} h(d_1)\overline{h(d_2)} \Big|,$$

and it is now transparent that $\mathsf{E}_h^{(iv)}$ also obeys the bound obtained for $\mathsf{E}_h^{(ii)}$. On collecting together, this shows that

$$\mathsf{E}_h(n;u) \ll \mathsf{M}_{h;2}(n) + (\log\log 9n)^2 \mathrm{B}_h(n)^2 + \frac{\mathrm{d}(n)^2}{(\log 9n)^3}$$

holds uniformly in u.

With this preparatory inequality at hand, the main argument begins with (105), showing that

$$\sum_p \frac{\log p}{p} \mathsf{N}_{h;q}(n;2,p) = \int_0^{\log n} |\mathrm{B}_h(n,u)|^{q-2} \mathsf{E}_h(n;u)\,\mathrm{d}u.$$

The bound for $\mathsf{E}_h(n;u)$ yields

$$\sum_p \frac{\log p}{p} \mathsf{N}_{h;q}(n;2,p) \ll \mathsf{M}_{h;q-2}\left(\mathsf{M}_{h;2}(n) + (\log\log 9n)^2 \mathrm{B}_h(n)^2 + \frac{\mathrm{d}(n)^2}{(\log 9n)^3}\right),$$

and by Hölder's inequality,

$$\mathsf{M}_{h,q-2}(n) \le \mathsf{M}_{h;2}^{2/(q-2)}(n) \mathsf{M}_{h;q}^{(q-4)/(q-2}(n).$$

The case $j = 2$ of the lemma now follows from (96).

It remains to consider the case $j = q-2$. In contrast to the problem considered in Lemma 8, a treatment rather different form the above is now required, calling for a detailed account. The obvious change of variable in (105) gives

$$\mathsf{N}_{h;q}(n;q-2,p) = \int_0^{\log(n/p)} |\mathrm{B}_h(n;u+\log p)|^2 |\mathrm{B}_h(n;u)|^{q-2}\,\mathrm{d}u.$$

Thus, the point of departure now is the formula

$$\sum_p \frac{\log p}{p} \mathsf{N}_{h;q}(n;q-2,p) = \int_0^{\log \frac{1}{2}n} \mathrm{B}_h(n;u)^{q-2} \mathsf{D}_h(n;u)\,\mathrm{d}u, \tag{107}$$

in which

$$\mathsf{D}_h(n;u) = \sum_{p \le n\mathrm{e}^{-u}} \frac{\log p}{p} |\mathrm{B}_h(n;u+\log p)|^2.$$

By (87), one has

$$\mathsf{D}_h(n;u) = \sum_{d_1|n, d_2|n} h(d_1)\overline{h(d_2)} \sum_{p \le n\mathrm{e}^{-u}} \frac{\log p}{p}.$$

Let D_h' be the subsum of $\mathsf{D}_h(n,u)$ where $d^+\mathrm{e}^{-u} \geq \gamma^{-1}$. If $\mathrm{e}^u > n\gamma$, then this condition demands that $d^+ > n$, which is impossible. Hence, $\mathsf{D}_h' = 0$ in this case. Therefore, we now suppose that

$$\mathrm{e}^u \leq n\gamma, \tag{108}$$

in which case (54) yields

$$\sum_{\mathrm{e}^{-u}d^+ < p \leq \mathrm{e}^{-u}n} \frac{\log p}{p} = \log \frac{n}{d^+} + O\big((\log 9n)^{-3}\big),$$

and a now familiar argument produces the formula

$$\mathsf{D}_h' = \sum_{\substack{d_1 | n, d_2 | n \\ d^+ > \mathrm{e}^u/\gamma}} h(d_1)\overline{h(d_2)} \log \frac{n}{d^+} + O\left(\frac{\mathrm{d}(n)^2}{(\log 9n)^3}\right).$$

One may flip the condition that $d^+ > \mathrm{e}^u/\gamma$ into the complementary one, observing the identity (99). This gives

$$\mathsf{D}_h' = -\mathsf{M}_{h;2}(n) + \sum_{\substack{d_1 | n, d_2 | n \\ d^+ \leq \mathrm{e}^u/\gamma}} h(d_1)\overline{h(d_2)} \log \frac{d^+}{n} + O\left(\frac{\mathrm{d}(n)^2}{(\log 9n)^3}\right).$$

The obvious formula

$$\log \frac{d}{n} = \log \frac{d^+\gamma}{\mathrm{e}^u} - \log \frac{n\gamma}{\mathrm{e}^u}$$

and (98) again, with $\tau = \mathrm{e}^u/\gamma$, provide

$$D_h' = -\log \frac{n\gamma}{\mathrm{e}^u} \sum_{\substack{d_1 | n, d_2 | n \\ d^+ \leq \mathrm{e}^u/\gamma}} h(d_1)\overline{h(d_2)} + O\left(\frac{\mathrm{d}(n)^2}{(\log 9n)^3}\right).$$

Now consider the remaining portion of the sum $\mathsf{D}_h(n;u)$, where $d^+ \leq \mathrm{e}^u/\gamma$, and denote this subsum by D_h'', so that $\mathsf{D}_h(n;u) = \mathsf{D}_h' + \mathsf{D}_h''$. In the sum defining D_h'', one may restore the terms with $p \leq \mathrm{e}^{-u}d^+$ in the sum over p, at the cost of an error

$$\sum_{p \leq \gamma^{-1}} \frac{\log p}{p} \Bigg| \sum_{\substack{d_1 | n, d_2 | n \\ d^+ \leq \mathrm{e}^u/\gamma}} h(d_1)\overline{h(d_2)} \Bigg| \ll \mathrm{B}_h(n)^2 (\log\log 9n)^2.$$

This shows that

$$\mathsf{D}_h'' = \sum_{\substack{d_1 | n, d_2 | n \\ d^+ \leq \mathrm{e}^u/\gamma}} h(d_1)\overline{h(d_2)} \sum_{p \leq \mathrm{e}^{-u}d^+} \frac{\log p}{p} + O\big(\mathrm{B}_h(n)^2 (\log\log 9n)^2\big). \tag{109}$$

The sum over p may be evaluated by (54). In view of the currently active assumption (108), it follows that

$$\mathsf{D}_h'' = \left(\log \frac{n}{\mathrm{e}^u}\right) \sum_{\substack{d_1|n, d_2|n \\ d^+ \leq \mathrm{e}^u/\gamma}} h(d_1)\overline{h(d_2)} + O\left(\mathrm{B}_h(n)^2(\log\log 9n)^2 + \frac{\mathrm{d}(n)^2}{(\log 9n)^3}\right).$$

Now sum D_h' and D_h''. The terms that were kept explicit largely cancel out, leaving the contribution

$$(\log \gamma^{-1}) \sum_{\substack{d_1|n, d_2|n \\ d^+ \leq \mathrm{e}^u/\gamma}} h(d_1)\overline{h(d_2)} \ll \mathrm{B}_h(n)^2(\log\log 9n)^2,$$

and one finds that

$$\mathsf{D}_h(n;u) \ll \mathrm{B}_h(n)^2(\log\log 9n)^2 + \frac{\mathrm{d}(n)^2}{(\log 9n)^3}. \tag{110}$$

Although this final estimate has been verified so far only for all u satisfying (108), it remains valid for all $u \leq \log n$. In fact, it has been noted earlier that whenever $\mathrm{e}^u > n\gamma$, then $\mathsf{D}_h' = 0$, and one also readily verifies that the innermost sum over p in (109) does not exceed $O(\log \gamma^{-1})$, so that in this case one has $\mathsf{D}_h'' \ll \mathrm{B}_h(n)^2(\log\log 9n)^2$, confirming (110) for the previously missing range of u. If one now uses (110) in (107) and then proceeds as in the final steps of the estimation in the case $j = 2$, then one arrives at the claim of Lemma 16 in the case $j = q-2$. This completes the proof.

2.4 A Roadmap for the Way Home

The principal difficulties in the proof of Theorem 2 have now been overcome, and in the remaining steps, one may follow the pattern of the proof of Theorem 1 very closely. Therefore, only very brief commentary is offered beyond this point, leaving most of the details to the reader. The endgame begins with an analogue of (68). One now studies the Dirichlet series

$$\mathsf{L}_h(\sigma) = \mathsf{L}_h(\sigma; \rho, q, t) = \sum_{n=1}^{\infty} \mu(n)^2 \rho(n) \mathsf{M}_{h;q}(n)^{t/q} n^{-\sigma},$$

which converges absolutely for $\sigma > 1$, as is evident from (96). Recalling (106), the obvious analogon of (69) for $-\mathsf{L}_h'(\sigma)$ remains valid. From (86), one finds that for $p \nmid n$, one has

$$\mathrm{B}_h(pn;u) = h(p)\mathrm{B}_h(n;u - \log p) + \mathrm{B}_h(n;u),$$

whence

$$|\mathrm{B}_h(pn;u)| \leq |\mathrm{B}_h(n;u-\log p)| + |\mathrm{B}_h(n;u)|.$$

A version of (70), with M, N in place of M, N now follows in the same way as (70) was proved, and one then derives an analogue of (71), applying Lemma 16 rather than Lemma 8. The definition of Γ in (72) needs appropriate adjustment, and in (71), one should now read $\log n$ for the factor v that occurs together with $\mathrm{d}(n)^2$, and in the following argument, this factor may be incorporated in the adjusted version of J_h. This logarithm presents little difficulty, and one readily confirms an appropriate analogon of Lemma 10, which may then be fed into the proof of Lemma 11. Thus, one now estimates the sum

$$\sum_{n\leq x} \frac{\mu(n)^2\rho(n)\mathrm{B}_h(n)^t}{n}$$

by first applying Lemma 14. This produces two sums of which one is the sum estimated in Lemma 11, and the other contains the moment $\mathsf{M}_{h;q}(n)^{t/q}$. The latter may be controlled by the bound for $\mathsf{L}_h(\sigma)$ provided by the adjusted version of Lemma 10. One then finds that the claim in Lemma 11 remains valid with $\mathrm{B}_h(n)$ in place of $\Delta_h(n)$. Equipped with this bound, one first observes that for coprime numbers n, m, one has

$$|\mathrm{B}_h(nm;u)| \leq \sum_{d|n} |\mathrm{B}_h(m;u-\log d)|,$$

as is evident from (86). This implies that

$$\mathrm{B}_h(nm) \leq \mathrm{d}(n)\mathrm{B}_h(m),$$

which may replace (83) in the work of Sect. 1.6 to complete the proof of Theorem 2.

References

1. S. Daniel. On the divisor-sum problem for binary forms. *J. Reine Angew. Math.* 507 (1999), 107–129.
2. P. Erdős, I. Kátai. Non-complete sums of multiplicative functions. *Periodica Mathematica Hungarica* 1 (1971), 209–212.
3. R.R. Hall, G. Tenenbaum. Divisors. *Cambridge Tracts in Mathematics*, 90. Cambridge University Press, Cambridge, 1988.
4. C. Hooley. On a new technique and its applications to the theory of numbers. *Proc. London Math. Soc.* (3) 38 (1979), 115–151.
5. H. Maier. On the Möbius function. *Trans. Amer. Math. Soc.* 301 (1987), 649–664.
6. H. Maier, G. Tenenbaum. On the normal concentration of divisors, II. *Math. Proc. Cambridge Philos. Soc.* 147 (2009), 513–540.
7. G. Tenenbaum. Sur la concentration moyenne des diviseurs. *Comment. Math. Helv.* 60 (1985), 411–428.

8. G. Tenenbaum. Fonctions Δ de Hooley et applications. Séminaire de théorie des nombres, Paris 1984–85, 225–239, *Progr. Math.*, 63, Birkhäuser Boston, Boston, MA, 1986.
9. G. Tenenbaum. Sur une question d'Erdös et Schinzel. II. *Invent. Math.* 99 (1990), 215–224.
10. R.C. Vaughan. On Waring's problem for cubes. *J. reine angew. Math.* 365 (1986), 122–170.
11. R.C. Vaughan. On Waring's problem for smaller exponents. II. *Mathematika* 33 (1986), 6–22.

Coefficients of the n-Fold Theta Function and Weyl Group Multiple Dirichlet Series

Benjamin Brubaker, Daniel Bump, Solomon Friedberg, and Jeffrey Hoffstein

Abstract We establish a link between certain Whittaker coefficients of the generalized metaplectic theta functions, first studied by Kazhdan and Patterson [Kazhdan and Patterson, Metaplectic forms, Inst. Hautes Études Sci. Publ. Math., (59): 35–142, 1984], and the coefficients of stable Weyl group multiple Dirichlet series defined in [Brubaker, Bump, Friedberg, Weyl group multiple Dirichlet series. II. The stable case. Invent. Math., 165(2):325–355, 2006]. The generalized theta functions are the residues of Eisenstein series on a metaplectic n-fold cover of the general linear group. For n sufficiently large, we consider *different* Whittaker coefficients for such a theta function which lie in the orbit of Hecke operators at a given prime p. These are shown to be equal (up to an explicit constant) to the p-power supported coefficients of a Weyl group multiple Dirichlet series (MDS). These MDS coefficients are described in terms of the underlying root system; they have also recently been identified as the values of a p-adic Whittaker function attached to an unramified principal series representation on the metaplectic cover of the general linear group.

B. Brubaker (✉)
Department of Mathematics, MIT, Cambridge, MA 02139-4307, USA
e-mail: brubaker@math.mit.edu

D. Bump
Department of Mathematics, Stanford University, Stanford, CA 94305-2125, USA
e-mail: bump@math.stanford.edu

S. Friedberg
Department of Mathematics, Boston College, Chestnut Hill, MA 02467-3806, USA
e-mail: friedber@bc.edu

J. Hoffstein
Department of Mathematics, Brown University, Providence, RI 02912-9032, USA
e-mail: jhoff@math.brown.edu

V. Blomer and P. Mihăilescu (eds.), *Contributions in Analytic and Algebraic Number Theory*, Springer Proceedings in Mathematics 9, DOI 10.1007/978-1-4614-1219-9_4,

Dedicated to Professor Samuel J. Patterson in honor of his 60th birthday,

1 Introduction

This paper links the coefficients of two different Dirichlet series in several complex variables that arise in the study of automorphic forms on the metaplectic group. We begin with a brief discussion of the metaplectic group. Let F be a number field containing the group μ_{2n} of $2n$th roots of unity, and let F_v denote the completion of F at a place v. Let $\widetilde{G}_v$ denote an n-fold metaplectic cover of $\mathrm{GL}_{r+1}(F_v)$. Recall that $\widetilde{G}_v$ is a central extension of $\mathrm{GL}_{r+1}(F_v)$ by μ_n:

$$1 \longrightarrow \mu_n \longrightarrow \widetilde{G}_v \longrightarrow \mathrm{GL}_{r+1}(F_v) \longrightarrow 1.$$

This group is described by a 2-cocycle whose definition involves the nth power Hilbert symbol. See Matsumoto [17] or Kazhdan and Patterson [15] for this construction. The group $\widetilde{G}_v$ is generally not the F_v-points of an algebraic group. One may then take a suitable restricted direct product to define a global metaplectic cover $\widetilde{G}$ over $\mathrm{GL}_{r+1}(\mathbb{A}_F)$, the adelic points of the group GL_{r+1}. (The assumption that F contains μ_{2n} rather than μ_n is not necessary, but greatly simplifies the description of the cocycle and resulting formulas.)

Generalized theta series were introduced on the metaplectic covers of GL_2 by Kubota [16], and for GL_{r+1} in the visionary paper of Kazhdan and Patterson [15]. These remarkable automorphic forms are residues of the minimal parabolic Eisenstein series on the global metaplectic cover. They generalize classical theta functions of Jacobi and Siegel, which were shown by Weil to live on the metaplectic double covers of GL_2 and symplectic groups.

After Kubota introduced generalized theta series on the higher metaplectic covers of GL_2, Patterson and Heath-Brown [13] exploited the fact that when $n = 3$ their Fourier (Whittaker) coefficients are Gauss sums in order to settle the Kummer conjecture. Yet, it was found by Suzuki [19] that one could not so readily determine the coefficients of the theta function on the fourfold cover of GL_2. See Eckhardt and Patterson [10] for further discussion of this case. The difficulty in determining these coefficients is linked with the non-uniqueness of Whittaker models [9].

Thus, determining the Whittaker coefficients of generalized theta series was recognized as a fundamental question. The non-degenerate Whittaker coefficients on an n-fold cover of GL_{r+1} are non-zero only if $n \geq r+1$ [15]. Due to non-uniqueness of Whittaker models, their complete description is unavailable. Although they are thus mysterious, the partial information that is available is interesting indeed. They satisfy a periodicity property modulo nth powers, which is a generalization of the periodicity of the coefficients of the classical Jacobi theta function modulo squares. Moreover Kazhdan and Patterson found an action of the Weyl group on the coefficients modulo this periodicity in which each simple reflection adds or deletes a Gauss sum. This is an elegant formulation of the information that is available from Hecke theory. The non-uniqueness of Whittaker models when $n > r+2$ is a consequence of the fact that there is more than one free orbit in this Weyl group action. We review the definition of the generalized theta functions and expand on this discussion in Sect. 3.

More recently, Brubaker, Bump and Friedberg [6] have given an explicit description of the Whittaker coefficients of Borel Eisenstein series on the n-fold metaplectic cover of GL_{r+1}. In particular, they showed that the first non-degenerate Whittaker coefficient is a Dirichlet series in r complex variables (a "multiple Dirichlet series") that is roughly of the form

$$\sum_{c_1,\ldots,c_r} H(c_1,\ldots,c_r)|c_1|^{-2s_1}\ldots|c_r|^{-2s_r}. \tag{1}$$

Here, the sum runs over r-tuples of non-zero S-integers $\mathfrak{o}_S \subset F$ for a finite set of bad primes S and the coefficients H are sums of products of Gauss sums built with nth power residue symbols. The general expression for the coefficients H is best given in the language of crystal graphs, but this full description will not be needed here. Indeed, we will restrict our attention to cases where the degree of the cover n is at least $r+1$ (which is "stable" in the vocabulary of [3]), in which case the description of the coefficients simplifies considerably. In this situation, the coefficients supported at powers of a given prime p are in one-to-one correspondence with the Weyl group S_{r+1} of GL_{r+1}, and have a description in terms of the underlying root system [3]. Although we have described the series (1) in terms of global objects (Eisenstein series), let us also mention that the p-power supported terms are known to match the p-adic Whittaker function attached to the spherical vector for the associated principal series representation used to construct the Eisenstein series. This follows from combining the work of McNamara [18] with that of Brubaker, Bump and Friedberg [2, 6], or by combining [2, 18] and an unfolding argument of Friedberg and McNamara [11]. The precise definition of the coefficients H in the "stable" case will be reviewed in Sect. 2.

This paper establishes a link between *some* of the Whittaker coefficients of generalized theta functions and the coefficients of a stable Weyl group multiple Dirichlet series. Let us explain which coefficients are linked. We will show that, for $n > r+1$ fixed, the coefficients at p determined by Hecke theory are in one-to-one correspondence with the coefficients at p of the series (1). This is accomplished by comparing the two Weyl group actions – one on the Whittaker coefficients of generalized theta series found by Kazhdan and Patterson, and the other on the permutahedron supporting the stable multiple Dirichlet series coefficients. We know of no a priori reason for this link. On the one side, we have *different* Whittaker coefficients attached to a *residue* of an Eisenstein series. On the other hand, we have multiple Dirichlet series coefficients that contribute to the representation of a *single* Whittaker coefficient of the Eisenstein series itself. (More precisely, these contribute to the first non-degenerate coefficient.) For $n = r+1$, there is also a link, but this time to a multiple Dirichlet series coefficient attached to an Eisenstein series on the n-fold cover of GL_r, rather than on the n-fold cover of GL_{r+1}. Both comparison theorems for $n > r+1$ (Theorem 2) and $n = r+1$ (Theorem 3) are stated and proved in Sect. 4 of the paper. These theorems sharpen and extend the work of Kazhdan and Patterson (see [15], Theorems I.4.2 and II.2.3) on this connection.

2 Weyl Group Multiple Dirichlet Series

In [3], Brubaker, Bump, and Friedberg defined a *Weyl group multiple Dirichlet series* for any reduced root system and for n sufficiently large (depending on the root system). The requirement that n be sufficiently large is called stability, as the coefficients of the Dirichlet series are uniformly described Lie-theoretically for all such n. In this paper, we will be concerned with root systems of type A_r and in this case, the stability condition is satisfied if $n \geq r$.

As above, let F be a number field containing the group μ_{2n} of $2n$th roots of unity. Let S be a finite set of places of F containing the archimedean ones and those ramified over $\mathbb{Q}$ and that is large enough that the ring $\mathfrak{o}_S$ of S-integers in F is a principal ideal domain.

The multiple Dirichlet series coefficients are built from Gauss sums g_t, whose definition we now give. Let e be an additive character of $F_S = \prod_{v \in S} F_v$ that is trivial on $\mathfrak{o}_S$ but no larger fractional ideal. If $m, c \in \mathfrak{o}_S$ with $c \neq 0$ and if $t \geq 1$ is a rational integer, let

$$g_t(m,c) = \sum_{a \bmod^{\times} c} \left(\frac{a}{c}\right)^t e\left(\frac{am}{c}\right), \tag{2}$$

where $\left(\frac{a}{c}\right)$ is the nth power residue symbol and the sum is over a modulo c with $(a,c) = 1$ in $\mathfrak{o}_S$. For convenience, we let $g(m,c) = g_1(m,c)$. Let p be a fixed prime element of $\mathfrak{o}_S$, and q be the cardinality of $\mathfrak{o}_S / p\mathfrak{o}_S$. For brevity, we may sometimes write $g_t = g_t(1,p)$.

The multiple Dirichlet series of type A_r defined in [3] has the form

$$Z_\Psi(s_1,\ldots,s_r) = \sum H\Psi(c_1,\ldots,c_r) \mathbb{N}c_1^{-2s_1} \ldots \mathbb{N}c_r^{-2s_r}, \tag{3}$$

where the sum is over non-zero ideals c_i of $\mathfrak{o}_S$. Here, H and Ψ are functions defined when the c_i are non-zero elements of $\mathfrak{o}_S$, but their product is well defined over ideals, since H and Ψ behave in a coordinated way when c_i is multiplied by a unit. Thus the sum is essentially over ideals $c_i\mathfrak{o}_S$. However, we will want to consider H independently of Ψ, so for each prime $\mathfrak{p}$ of $\mathfrak{o}_S$ we fix a generator p of $\mathfrak{p}$, and only consider c_i which are products of powers of these fixed p's.

The function Ψ is chosen from a finite-dimensional vector space that is well understood and defined in [3] or [2], and we will not discuss it further here. The function H contains the key arithmetic information. It has a twisted multiplicativity, so that while Z_Ψ is not an Euler product, the specification of its coefficients is reduced to the case, where the c_i are powers of the same prime p. See [1, 3, 4] for further details.

To describe $H(p^{t_1},\ldots,p^{t_r})$, let Φ be the roots of A_r with Φ^+ (resp. Φ^-) the positive (resp. negative) roots. The Weyl group W acts on Φ. Let

$$\Phi(w) = \{\alpha \in \Phi^+ \mid w\alpha \in \Phi^-\}.$$

Also, let $\rho = \frac{1}{2}\sum_{\alpha\in\Phi^+}\alpha$ be the Weyl vector and let $\Sigma = \{\alpha_1,\ldots,\alpha_r\}$ denote the set of simple roots. Then as described in [3], we have:

- $H(p^{t_1},\ldots,p^{t_r}) \neq 0$ if and only if $\rho - w\rho = \sum_{i=1}^r t_i\alpha_i$ for some $w \in W$.
- If $\rho - w\rho = \sum_{i=1}^r t_i\alpha_i$, then

$$H(p^{t_1},\ldots,p^{t_r}) = \prod_{\alpha\in\Phi(w)} g\left(p^{d(\alpha)-1}, p^{d(\alpha)}\right) \tag{4}$$

with $d(\sum_{i=1}^r t_i\alpha_i) = \sum_{i=1}^r t_i$.

Thus in the stable case the Weyl group multiple Dirichlet series of type A_r has exactly $(r+1)!$ non-zero coefficients at each prime p. For motivation, more details, and generalizations to the case of smaller n (where there are additional nonzero coefficients), see [1–6].

3 Theta Functions

As in the introduction, $\widetilde{G}$ denotes an n-fold metaplectic cover of $\mathrm{GL}_{r+1}(\mathbb{A}_F)$. Suppose that $c = -1$ in the notation of [15] if $n = r+1$. Let $\theta_r^{(n)}$ denote the theta function on $\widetilde{G}$. This function is the normalized K-fixed vector in the space spanned by the residues at the rightmost poles of the minimal parabolic Eisenstein series on $\widetilde{G}$. Here, K denotes a suitable compact open subgroup. We will be concerned with the Whittaker coefficients of this vector, when they exist. By Hecke theory, these are determined by the values of these coefficients at prime power indices, or equivalently by the values of the local Whittaker functions for the exceptional theta representations $\Theta_r^{(n)}$, in the sense of Kazhdan and Patterson [15]. We now pass to a fixed completion of F at a good finite prime. In Sect. I.3 of [15], these authors have shown:

1. The representation $\Theta_r^{(n)}$ has a unique Whittaker model if and only if $n = r+1$ or $n = r+2$.
2. The representation $\Theta_r^{(n)}$ does not have a Whittaker model if $n \leq r$.
3. The representation $\Theta_r^{(n)}$ has a finite number of independent non-zero Whittaker models if $n > r+2$.

4. When the Whittaker model for $\Theta_r^{(n)}$ is non-zero, it is completely determined by the values of the associated Whittaker function on diagonal matrices of the form

$$\begin{pmatrix} \varpi^{f_1} & & & \\ & \varpi^{f_2} & & \\ & & \cdots & \\ & & & \varpi^{f_{r+1}} \end{pmatrix}$$

with $0 \leq f_i - f_{i+1} \leq n-1$ for $1 \leq i \leq r$.

The reason that this last holds is that the remaining values are determined by Kazhdan and Patterson's Periodicity Theorem. This states that shifting one of the $f_i - f_{i+1}$ by a multiple of n multiplies the Whittaker value by a specific power of q.

Suppose $n \geq r+1$. Fix a prime element p of $\mathfrak{o}_S$. Let $\tau_{n,r}(k_1,\ldots,k_r)$ be the $(p^{k_1},\ldots,p^{k_r})$th Whittaker coefficient of $\theta_r^{(n)}$. This coefficient is obtained by integrating against the character

$$e_U(u) = e\left(\sum_{i=1}^{r} p^{k_i} u_{i,i+1}\right)$$

of the subgroup U of upper triangular unipotents of GL_{r+1}, which is embedded in $\widetilde{G}$ via the trivial section.

Kazhdan and Patterson observed that Hecke theory may be used to compute all these Whittaker coefficients in the unique model case, and a subset of the coefficients in general. (See also Bump and Hoffstein [8] and Hoffstein [14].) We shall now review their description.

Let W denote the Weyl group for the root system A_r, isomorphic to the symmetric group S_{r+1}. In Sect. I.3 of [15], Kazhdan and Patterson define an action of W on the weight lattice (identified with $\mathbb{Z}^{r+1}$) by the formula

$$w[\mathbf{f}] = w(\mathbf{f} - \rho) + \rho,$$

where $\mathbf{f} = (f_1,\ldots,f_{r+1})$, the Weyl vector $\rho = (r, r-1,\ldots,0)$, and

$$w(\mathbf{f}) = (f_{w^{-1}(1)},\ldots,f_{w^{-1}(r+1)}).$$

This action of W on $\mathbb{Z}^{r+1}$ may then be projected down to $(\mathbb{Z}/n\mathbb{Z})^{r+1}$.

Because we prefer to use coordinates on the root lattice, we will reformulate this action in the language of the previous section. It suffices to define it for simple reflections σ_i, which generate W. Let $\mathbf{K}_r = \{\mathbf{k} = (k_1,\ldots,k_r) \mid 0 \leq k_j < n \text{ for all } j\}$.

Table 1 The orbit of $(0,0)$ under S_3

Element σ of S_3	$\sigma((0,0))$
e	$(0,0)$
σ_1	$(n-2,1)$
σ_2	$(1,n-2)$
$\sigma_1\sigma_2$	$(n-3,0)$
$\sigma_2\sigma_1$	$(0,n-3)$
$\sigma_1\sigma_2\sigma_1$	$(n-2,n-2)$

Table 2 The orbit of $(0,0,0)$ under S_4

Element σ of S_4	$\sigma((0,0,0))$	Element σ of S_4	$\sigma((0,0,0))$
e	$(0,0,0)$	$\sigma_3\sigma_1\sigma_2$	$(n-3,2,n-3)$
σ_1	$(n-2,1,0)$	$\sigma_2\sigma_3\sigma_2$	$(2,n-2,n-2)$
σ_2	$(1,n-2,1)$	$\sigma_1\sigma_2\sigma_3$	$(n-4,0,0)$
σ_3	$(0,1,n-2)$	$\sigma_1\sigma_3\sigma_2\sigma_1$	$(n-2,1,n-4)$
$\sigma_1\sigma_2$	$(n-3,0,1)$	$\sigma_2\sigma_3\sigma_2\sigma_1$	$(1,n-2,n-3)$
$\sigma_2\sigma_1$	$(0,n-3,2)$	$\sigma_1\sigma_2\sigma_3\sigma_1$	$(n-3,n-2,1)$
$\sigma_1\sigma_3$	$(n-2,2,n-2)$	$\sigma_2\sigma_3\sigma_1\sigma_2$	$(0,n-4,0)$
$\sigma_2\sigma_3$	$(2,n-3,0)$	$\sigma_1\sigma_2\sigma_3\sigma_2$	$(n-4,1,n-2)$
$\sigma_3\sigma_2$	$(1,0,n-3)$	$\sigma_2\sigma_1\sigma_3\sigma_2\sigma_1$	$(0,n-3,n-2)$
$\sigma_1\sigma_2\sigma_1$	$(n-2,n-2,2)$	$\sigma_1\sigma_2\sigma_3\sigma_2\sigma_1$	$(n-3,0,n-3)$
$\sigma_3\sigma_2\sigma_1$	$(0,0,n-4)$	$\sigma_1\sigma_2\sigma_3\sigma_1\sigma_2$	$(n-2,n-3,0)$
$\sigma_2\sigma_3\sigma_1$	$(1,n-4,1)$	$\sigma_1\sigma_2\sigma_1\sigma_3\sigma_2\sigma_1$	$(n-2,n-2,n-2)$

Then the action of σ_i for any $i = 1,\dots,r$ on multi-indices $\mathbf{k} \in \mathbf{K}_r$ is given by $\sigma_i(\mathbf{k}) = \mathbf{m}$ with

$$
\begin{aligned}
m_{i-1} &= \begin{cases} 1+k_i+k_{i-1} & \text{if } 1+k_i+k_{i-1} < n \\ 1+k_i+k_{i-1}-n & \text{if } 1+k_i+k_{i-1} \geq n \end{cases} \\
m_i &= \begin{cases} n-2-k_i & \text{if } k_i < n-1 \\ 2n-2-k_i & \text{if } k_i = n-1 \end{cases} \\
m_{i+1} &= \begin{cases} 1+k_i+k_{i+1} & \text{if } 1+k_i+k_{i+1} < n \\ 1+k_i+k_{i+1}-n & \text{if } 1+k_i+k_{i+1} \geq n \end{cases} \\
m_j &= k_j \qquad \text{if } j \neq i-1, i, i+1.
\end{aligned} \tag{5}
$$

In these formulas, we take $k_0 = k_{r+1} = 0$. It is a simple exercise to verify that this matches the action on the weight lattice described above.

To illustrate, the orbit of the origin when $r = 2$ and $n \geq 3$ is given in Table 1, and the orbit of the origin when $r = 3$ and $n \geq 4$ is given in Table 2. We shall also show that the stabilizer of the origin is trivial for $n > r+1$. However, this fails for $n = r+1$, as one sees immediately in these two examples.

The action above is essentially that corresponding to the action of the Hecke operators. Since $\theta_r^{(n)}$ is an eigenfunction of these operators, one can deduce the following relation.

Theorem 1. *Suppose that* $0 \le k_j < n$ *for* $1 \le j \le r$, $k_i \not\equiv -1 \bmod n$, *and* $\sigma_i(\mathbf{k}) = \mathbf{m}$. *Then*

$$\tau_{n,r}(\mathbf{m}) = q^{i-r/2-1+\delta(i,r,\mathbf{k})} g_{1+k_i} \tau_{n,r}(\mathbf{k}),$$

where

$$\delta(i,r,\mathbf{k}) = \begin{cases} -(i-1)(r-i+2)/2 & \text{if } 1+k_i+k_{i-1} \ge n \\ 0 & \text{otherwise} \end{cases} + \begin{cases} (i+1)(r-i)/2 & \text{if } 1+k_i+k_{i+1} < n \\ 0 & \text{otherwise.} \end{cases}$$

Here, we have used the Gauss sum g_t as defined in (2). The result follows from Proposition 5.3 of [14] and the periodicity property of the Fourier coefficients $\tau_{n,r}$, given in Proposition 5.1 there. (Note that $G_{1+c_i}(m_i, p)$ in [14], Proposition 5.3, is normalized to have absolute value 1, while g_{1+k_i} has absolute value $q^{1/2}$.) See also [8], and Corollary I.3.4 of [15].

4 A Link Between Theta Coefficients and Weyl Group Multiple Dirichlet Series

To give a link between the Whittaker coefficients of the generalized theta functions that are determined by Hecke theory and the Weyl group multiple Dirichlet series, we begin by linking the action of (5) to roots. Suppose that $n \ge r+1$.

Proposition 1. *Let* $w \in W$, *and suppose that* $w((0,\ldots,0)) = \mathbf{k}$. *Then* $1 + k_i \equiv d(w^{-1}\alpha_i) \bmod n$ *for each* i, $1 \le i \le r$.

Proof. We prove the Proposition by induction on $\ell(w)$, the length of w as a reduced word composed of simple reflections σ_i. The case $w = 1$ is clear. Suppose that $w((0,\ldots,0)) = \mathbf{k}$ and $1 + k_i \equiv d(w^{-1}\alpha_i) \bmod n$ for each i. Choose $\sigma_j \in W$ such that $\ell(\sigma_j w) = \ell(w) + 1$. If $\sigma_j w((0,\ldots,0)) = \mathbf{m}$, then $\mathbf{m} = \sigma_j(\mathbf{k})$, so by (5)

$$m_i \equiv \begin{cases} -2-k_i & j = i \\ 1+k_i+k_j & j = i+1 \text{ or } j = i-1 \\ k_i & \text{otherwise} \end{cases} \tag{6}$$

modulo n. On the other hand, we have

$$\begin{aligned} (\sigma_j w)^{-1}(\alpha_i) &= w^{-1}(\sigma_j(\alpha_i)) \\ &= \begin{cases} w^{-1}(-\alpha_i) & j = i \\ w^{-1}(\alpha_i + \alpha_j) & j = i+1 \text{ or } j = i-1 \\ w^{-1}(\alpha_i) & \text{otherwise.} \end{cases} \end{aligned}$$

Hence,

$$d((\sigma_j w)^{-1}(\alpha_i)) = \begin{cases} -d(w^{-1}(\alpha_i)) & j=i \\ d(w^{-1}(\alpha_i))+d(w^{-1}(\alpha_j)) & j=i+1 \text{ or } j=i-1 \\ d(w^{-1}(\alpha_i)) & \text{otherwise.} \end{cases}$$

Using the inductive hypothesis, we see that modulo n

$$d((\sigma_j w)^{-1}(\alpha_i)) \equiv \begin{cases} -(1+k_i) & j=i \\ 2+k_i+k_j & j=i+1 \text{ or } j=i-1 \\ 1+k_i & \text{otherwise} \end{cases}$$

$$= \begin{cases} 1+(-2-k_i) & j-i \\ 1+(1+k_i+k_j) & j=i+1 \text{ or } j=i-1 \\ 1+k_i & \text{otherwise.} \end{cases}$$

Comparing this to (6), we see that the Proposition holds.

Corollary 1. *Let $w \in W$, and suppose that $w((0,\ldots,0)) = \mathbf{k}$. Then for all i, $1 \le i \le r$, $k_i \not\equiv -1 \bmod n$.*

Proof. Since $w^{-1}\alpha_i$ is a root, we have $d(w^{-1}\alpha_i) \neq 0$. Moreover, from the explicit description of the roots of type A_r, for any root $\beta \in \Phi$, we have $-r \le d(\beta) \le r$. If $k_i \equiv -1 \bmod n$, Proposition 1 would imply that $d(w^{-1}\alpha_i) \equiv 0 \bmod n$, which is impossible as $n \ge r+1$ and $d(w^{-1}\alpha_i) \neq 0$.

Corollary 2. *Suppose that $n > r+1$. Then the stabilizer in W of $(0,\ldots,0)$ is trivial. Thus, the orbit of the origin has cardinality $(r+1)!$, and every point in the orbit may be described uniquely as $w((0,\ldots,0))$ for some $w \in W$.*

Proof. Let $\sigma \in W$ and suppose that $\sigma((0,\ldots,0)) = (0,\ldots,0)$. By Proposition 1, $d(\sigma^{-1}\alpha_i) \equiv 1 \bmod n$ for all i, $1 \le i \le r$. But as noted above, $-r \le d(\sigma^{-1}\alpha_i) \le r$. Since $n > r+1$, the congruence can only hold if $d(\sigma^{-1}\alpha_i) = 1$ for all i, $1 \le i \le r$. Thus, $\sigma^{-1}(\alpha_i) \in \Phi^+$ for all i. This implies that $\sigma^{-1}(\Phi^+) \subset \Phi^+$, which is true only if σ is the identity element.

Note that Corollary 2 does not remain valid if $n = r+1$; it is possible that there exists an i for which $d(\sigma^{-1}\alpha_i)$ is $-r$ and not 1. This occurs, for example, when $r = 2$, $\sigma = \sigma_1\sigma_2$, and $i = 1$. More generally, see Lemma 1 below.

We may now establish a link between the Whittaker coefficients of the generalized theta function that are determined by Hecke theory and the Weyl group multiple Dirichlet series.

Theorem 2. *Suppose that $n > r+1$. Let $w \in W$, and set $w((0,\ldots,0)) = \mathbf{k}$, $\rho - w\rho = \sum t_i\alpha_i$. Then*

$$\tau_{n,r}(\mathbf{k}) = q^{\eta(w,n,r,\mathbf{k})} H(p^{t_1},\ldots,p^{t_r}),$$

where the function $\eta(w,n,r,\mathbf{k})$ is described in (9) below.

Remark 1. We should emphasize that for given $w \in W$ the coordinates k_i, which are determined from the equation $w((0,\dots,0)) = \mathbf{k}$, are not the same as the coordinates t_i, which are determined from the equation $\rho - w\rho = \sum t_i \alpha_i$. For example, on A_2 we have $\sigma_1((0,0)) = (n-2,1)$ while $\rho - \sigma_1(\rho) = \alpha_1$, so for σ_1, $(t_1,t_2) = (1,0)$. Thus in this case Theorem 2 asserts that for $n > 3$, $\tau_{n,2}(p^{n-2},p)$ is equal, up to a power of q, to $H(p,1)$. (From (4), $H(p,1) = g(1,p)$.)

Coincidentally, on A_2 with $n = 4$ (a unique model case), as one runs over all $w \in W$, one obtains the same 6 integer lattice points for the (k_1,k_2) and the (t_1,t_2) (albeit with some of those lattice point attached to different Weyl group elements for the two parametrizations), but this phenomenon does not persist to higher rank.

Proof. We prove this by induction on the length of w. If w is the identity, the result is clear (with $\eta(e,n,r,\mathbf{k}) = 0$). Suppose that the result is proved for w and that $\ell(\sigma_i w) = \ell(w) + 1$. Let $\sigma_i(\mathbf{k}) = \mathbf{m}$. By Corollary 1, the hypothesis of Theorem 1 is satisfied. Thus by this result, we have

$$\tau_{n,r}(\mathbf{m}) = q^{i-r/2-1+\delta(i,r,\mathbf{k})} g_{1+k_i} \tau_{n,r}(\mathbf{k}).$$

By Proposition 1, $g_{1+k_i} = g_{d(w^{-1}\alpha_i)}$. Moreover, under the assumption that $\ell(\sigma_i w) = \ell(w) + 1$, it follows that $w^{-1}\alpha_i \in \Phi^+$, so $d(w^{-1}\alpha_i) > 0$. (See, for example, Bump [7], Propositions 21.2 and 21.10.). Thus, by elementary properties of Gauss sums,

$$g_{d(w^{-1}\alpha_i)} = q^{1-d(w^{-1}\alpha_i)} g(p^{d(w^{-1}\alpha_i)-1}, p^{d(w^{-1}\alpha_i)}).$$

So we arrive at the formula

$$\tau_{n,r}(\mathbf{m}) = q^{i-r/2+\delta(i,r,\mathbf{k})-d(w^{-1}\alpha_i)} g(p^{d(w^{-1}\alpha_i)-1}, p^{d(w^{-1}\alpha_i)}) \tau_{n,r}(\mathbf{k}). \tag{7}$$

On the other hand, it is well known (see, for example, Bump [7], Proposition 21.10) that

$$\Phi(\sigma_i w) = \Phi(w) \cup \{w^{-1}\alpha_i\}.$$

Thus, (4) implies that

$$H(p^{u_1},\dots,p^{u_r}) = g(p^{d(w^{-1}\alpha_i)-1}, p^{d(w^{-1}\alpha_i)}) H(p^{t_1},\dots,p^{t_r}), \tag{8}$$

where $\rho - w(\rho) = \sum_i t_i \alpha_i$ and $\rho - \sigma_i w(\rho) = \sum_i u_i \alpha_i$. Comparing (7) and (8), the theorem follows.

To give the precise power of q, suppose that $w = \sigma_{j_c}\sigma_{j_{c-1}}\cdots\sigma_{j_1}$ is a reduced word for w, so $c = \ell(w)$. Let $\mathbf{k}^{(0)} = (0,\dots,0)$ and $\sigma_{j_i}(\mathbf{k}^{(i-1)}) = \mathbf{k}^{(i)}$, $1 \le i \le c$. Also let $\tau_1 = 1$ and $\tau_t = \sigma_{j_{t-1}}\sigma_{j_{t-2}}\cdots\sigma_{j_1}$ for $1 < t \le c$. Then applying (7) repeatedly, we find that

$$q^{\eta(w,n,r,\mathbf{k})} = q^{-r\ell(w)/2} \prod_{t=1}^{\ell(w)} q^{j_t+\delta(j_t,r,\mathbf{k}^{(t-1)})-d(\tau_t^{-1}\alpha_{j_t})}. \tag{9}$$

Next we turn to the case $n = r+1$. This equality implies that the Whittaker model of the theta representation is unique (see Kazhdan-Patterson [15], Corollary I.3.6 for the local uniqueness and Theorem II.2.5 for its global realization). To describe the corresponding Whittaker coefficients in terms of multiple Dirichlet series, we first describe the orbit of the origin under W. As noted above, the stabilizer of the origin is non-trivial. Indeed, we have

Lemma 1. *Suppose $n = r+1$. Then $\sigma_1\sigma_2\cdots\sigma_r((0,\ldots,0)) = (0,\ldots,0)$.*

Proof. The proof is a straightforward calculation, left to the reader.

Since the stabilizer of the origin is non-trivial, let us restrict the action of W on r-tuples to the subgroup generated by the transpositions σ_i, $1 \le i < r$. We will denote this group $\mathfrak{S}_r$; note that $\mathfrak{S}_r$ is isomorphic to the symmetric group S_r, but the action of $\mathfrak{S}_r$ on r-tuples is *not* the standard permutation action.

Lemma 2. *Suppose $n = r+1$. Then the stabilizer in $\mathfrak{S}_r$ of $(0,\ldots,0)$ is trivial.*

Proof. In this proof (and in the proof of Theorem 3 below), we write W_r instead of W for the Weyl group of type A_r. W_r acts on $\mathbf{K}_r$ by the action given in (5). Observe that under the projection π from $\mathbf{K}_r$ to $\mathbf{K}_{r-1}$ obtained by forgetting the last coordinate, the action of $\mathfrak{S}_r$ on $\mathbf{K}_r$ restricts to the action of the Weyl group W_{r-1} on $\mathbf{K}_{r-1}$. Indeed, this is true since the actions on the first $r-1$ entries are the same; note that changing the rth entry of an element of $\mathbf{K}_r$ does not affect its image under $\pi\circ\sigma_i$ for $\sigma_i \in \mathfrak{S}_r$. Then the Lemma follows at once from Corollary 2, which applies as $n > (r-1)+1$.

Combining these, we may describe the orbit of the origin.

Proposition 2. *Suppose $n = r+1$. Then the stabilizer in W of the origin has order $r+1$ and is the group generated by the element $\sigma_1\sigma_2\cdots\sigma_r$. The orbit of the origin under W has order $r!$, and every point in the orbit may be described uniquely as $w((0,\ldots,0))$ for some $w \in \mathfrak{S}_r$.*

Proof. Since $\sigma_1\sigma_2\cdots\sigma_r$ has order $r+1$, the stabilizer $W^{(0,\ldots,0)}$ of the origin in W has order at least $r+1$. Hence, $[W : W^{(0,\ldots,0)}] \le r!$. But by Lemma 2, the image of W has order at least $r!$ Since the cardinality of this image is exactly $[W : W^{(0,\ldots,0)}]$, equality must obtain, and the Proposition follows.

Finally, we give the analogue of Theorem 2 when $n = r+1$. The link is once again between theta Whittaker coefficients and stable Weyl group multiple Dirichlet series coefficients, but this time the latter are of type A_{r-1} rather than type A_r.

Theorem 3. *Suppose that $n = r+1$. Let $w \in \mathfrak{S}_r$, and set $w((0,\ldots,0)) = \mathbf{k}$, $\rho - w\rho = \sum t_i\alpha_i$. Then*

$$\tau_{r+1,r}(\mathbf{k}) = q^{\eta(w,r+1,r,\mathbf{k})} H(p^{t_1},\ldots,p^{t_{r-1}}),$$

where the coefficient H is the coefficient of the type A_{r-1} multiple Dirichlet series, and the function $\eta(w,r+1,r,\mathbf{k})$ is given by (9) above.

Remark 2. Note that since $n > r-1$, the coefficients H are stable, and account for the full set of non-zero Weyl group multiple Dirichlet series coefficients for A_{r-1}. See [3]. Also, if $w \in \mathfrak{S}_r$ and $\rho - w\rho = \sum_{i=1}^{r} t_i \alpha_i$, then necessarily $t_r = 0$, so the restriction to $(r-1)$-tuples in the right-hand side of the Theorem is natural. In addition, one can check that

$$\eta(\sigma_1\sigma_2\cdots\sigma_r, r+1, r, (0,\ldots,0)) = 0,$$

and that one can use any $w \in W$ to reach $\mathbf{k}$ in the orbit of $(0,\ldots,0)$ in order to compute the coefficient $\tau_{r+1,r}(\mathbf{k})$. (Doing so one obtains each coefficient $r+1$ times.)

Proof. The Weyl group of type A_r, W_r, acts on its root system Φ and on $\mathbf{K}_r$. These actions each restrict: the subgroup $\mathfrak{S}_r$ acts on

$$\Phi_{r-1} = \left\{ \alpha \in \Phi \mid \alpha = \sum_{i=1}^{r-1} m_i \alpha_i \text{ for some } m_i \in \mathbb{Z} \right\},$$

and, as noted in the proof of Lemma 2 above, it also acts on $\mathbf{K}_{r-1}$. These actions are each compatible with the isomorphism $\mathfrak{S}_r \cong W_{r-1}$. Thus, we may follow the argument given in the proof of Theorem 2. However, in that case we obtain the $r!$ stable coefficients of the type A_{r-1} Weyl group multiple Dirichlet series. (Note that these coefficients are a subset of the $(r+1)!$ stable coefficients of type A_r.)

In concluding, we note that one can ask whether theorems that are similar to Theorems 2 and 3 hold for metaplectic covers of the adelic points other reductive groups. The theory of theta functions, that is, residues of Eisenstein series on metaplectic covers, is not yet fully established when the underlying group in question is not a general linear group. However, we do expect that it can be developed using methods similar to those of [15], and that the link between the stable Weyl group multiple Dirichlet series and the Whittaker coefficients determined by Hecke theory persists. Indeed, Brubaker and Friedberg have carried out computations of Hecke operators on the four and fivefold covers of $GSp(4)$, following the approach of Goetze [12]. Under reasonable hypotheses about the periodicity relation (which should vary depending on root lengths for simple roots) for theta coefficients for those groups, such a link once again holds in those cases.

Acknowledgements This work was supported by NSF grants DMS-0844185, DMS-1001079 and DMS-1001326, NSF FRG grants DMS-0652817, DMS-0652609, and DMS-0652312, and by NSA grant H98230-10-1-0183.

References

1. B. Brubaker, D. Bump, G. Chinta, S. Friedberg, and J. Hoffstein. Weyl group multiple Dirichlet series I. In: *Multiple Dirichlet Series, Automorphic Forms, and Analytic Number Theory*, Proc. Sympos. Pure Math., 75. Amer. Math. Soc., Providence, R.I., 2006, pp. 91–114.
2. B. Brubaker, D. Bump, and S. Friedberg. Twisted Weyl group multiple Dirichlet series: the stable case. In: *Eisenstein series and applications (Gan, Kudla and Tschinkel, eds.)*. Birkhäuser Progress in Mathematics 258, 2008, pp. 1–26.
3. B. Brubaker, D. Bump, and S. Friedberg. Weyl group multiple Dirichlet series. II. The stable case. *Invent. Math.*, 165(2):325–355, 2006.
4. B. Brubaker, D. Bump, S. Friedberg, and J. Hoffstein. Weyl group multiple Dirichlet series III: Eisenstein series and twisted unstable A_r. *Annals of Mathematics*, 166(1):293–316, 2007.
5. B. Brubaker, D. Bump, and S. Friedberg. Weyl Group Multiple Dirichlet Series: Type A Combinatorial Theory. *Annals of Mathematics Studies*, Volume 175, Princeton Univ. Press, 2011
6. B. Brubaker, D. Bump, and S. Friedberg. Weyl group multiple Dirichlet series, Eisenstein series and crystal bases. *Annals of Mathematics*, 173(2):1081–1120, 2011.
7. D. Bump. Lie Groups, Graduate Texts in Mathematics 225. Springer-Verlag, New York, 2004.
8. D. Bump and J. Hoffstein. On Shimura's correspondence. *Duke Math. J.*, 55(3):661–691, 1987.
9. P. Deligne. Sommes de Gauss cubiques et revêtements de SL(2) [d'après S. J. Patterson]. *Sém. Bourbaki (1978/79), Exposé No. 539*. In: Springer Lecture Notes 770, 1980, pp. 244–277.
10. C. Eckhardt and S. J. Patterson. On the Fourier coefficients of biquadratic theta series. *Proc. London Math. Soc. (3)*, 64(2):225–264, 1992.
11. S. Friedberg and P. McNamara. in preparation.
12. T. Goetze. Euler products associated to metaplectic automorphic forms on the 3-fold cover of $GSp(4)$. *Trans. A.M.S.*, 350(3):975–1011, 1998.
13. D. R. Heath-Brown and S. J. Patterson. The distribution of Kummer sums at prime arguments. *J. Reine Angew. Math.*, 310:111–130, 1979.
14. J. Hoffstein. Eisenstein series and theta functions on the metaplectic group. In: *Theta functions: from the classical to the modern (M. Ram Murty, ed.)*. CRM Proc. Lect. Notes 1, 1993, pp. 65–104.
15. D. A. Kazhdan and S. J. Patterson. Metaplectic forms. *Inst. Hautes Études Sci. Publ. Math.*, (59):35–142, 1984.
16. T. Kubota. Some results concerning reciprocity law and real analytic automorphic functions. 1969 Number Theory Institute (Proc. Sympos. Pure Math., Vol. XX, State Univ. New York, Stony Brook, N.Y., 1969), pp. 382–395. Amer. Math. Soc., Providence, R.I., 1971.
17. H. Matsumoto. Sur les sous-groupes arithmétiques des groupes semi-simples déployés. *Ann. Sci. École Norm. Sup.*, (4)2:1–62, 1969.
18. P. McNamara. Metaplectic Whittaker functions and crystal bases. *Duke Math. J.*, 156(1): 1–31, 2011.
19. T. Suzuki. Some results on the coefficients of the biquadratic theta series. *J. Reine Angew. Math.*, 340:70–117, 1983.

Towards the Trace Formula for Convex-Cocompact Groups

Ulrich Bunke and Martin Olbrich

Abstract We develop a general representation theoretic framework for trace formulas for quotients of rank one simple Lie groups by convex-cocompact discrete subgroups. We further discuss regularized traces of resolvents with applications to Selberg-type zeta functions.

Dedicated to Samuel J. Patterson on the occasion of his 60th birthday.

1 Introduction

In this paper, we develop a part of the harmonic analysis associated with a convex cocompact subgroup Γ of a semisimple Lie group G of real rank one that could play the same role as the trace formula in the case of cocompact groups or groups of finite covolume. In these classical situations, a smooth, compactly supported, and K-finite function f on G acts by right convolution $R_\Gamma(f)$ on the Hilbert space $L^2(\Gamma\backslash G)$. The trace formula is an expression of the trace of the restriction of this operator to the discrete subspace in terms of the function f and its Fourier transform $\hat{f}$. The part involving f is called the geometric side and usually written as a sum of orbital integrals. Depending on the applications one has in mind, one might prefer to express some of the orbital integrals (as the identity contribution for instance)

U. Bunke (✉)
NWF I – Mathematik, Universität Regensburg, 93040 Regensburg, Germany
e-mail: ulrich.bunke@mathematik.uni-regensburg.de

M. Olbrich
Université du Luxembourg, 6, rue Richard Coudenhove-Kalergi, 1359 Luxembourg
e-mail: martin.olbrich@uni.lu

V. Blomer and P. Mihăilescu (eds.), *Contributions in Analytic and Algebraic Number Theory*, Springer Proceedings in Mathematics 9, DOI 10.1007/978-1-4614-1219-9_5,

by $\hat{f}$, instead. The Fourier transform certainly enters the trace formula in the case of non-cocompact subgroups via the contribution of the scattering matrix.

In this paper, we assume that Γ is a convex cocompact subgroup of G. Let X be the symmetric space of G and ∂X be its geodesic boundary. If $\Gamma \subset G$ is a discrete torsion-free subgroup, then there is a Γ-invariant disjoint decomposition $\partial X = \Omega \cup \Lambda$, where Λ is the limit set of Γ. Here, we call Γ convex-cocompact if $\Gamma \backslash X \cup \Omega$ is a compact manifold with non-empty boundary. In particular, $\Gamma \backslash G$ has infinite volume. We assume in addition that G is different from the exceptional rank one group F_4^{-20} since the necessary spectral and scattering theoretic results are not yet available in this case (see [4]).

Since the discrete spectrum of $L^2(\Gamma \backslash G)$ is rather sparse – even empty in some cases – we take the point of view that the contribution of the scattering matrix is essentially (up to the contribution of the discrete spectrum) the Fourier transform of the geometric side of the trace formula.

Thus, our starting point is the geometric side. It is a distribution Ψ on G given as a sum of suitably normalized orbital integrals associated with the hyperbolic conjugacy classes of Γ

$$\Psi(f) := \sum_{\gamma \in \widetilde{C\Gamma}} \operatorname{vol}(\Gamma_\gamma \backslash G_\gamma) \theta_\gamma(f)$$

(see Definition 1). Note that our definition of the geometric side does not contain any contribution of the identity element of Γ. In fact, the usual identity term would be infinite by the infinite volume of $\Gamma \backslash G$. The objective of the trace formula for convex cocompact Γ is an explicit expression for the Fourier transform of Ψ. We are looking for a "measure" Φ on the unitary dual $\hat{G}$ such that

$$\Psi(f) = \int_{\hat{G}} \theta_\pi(f) \Phi(\mathrm{d}\pi), \tag{1}$$

where $\theta_\pi(f) := \operatorname{Tr} \hat{f}(\pi)$ is the character of π. In this paper we will formulate a precise conjecture about Φ, but we are not able to prove the formula (1) in the general case. We conjecture that

$$\begin{aligned}\int_{\hat{G}} \theta_\pi(f) \Phi(\mathrm{d}\pi) = & \sum_{\sigma \in \hat{M}} \frac{1}{4\pi} \int_{ia^*} L_\Gamma(\pi^{\sigma,\lambda}) \theta_{\pi^{\sigma,\lambda}}(f) \mathrm{d}\lambda \\ & + \sum_{\sigma \in \hat{M}, \sigma \cong \sigma^w, \pi^{\sigma,0} \text{red.}} \sum_{\varepsilon \in \{+,-\}} \tilde{N}_\Gamma(\pi^{\sigma,\varepsilon}) \theta_{\pi^{\sigma,\varepsilon}}(f) \\ & + \sum_{\sigma \in \hat{M}, \sigma \cong \sigma^w, \pi^{\sigma,0} \text{ irred.}} \tilde{N}_\Gamma(\pi^{\sigma,0}) \theta_{\pi^{\sigma,0}}(f) \\ & + \sum_{\pi \in \hat{G}_c \cup \hat{G}_d} N_\Gamma(\pi) \theta_\pi(f). \end{aligned} \tag{2}$$

For the precise notation, we refer to Conjecture 1. The first three terms on the right-hand side are the contributions of the continuous spectrum. The number

$L_\Gamma(\pi^{\sigma,\lambda})$ in the first term can be considered as the logarithmic derivative of a kind of regularized determinant of the scattering matrix. The irreducible components of the unitary principal series of G with zero parameter (e.g. of the limits of discrete series representations) contribute with additional terms. The last sum is the contribution of the discrete spectrum, which comes from complementary series – the non-tempered unitary representations of G – and some unitary principal series with zero parameter (together denoted by $\hat{G}_{\mathrm{c}}$ here) and the discrete series $\hat{G}_{\mathrm{d}}$ of G. Note that other unitary principal series do not contribute to the discrete spectrum [4]. If $\pi \in \hat{G}_{\mathrm{c}}$, then the integer $N_\Gamma(\pi)$ denotes its multiplicity in $L^2(\Gamma\backslash G)$.

The multiplicity of a discrete series representation π in $L^2(\Gamma\backslash G)$ is infinite. In this case, the number $N_\Gamma(\pi)$ is a regularized multiplicity and a priory a real number. But by Proposition 5 $N_\Gamma(\pi) = \Phi(\{\pi\})$ is an integer, too. In Proposition 3, we show that it is zero for integrable discrete series representations π. It is an interesting problem for future research to study this number for non-integrable discrete series in detail.

The unitary dual $\hat{G}$ has a natural topology. Now observe that the intersection of the support of $\hat{f}$ and the support of the Plancherel measure of $L^2(\Gamma\backslash G)$ is the spectrum of $R_\Gamma(f)$. The Fourier transform of a compactly supported function f on G is never compactly supported on $\hat{G}$. In order to do our computations, we have to approximate $R_\Gamma(f)$ by operators, which have compact spectrum. The missing piece for the proof of (1) is some estimate, which eventually allows for dropping the cut-off.[1] However, Conjecture 1 can easily be verified in case that Γ has a negative critical exponent (see the Remark 1).

In this paper, we will prove a formula which is similar to (1), but where Ψ has a different interpretation. Let $R(f)$ denote the right-convolution operator on $L^2(G)$ induced by f. Then both, $R(f)$ and $R_\Gamma(f)$, have smooth integral kernels $K_{R(f)}$, $K_{R_\Gamma(f)}$, and, by Lemma 1, the value $\Psi(f)$ is nothing else than the integral

$$\Psi'(f) := \int_{\Gamma\backslash G} (K_{R_\Gamma(f)}(g,g) - K_{R(f)}(g,g))\mu_G(\mathrm{d}g).$$

We will show that Ψ' can be applied to functions with compactly supported Fourier transform, and our main Theorem 2 is a formula

$$\Psi'(f) = \int_{\hat{G}} \theta_\pi(f)\Phi(\mathrm{d}\pi). \qquad (3)$$

together with the explicit expression (2) for Φ.

A related regularized trace formula for the scalar wave operator on real hyperbolic spaces has been obtained in [9].

---

[1] The problem cannot be solved by just looking at a space of functions f having not necessarily compact support to which Ψ can be applied such that Lemma 1 remains valid. If the critical exponent of Γ is positive, any reasonable space of this kind is contained in some $L^p(G)$ for $p < 2$. The only L^p-functions f on G ($p < 2$) with compactly supported Fourier transform are linear combinations of matrix coefficients of certain discrete series representations. They satisfy $\Psi(f) = 0$, compare Proposition 3.

There are interesting operators with non-compact spectrum to which Ψ' can be applied. Let $K \subset G$ be a maximal compact subgroup and C be the Casimir operator of G. We fix a K-type τ. For a representation V of G, we let $V(\tau)$ denote the τ-isotypical component. We consider the resolvent $(z-C)^{-1}$ on $L^2(\Gamma\backslash G)(\tau)$ and $L^2(G)(\tau)$ if z is not in the spectrum. Let $K_\Gamma(z)$ and $K(z)$ denote the corresponding integral kernels. The difference $K_\Gamma(z) - K(z)$ is smooth on the diagonal and goes into Ψ' if $\mathrm{Re}(z) \ll 0$. The study of this regularized resolvent trace, which is closely related to Selberg-type zeta functions, is the aim of the long final section of the paper. It is quite independent of the previous ones, only some computations of Sect. 4.4 are used.

The analysis of $\Psi'(f)$ for functions on G with compactly supported Fourier transform $\hat{f}$ is based on the Plancherel theorem for $L^2(\Gamma\backslash G)$, which has been shown in [4]. In the case of the resolvent kernel, we also use the hyperfunction boundary value theory [18] for eigenfunctions of the Casimir operator. This additional information provides the asymptotic behavior of the resolvent kernel avoiding the use of unknown estimates on the growth of Eisenstein series with respect to the spectral parameter. Note that we do not prove Formula (2) for resolvent kernels. What we obtain is a functional equation for the meromorphic continuation of resolvent traces in the spectral parameter z (Proposition 4), compare (4) below. This functional equation is compatible with Conjecture 1.

Considering traces of resolvents provides the link between the continuous part of Φ and the Selberg zeta functions $Z_S(\sigma,\lambda)$ associated to Γ. In fact, using the analysis of the resolvent traces, we can show the meromorphic continuation of the logarithmic derivative and the functional equation (Theorem 3) of the Selberg zeta functions (By other methods, we a priori know that the Selberg zeta functions are meromorphic, see below.). The basic identity is

$$\frac{Z_S'(-\lambda,\sigma)}{Z_S(-\lambda,\sigma)} + \frac{Z_S'(\lambda,\sigma)}{Z_S(\lambda,\sigma)} = L_\Gamma(\pi^{\sigma,\lambda}). \tag{4}$$

shown in the proof of Theorem 3. In particular, we obtain the following description of the singularities of the Selberg zeta function

$$\mathrm{ord}_{\lambda=\mu} Z_S(\lambda,\sigma) = \begin{cases} n_{\mu,\sigma} & \mathrm{Re}(\mu) > 0, \\ \mathrm{res}_{\lambda=\mu} L_\Gamma(\pi^{\sigma,\lambda}) + n_{-\mu,\sigma} & \mathrm{Re}(\mu) < 0, \end{cases}$$

where the integers $n_{\mu,\sigma}$ are related to the multiplicities of complementary series representations, and the numbers $\mathrm{res}_{\lambda=\mu} L_\Gamma(\pi^{\sigma,\lambda})$ can be expressed in terms of dimensions of spaces of invariant distributions supported on the limit set (see Corollary 2 for a more precise explanation). Our work extends previous results [11] in the two-dimensional case and [20] in the spherical case of $G = SO(1,n)$, $n \geq 2$.

In [8], using different methods in the real hyperbolic and spherical case, the right-hand side of the functional equation for the Selberg zeta function has been expressed in terms of a regularized scattering determinant. It is an interesting problem to get a similar result in the framework of this paper. This could help to understand the singularities of the Selberg zeta function at the negative integral points. In [10], the authors consider Selberg zeta functions associated with spinor representations on odd-dimensional hyperbolic spaces and identify the special value at the symmetry point with regularized η-invariants.

Using symbolic dynamics of the geodesic flow and the thermodynamic formalism, one can show that the Selberg zeta functions themselves are meromorphic functions of finite order [20]. This gives information about the growth of Φ and on the counting function of resonances. In particular, it shows that Φ can be applied to Schwartz functions like the Fourier transform $\hat{f}$ of a K-finite smooth function f of compact support on G. For the relation between the growth, resonance counting, and the Hausdorff dimension of the limit set, see also [12].

2 The Distribution Ψ

2.1 *Invariant Distributions*

Let G be a semisimple Lie group. We fix once and for all a Haar measure μ_G on G. In this subsection, we describe two sorts of conjugation invariant distributions on G, namely orbital integrals and characters of irreducible representations.

Let $\gamma \in G$ be a semisimple element. The orbit $\mathscr{O}_\gamma := \{g\gamma g^{-1} \,|\, g \in G\}$ of γ under conjugation by G is a submanifold of G, which can be identified with $G_\gamma \backslash G$, where G_γ denotes the centralizer of γ. The inclusion $i_\gamma : G_\gamma \backslash G \cong \mathscr{O}_\gamma \hookrightarrow G$ is a proper map. Therefore, the pull-back by i_γ is a continuous map

$$i_\gamma^* : C_c^\infty(G) \to C_c^\infty(G_\gamma \backslash G).$$

If we choose a Haar measure μ_{G_γ} on G_γ, then we obtain an induced measure $\mu_{G_\gamma \backslash G}$ on $G_\gamma \backslash G$ such that

$$\int_G f(g)\mu_G(dg) = \int_{G_\gamma \backslash G} \int_{G_\gamma} f(hg)\mu_{G_\gamma}(dh)\mu_{G_\gamma \backslash G}(dg).$$

The orbital integral θ_γ associated with γ and the choice of the Haar measure μ_{G_γ} is, by definition, the composition of i_γ^* and the measure $\mu_{G_\gamma \backslash G}$, i.e.

$$\theta_\gamma(f) := \mu_{G_\gamma \backslash G} \circ i_\gamma^*(f) = \int_{G_\gamma \backslash G} f(g\gamma g^{-1})\mu_{G_\gamma \backslash G}(dg).$$

We now recall the character θ_π of an irreducible admissible representation π of G, on a Hilbert space V_π, see e.g. [21]. If $f \in C_c^\infty(G)$, then

$$\pi(f) := \int_G f(g)\pi(g)\mu_G(dg)$$

is a trace class operator on V_π. The character θ_π is the distribution on G given by

$$\theta_\pi(f) := \operatorname{Tr}\pi(f).$$

2.2 An Invariant Distribution Associated to Γ

Let G be a semisimple linear connected Lie group of real rank one. We consider a torsion-free discrete convex-cocompact, non-cocompact subgroup $\Gamma \subset G$ (see [4], Sect. 2). Let $\tilde{\mathcal{O}}_\Gamma$ denote the disjoint union of manifolds $G_\gamma\backslash G$, where γ runs over a set $\widetilde{C\Gamma}$ of representatives of the set $C\Gamma \setminus \{[1]\}$ of non-trivial conjugacy classes of Γ:

$$\tilde{\mathcal{O}}_\Gamma = \bigsqcup_{\gamma\in\widetilde{C\Gamma}} G_\gamma\backslash G.$$

The natural map $i_\Gamma : \tilde{\mathcal{O}}_\Gamma \to G$ is proper, and we obtain a continuous map

$$i_\Gamma^* : C_c^\infty(G) \to C_c^\infty(\tilde{\mathcal{O}}_\Gamma).$$

For each $\gamma \in \widetilde{C\Gamma}$, we fix a Haar measure μ_{G_γ}. Then we define a measure μ_Γ on $\tilde{\mathcal{O}}_\Gamma$ such that its restriction to $G_\gamma\backslash G$ is $\operatorname{vol}(\Gamma_\gamma\backslash G_\gamma)\mu_{G_\gamma\backslash G}$. Note that this measure only depends on the Haar measure μ_G and not on the choices of μ_{G_γ}.

Definition 1. The geometric side of the trace formula is the distribution Ψ on G given by

$$\Psi := \mu_\Gamma \circ i_\Gamma^*.$$

In terms of orbital integrals, we can write

$$\Psi(f) = \sum_{\gamma\in\widetilde{C\Gamma}} \operatorname{vol}(\Gamma_\gamma\backslash G_\gamma)\theta_\gamma(f).$$

Note that this distribution is in fact a measure, invariant under conjugation, and it only depends on Γ and the Haar measure μ_G.

2.3 The Fourier Inversion Formula

Let $\hat{G}$ denote the unitary dual of G. This is the set of equivalence classes of irreducible unitary representations of G equipped with a natural structure of a measurable space. For $\pi \in \hat{G}$, the operator $\pi(f)$ is, by definition, the value of the Fourier tranform of f at π, which we will also denote by $\hat{f}(\pi)$.

It is a consequence of the Plancherel theorem for G [13,22] that there is a measure p on $\hat{G}$ such that for any $f \in C_c^\infty(G)$ and $g \in G$ we have

$$f(g) = \int_{\hat{G}} \operatorname{Tr} \pi(g)^{-1} \hat{f}(\pi)\, p(\mathrm{d}\pi).$$

Note that $p(\mathrm{d}\pi)$ depends on the choice of the Haar measure μ_G. Later in this paper, we will state a more explicit version of the Plancherel theorem.

2.4 The Fourier Transform of Ψ

The contents of a trace formula for convex-cocompact groups Γ would be an expression of $\Psi(f)$ in terms of the Fourier transform $\hat{f}$. In other words, we are interested in the Fourier transform of the distribution Ψ. Since Ψ is invariant, this expression should only involve the characters $\theta_\pi(f) = \operatorname{Tr}\hat{f}(\pi)$. Thus, there should exist a certain measure Φ on $\hat{G}$ such that the following equality holds true for all $f \in C_c^\infty(G)$:

$$\Psi(f) = \int_{\hat{G}} \theta_\pi(f) \Phi(\mathrm{d}\pi).$$

Note that there is a Paley–Wiener theorem for G, which characterizes the range of the Fourier transform as a certain Paley–Wiener space. A priori, Φ is a functional on this Paley–Wiener space, and it would be a non-trivial statement that this functional is in fact induced by a measure on $\hat{G}$.

2.5 The Distribution Ψ as a Regularized Trace

In this paper, we will not compute the Fourier transform Φ of Ψ in the sense of Sect. 2.4. Rather we will compute the candidate for Φ using a different interpretation of Ψ.

Let R denote the right-regular representation of G on $L^2(G)$. It extends to the convolution algebra $L^1(G)$ by the formula

$$R(f) = \int_G f(g) R(g) \mu_G(\mathrm{d}g).$$

If $f \in C_c^\infty(G)$, then $R(f)$ is an integral operator with smooth integral kernel $K_{R(f)}(g,h) = f(g^{-1}h)$. In a similar manner, we have a unitary right-regular representation R_Γ of G on the Hilbert space $L^2(\Gamma\backslash G)$, which can be extended to $L^1(G)$ using the formula

$$R_\Gamma(f) = \int_G f(g) R_\Gamma(g) \mu_G(\mathrm{d}g).$$

If $f \in C_c^\infty(G)$, then $R_\Gamma(f)$ is an integral operator with smooth kernel

$$K_{R_\Gamma(f)}(g,h) = \sum_{\gamma \in G} f(g^{-1}\gamma h). \tag{5}$$

Indeed, for $\phi \in L^2(\Gamma\backslash G)$ we have

$$\begin{aligned} R_\Gamma(f)\phi(g) &= \int_G \phi(gh)f(h)\mu_G(\mathrm{d}h) \\ &= \int_G \phi(h)f(g^{-1}h)\mu_G(\mathrm{d}h) \\ &= \int_{\Gamma\backslash G} \sum_{\gamma\in\Gamma} \phi(h)f(g^{-1}\gamma h)\mu_G(\mathrm{d}h). \end{aligned}$$

Lemma 1. *For $f \in C_c^\infty(G)$ we have*

$$\Psi(f) = \int_{\Gamma\backslash G} [K_{R_\Gamma(f)}(\Gamma g, \Gamma g) - K_{R(f)}(g,g)]\mu_G(\mathrm{d}g).$$

Proof. We compute

$$\begin{aligned} \Psi(f) &= \sum_{\gamma\in\widetilde{C\Gamma}} \mathrm{vol}(\Gamma_\gamma\backslash G_\gamma)\theta_\gamma(f) \\ &= \sum_{\gamma\in\widetilde{C\Gamma}} \mathrm{vol}(\Gamma_\gamma\backslash G_\gamma) \int_{G_\gamma\backslash G} f(g^{-1}\gamma g)\mu_{G_\gamma\backslash G}(\mathrm{d}g) \\ &= \sum_{\gamma\in\widetilde{C\Gamma}} \int_{\Gamma_\gamma\backslash G} f(g^{-1}\gamma g)\mu_G(\mathrm{d}g) \\ &= \int_{\Gamma\backslash G} \sum_{\gamma\in\widetilde{C\Gamma}} \sum_{h\in\Gamma_\gamma\backslash\Gamma} f(g^{-1}h^{-1}\gamma hg)\mu_G(\mathrm{d}g) \\ &= \int_{\Gamma\backslash G} \sum_{1\neq\gamma\in\Gamma} f(g^{-1}\gamma g)\mu_G(\mathrm{d}g) \\ &= \int_{\Gamma\backslash G} [K_{R_\Gamma(f)}(g,g) - K_{R(f)}(g,g)]\mu_G(\mathrm{d}g). \end{aligned}$$

□

Of course, Formula (5) and Lemma 1 remain valid if we replace the condition of compact support by certain weaker decay properties. We omit the discussion of possible such replacements at this point since we do not need it for the main line of argument of this paper.

The expression of $\Psi(f)$ in terms of the integral kernels of $R(f)$ and $R_\Gamma(f)$ can be used to define Ψ on other classes of functions or even on certain distributions.

Using the Plancherel theorems for $L^2(G)$ and $L^2(\Gamma\backslash G)$, the right-regular representations R and R_Γ can be extended. If f is K-finite and $\hat{f}$ is smooth (in a sense to be made precise) and has also compact support, then we will see that $g \mapsto [K_{R_\Gamma(f)}(g,g) - K_{R(f)}(g,g)]$ belongs to $L^1(\Gamma\backslash G)$, and thus

$$\Psi'(f) := \int_{\Gamma\backslash G} [K_{R_\Gamma(f)}(g,g) - K_{R(f)}(g,g)]\mu_G(\mathrm{d}g)$$

is well defined. The main result of this paper is an expression of $\Psi'(f)$ in terms of $\hat{f}$ for those functions.

As mentioned in the introduction, we are going to apply Ψ' to the difference of distribution kernels of the resolvents $(z-C)^{-1}$ of the Casimir operator restricted to a K-type of $L^2(\Gamma\backslash G)$ and $L^2(G)$, respectively. In this example, the single kernels are not smooth, but their difference is so on the diagonal of $\Gamma\backslash G$. Strictly speaking, the integral defining Ψ' exists for $\mathrm{Re}(z) \ll 0$. For other values of z, we introduce a truncated version Ψ'_R, $R > 0$, and we define the value of Ψ' as the constant term of the asymptotic expansion of Ψ'_R as $R \to \infty$. It seems to be an interesting problem to characterize the functions of C (restricted to a K-type) with the property that Ψ'_R (applied to the corresponding distribution kernels) admits such an asymptotic expansion.

Given a discrete series representation π of G, we can consider the corresponding isotypic components of $L^2(G)$ and $L^2(\Gamma\backslash G)$. If we further consider a K-type of π, then the projections onto these components have smooth integral kernels. As a by-product of the investigation of the resolvents, we can show that Ψ' can be applied to these integral kernels and that its values are integers.

3 The Plancherel Theorem and Integral Kernels

3.1 *The Plancherel Theorems for $L^2(G)$ and $L^2(\Gamma\backslash G)$: Support of Plancherel Measures*

From now on, it is our standing assumption that G has real rank one, $G \neq F_4^{-20}$. We start with describing the rough structure of the unitary dual $\hat{G}$. First, there is a countable family of square integrable unitary representations, the discrete series $\hat{G}_{\mathrm{d}}$. The discretely decomposable subspace $L^2(G)_{\mathrm{d}} \subset L^2(G)$ is composed out of these representations each occurring with infinite multiplicity.

The orthogonal complement $L^2(G)_{\mathrm{ac}}$ of $L^2(G)_{\mathrm{d}}$ is given by a countable direct sum of direct integrals of unitary principal series representations. We are going to describe their parametrization. Let $G = KAN$ be an Iwasawa decomposition of G. The abelian group A is isomorphic to the multiplicative group $\mathbb{R}^+$. Let $\mathfrak{a}$ and $\mathfrak{n}$ denote the Lie algebras of A and N. Then $\dim_{\mathbb{R}}(\mathfrak{a}) = 1$ and the roots of $(\mathfrak{a}, \mathfrak{n})$ fix an order on $\mathfrak{a}$. Let $M = Z_K(A)$ denote the centralizer of A in K. The unitary principal series representations $\pi^{\sigma,\lambda}$ of G are parametrized by the set $(\sigma, \lambda) \in \hat{M} \times i\mathfrak{a}^*$. Let W

denote the Weyl group $N_K(A)/M$, where $N_K(A)$ denotes the normalizer of A in K. It is isomorphic to $\mathbb{Z}/2$, and we can choose a representative of the non-trivial element $w \in N_K(A)$ such that $w^{-1} = w$. One knows that $\pi^{\sigma,\lambda}$ is equivalent to $\pi^{\sigma^w,-\lambda}$, where σ^w denotes the Weyl conjugate representation of σ given by $\sigma^w(m) := \sigma(m^w)$. For $\lambda \neq 0$, the representation $\pi^{\sigma,\lambda}$ is irreducible. If σ is equivalent to σ^w, i.e. σ is Weyl invariant, then it may happen that $\pi^{\sigma,0}$ is reducible. In this case, it decomposes into a sum $\pi^{\sigma,+} \oplus \pi^{\sigma,-}$ of limits of discrete series representations.

The set of equivalence classes of unitary representations of G, which we have listed above, is the set of tempered representations. We refer to Sect. 8 of [4] for a discussion of the notion of temperedness for $L^2(G)$ and $L^2(\Gamma\backslash G)$.

Of course, the Plancherel theorem for $L^2(G)$ has been known explicitly for a long time [13]. The Plancherel measure p is supported on the set of tempered representations (compare [2]). In particular, it is absolutely continuous with respect to the Lebesgue measure on $i\mathfrak{a}$. Thus, we can neglect the point $\lambda = 0$. Then $L^2(G)_{\mathrm{ac}}$ decomposes as a direct integral of unitary principal series representations over $\hat{M} \times i\mathfrak{a}_+^*$ with infinite multiplicity, and the Plancherel measure has full support. Note that the multiplicity space of the representation π can be realized as V_π^*.

By $\hat{G}_{\mathrm{ac}}$ we denote the set of irreducible unitary principal series representations $\pi^{\sigma,\lambda}$, $\lambda \neq 0$. The remaining unitary representations $\hat{G}_{\mathrm{c}} = \hat{G} \setminus (\hat{G}_{\mathrm{d}} \cup \hat{G}_{\mathrm{ac}})$ can be realized as subspaces of principal series representations $\pi^{\sigma,\lambda}$ with $\lambda \in \mathfrak{a}_+^* \cup \{0\}$. The case of limits of discrete series $\hat{G}_{\mathrm{ld}}$ (in this case $\lambda = 0$) was mentioned above. The representations with parameter $\lambda > 0$ are not tempered and belong to the complementary series $\hat{G}_{\mathrm{c}}$.

In [4], we studied the Plancherel theorem for $L^2(\Gamma\backslash G)$. Let us recall its rough structure. The support of the corresponding Plancherel measure p_Γ is the union of $\hat{G}_{\mathrm{d}}$, $\hat{G}_{\mathrm{ac}}$, and a countable subset of $\hat{G}_{\mathrm{c}}$. $L^2(\Gamma\backslash G)$ decomposes into sum of subspaces $L^2(\Gamma\backslash G)_{\mathrm{cusp}}$, $L^2(\Gamma\backslash G)_{\mathrm{ac}}$, and $L^2(\Gamma\backslash G)_{\mathrm{c}}$. Here, $L^2(\Gamma\backslash G)_{\mathrm{cusp}}$ is discretely decomposable into representations of the discrete series, each occurring with infinite multiplicity, $L^2(\Gamma\backslash G)_{\mathrm{c}}$ is discretely decomposable into representations belonging to $\hat{G}_{\mathrm{c}}$, each occurring with finite multiplicity, and $L^2(\Gamma\backslash G)_{\mathrm{ac}}$ is a direct integral of unitary principal series representations with infinite multiplicity over the parameter set $\hat{M} \times i\mathfrak{a}_+^*$. On this set the Plancherel measure p_Γ is absolutely continuous to the Lebesgue measure and has full support. The multiplicity space M_π can be realized as a subspace of the Γ-invariant distribution vectors of $V_{\tilde{\pi}}$, i.e., $M_\pi \subset {}^\Gamma V_{\tilde{\pi},-\infty}$, where $\tilde{\pi}$ denotes the dual representation of π. For $\pi \in \hat{G}_{\mathrm{ac}}$, we are going to describe M_π explicitly in Sect. 3.3.

3.2 Extension of R and R_Γ

The Plancherel theorem for G provides a G-equivariant unitary equivalence

$$U : L^2(G) \xrightarrow{\sim} \int_{\hat{G}} V_\pi^* \hat{\otimes} V_\pi \, p(d\pi), \tag{6}$$

where G acts on $L^2(G)$ by the right-regular representation R, and the action on the direct integral is given by $g \mapsto \{\pi \mapsto \mathrm{id}_{V_\pi^*} \otimes \pi(g)\}$. We can identify $V_\pi^* \hat{\otimes} V_\pi$ with the space of Hilbert–Schmidt operators on V_π. For $\phi \in C_c^\infty(G)$ we set

$$U(\phi) := \left\{\pi \mapsto \hat{\tilde{\phi}}(\pi)\right\},$$

where $\tilde{\phi}(g) := \phi(g^{-1})$. This fixes the normalization of the Plancherel measure p.

The inverse transformation maps the family $\pi \mapsto h(\pi)$ to the function

$$\check{h}(g) := U^{-1}(h)(g) = \int_{\hat{G}} \mathrm{Tr}\, \pi(g) h(\pi) p(\mathrm{d}\pi). \tag{7}$$

If $f \in L^1(G)$, then $UR(f)U^{-1}$ is given by $\{\pi \mapsto \mathrm{id}_{V_\pi^*} \otimes \hat{f}(\pi)\}$.

The Plancherel theorem for $\Gamma\backslash G$ provides a G-equivariant unitary equivalence

$$U_\Gamma : L^2(\Gamma\backslash G) \xrightarrow{\sim} \int_{\hat{G}} M_\pi \hat{\otimes} V_\pi \, p_\Gamma(\mathrm{d}\pi), \tag{8}$$

where G acts on $L^2(\Gamma\backslash G)$ by the right-regular representation R_Γ, and the representation of G on the direct integral is given by $g \mapsto \{\pi \mapsto \mathrm{id}_{M_\pi} \otimes \pi(g)\}$. Again, if $f \in L^1(G)$, than $U_\Gamma R_\Gamma(f) U_\Gamma^{-1}$ is given by $\{\pi \mapsto \mathrm{id}_{M_\pi} \otimes \hat{f}(\pi)\}$. In order to write down an explicit formula for U_Γ, we first identify M_π^* with $M_{\tilde{\pi}}$ and embed $M_\pi \hat{\otimes} V_\pi$ into $\mathrm{Hom}(M_{\tilde{\pi}}, V_\pi)$. For $\phi \in C_c^\infty(\Gamma\backslash G)$ we define

$$U_\Gamma(\phi)(\pi) := \left\{ M_{\tilde{\pi}} \ni v \mapsto \int_{\Gamma\backslash G} \phi(g) \pi(g^{-1}) v \, \mu_G(\mathrm{d}g) \in V_\pi \right\}.$$

This fixes the normalization of p_Γ.

Let now h be a function on $\mathrm{supp}(p)$ such that $h(\pi)$ is a bounded operator on V_π for almost all $\pi \in \mathrm{supp}(p)$. If h is measurable[2] and essentially bounded, then it acts on the direct integral (6) by $\{\pi \mapsto \mathrm{id}_{V_\pi^*} \otimes h(\pi)\}$ and thus defines a bounded operator

$$U^{-1}\{\pi \mapsto \mathrm{id}_{V_\pi^*} \otimes h(\pi)\} U =: \check{R}(h)$$

on $L^2(G)$ commuting with the left-regular action of G.

In a similar manner, if h is a function on $\mathrm{supp}(p_\Gamma)$ such that $h(\pi)$ is a bounded operator on V_π for almost all $\pi \in \mathrm{supp}(p_\Gamma)$, and h is essentially bounded, then it acts on the direct integral (8) by $\{\pi \mapsto \mathrm{id}_{M_\pi} \otimes h(\pi)\}$ and thus defines a bounded operator

$$U_\Gamma^{-1}\{\pi \mapsto \mathrm{id}_{M_\pi} \otimes h(\pi)\} U_\Gamma =: \check{R}_\Gamma(h)$$

on $L^2(\Gamma\backslash G)$.

---

[2]The notion of measurability is part of the structure of the direct integral, see e.g. [22], Chap. 14. In our case it just amounts to the measurable dependence on the inducing parameters discussed in the previous section.

Let us now assume that the $h(\pi)$ are of trace-class, and that $\int_{\hat{G}} \|h(\pi)\|_1 \, p(d\pi)$ is finite, where $\|.\|_1$ denotes the trace norm $\|A\|_1 = \operatorname{Tr}|A|$ for a trace class operator A on V_π. Then $\check{R}(h)$ is an integral operator with integral kernel $K_{\check{R}(h)}(g,k) = \check{h}(g^{-1}k)$. Indeed, for $\phi \in C_c^\infty(G)$ we have

$$\begin{aligned}(\check{R}(h)\phi)(g) &= \int_{\hat{G}} \operatorname{Tr}\pi(g)h(\pi)\hat{\hat{\phi}}(\pi)p(d\pi)\\ &= \int_{\hat{G}} \operatorname{Tr}\int_G \pi(g)h(\pi)\pi(k^{-1})\phi(k)\mu_G(dk)p(d\pi)\\ &= \int_G\int_{\hat{G}} \operatorname{Tr}\pi(k^{-1}g)h(\pi)p(d\pi)\phi(k)\mu_G(dk)\\ &= \int_G \check{h}(g^{-1}k)\phi(k)\mu_G(dk).\end{aligned}$$

We are looking for a similar formula for the integral kernel of $\check{R}_\Gamma(h)$ in Sect. 3.3.

3.3 *The Absolute Continuous Part of $L^2(\Gamma\backslash G)$: Integral Kernels for $\check{R}_\Gamma(h)$*

In this subsection, we describe in detail the Plancherel decomposition of $L^2(\Gamma\backslash G)_{ac}$. The goal is to exhibit a class of functions $\pi \mapsto h(\pi)$ with the property that $\check{R}_\Gamma(h)$ is an integral operator.

Let $P = MAN$ be a fixed parabolic subgroup. If $(\sigma,\lambda) \in \hat{M} \times \mathfrak{a}^*_{\mathbb{C}}$, then we define the representation σ_λ of P by $\sigma_\lambda(man) := \sigma(m)a^{\rho-\lambda}$, where $\rho \in \mathfrak{a}^*$ is given by $2\rho(H) = \operatorname{tr}\operatorname{ad}(H)_{|\mathfrak{n}}$, $H \in \mathfrak{a}$, and for $\lambda \in \mathfrak{a}^*_{\mathbb{C}}$ and $a = \exp(H) \in A$ we put $a^\lambda = e^{\lambda(H)}$. We realize the principal series representation $H^{\sigma,\lambda} := V_{\pi^{\sigma,\lambda}}$ as the subspace of those distributions in $C^{-\infty}(G \times_P V_{\sigma_\lambda})$ that are given by locally integrable functions f such that $f_{|K} \in L^2(K \times_M V_\sigma)$. Then $H^{\sigma,\lambda}_{\pm\infty} = C^{\pm\infty}(G \times_P V_{\sigma_\lambda})$ are the spaces of smooth (resp. distribution) vectors of $\pi^{\sigma,\lambda}$. By restriction to K, we obtain canonical isomorphisms $H^{\sigma,\lambda} \cong L^2(K \times_M V_\sigma)$. Since the space on the right-hand side does not depend on λ, it therefore makes sense to speak of smooth functions f on $i\mathfrak{a}^*$ such that $f(\lambda) \in H^{\sigma,\lambda}$.

Note that ∂X can be identified with G/P. Let $G/P = \Omega \cup \Lambda$ be the Γ-invariant decomposition of the space G/P into the (open) domain of discontinuity Ω and the (closed) limit set Λ. As a convex-cocompact subgroup Γ acts freely and cocompactly on Ω. We put $B := \Gamma\backslash\Omega$. Furthermore, we define the bundle $V_B(\sigma_\lambda) \to B$ by $V_B(\sigma_\lambda) := \Gamma\backslash(G \times_P V_{\sigma_\lambda})_{|\Omega}$. If $\lambda \in i\mathfrak{a}^*$, then we have a natural Hilbert space $L^2(B, V_B(\sigma_\lambda))$. Again, fixing a volume form on B, we obtain identifications of the spaces $L^2(B, V_B(\sigma_\lambda))$ with the fixed space $L^2(B, V_B(\sigma_0))$ so that it makes sense to speak of smooth functions f on $i\mathfrak{a}^*$ such that $f(\lambda) \in L^2(B, V_B(\sigma_\lambda))$. We refer to [4, Sect. 3] for more details.

By ${}^{\Gamma}H_{-\infty}^{\sigma,\lambda}$, we denote the space of Γ-invariant distribution vectors of the principal series representation. In [4], we defined a family of extension maps ext: $L^2(B,V_B(\sigma_\lambda)) \to {}^{\Gamma}H_{-\infty}^{\sigma,\lambda}$ depending meromorphically on λ. It has no pole for $\lambda \in i\mathfrak{a}^*$, $\lambda \neq 0$, and for these parameters it provides an explicit identification of the multiplicity space $M_{\pi^{\sigma,\lambda}} \subset {}^{\Gamma}H_{-\infty}^{\tilde{\sigma},-\lambda}$ with $L^2(B,V_B(\tilde{\sigma}_{-\lambda}))$.

The Plancherel measures p and p_Γ on $\{\sigma\} \times i\mathfrak{a}_+^*$ are given by $\frac{\dim(V_\sigma)}{2\pi\omega_X} p_\sigma(\lambda)\mathrm{d}\lambda$, where p_σ is a smooth symmetric function on $i\mathfrak{a}^*$ of polynomial growth (see [4], Lemma 5.5. (3)), and $\omega_X := \lim_{a\to\infty} a^{-2\rho}\mathrm{vol}_{G/K}(KaK)$ (see [4, Sect. 11]). Note that $d\lambda$ is the *real* Lebesgue measure on $i\mathfrak{a}$.

We now describe the embedding

$$U_\Gamma^{-1} : \frac{\dim(V_\sigma)}{2\pi\omega_X} \int_{\{\sigma\}\times i\mathfrak{a}_+^*} L^2(B,V_B(\tilde{\sigma}_{-\lambda})) \otimes H^{\sigma,\lambda} p_\sigma(\lambda)\mathrm{d}\lambda \to L^2(\Gamma\backslash G)_{\mathrm{ac}}.$$

If $v \otimes w \in L^2(B,V_B(\tilde{\sigma}_{-\lambda})) \otimes H_\infty^{\sigma,\lambda}$, then we define $\langle v\otimes w\rangle := \langle \mathrm{ext}(v), w\rangle$. Let ϕ be a smooth function of compact support on $i\mathfrak{a}_+^* \cup \{0\}$ such that $\phi(\lambda) \in L^2(B,V_B(\tilde{\sigma}_{-\lambda})) \otimes H_\infty^{\sigma,\lambda}$, then we have

$$U_\Gamma^{-1}(\phi)(g) = \frac{\dim(V_\sigma)}{2\pi\omega_X} \int_{i\mathfrak{a}_+^*} \langle (1\otimes \pi^{\sigma,\lambda}(g))\phi(\lambda)\rangle p_\sigma(\lambda)\mathrm{d}\lambda.$$

Note that ext may be singular at $\lambda = 0$. In this case, it has a first-order pole and $p_\sigma(0) = 0$ (see [4], Prop. 7.4) such that the integral is still well defined.

We now fix a K-type $\tau \in \hat{K}$. Let $H^{\sigma,\lambda}(\tau)$ denote the τ-isotypic component of $H^{\sigma,\lambda}$. By Frobenius reciprocity, we have a canonical identification

$$H^{\sigma,\lambda}(\tau) \overset{\alpha}{=} V_\tau \otimes \mathrm{Hom}_K(V_\tau, H^{\sigma,\lambda}) \overset{1\otimes\beta}{=} V_\tau \otimes \mathrm{Hom}_M(V_\tau, V_\sigma). \tag{9}$$

Here, $\alpha^{-1}(v\otimes U) := U(v)$ and $\beta(U)(v) := U(v)(1)$.
Any operator $A \in \mathrm{End}(V_\tau \otimes \mathrm{Hom}_M(V_\tau,V_\sigma))$ gives rise to a finite rank operator $F(A) := ((1\otimes\beta)\alpha)^{-1}A(1\otimes\beta)\alpha \in \mathrm{End}(H^{\sigma,\lambda})$, which is trivial on the orthogonal complement of $H^{\sigma,\lambda}(\tau)$.

Let q be a smooth function of compact support on $\hat{M} \times i\mathfrak{a}^*$ such that $q(\sigma,\lambda) \in \mathrm{End}(V_\tau \otimes \mathrm{Hom}_M(V_\tau,V_\sigma))$. We call q symmetric if it is compatible with the equivalences $J_{\sigma,\lambda}^w : H^{\sigma,\lambda} \to H^{\sigma^w,-\lambda}$, i.e. if $F(q(\sigma^w,-\lambda)) = J_{\sigma,\lambda}^w \circ F(q(\sigma,\lambda)) \circ (J_{\sigma,\lambda}^w)^{-1}$. If q is symmetric, then we define a function h_q on $\hat{G}$ with $h_q(\pi) \in \mathrm{End}(V_\pi)$ by $h_q(\pi^{\sigma,\lambda}) := F(q(\sigma,\lambda))$ for $(\sigma,\lambda) \in \hat{M} \times i\mathfrak{a}^*$, $\lambda \neq 0$, and by $h_q(\pi) = 0$ for all other representations.

Let $\pi_* : H_\infty^{\sigma,\lambda} \to L^2(B,V_B(\sigma_\lambda))$ denote the push-down map [4], which is the adjoint of the extension ext: $L^2(B,V_B(\tilde{\sigma}_{-\lambda})) \to H_{-\infty}^{\tilde{\sigma},-\lambda}$. The composition

$$\pi^{\sigma,\lambda}(g)h_q(\pi^{\sigma,\lambda})\pi^{\sigma,\lambda}(k^{-1})\mathrm{ext}\,\pi_*$$

is an operator of finite rank from $H_\infty^{\sigma,\lambda}$ to $H_\infty^{\sigma,\lambda}$. It is therefore nuclear and has a well-defined trace.

Lemma 2. *The operator $\check{R}_\Gamma(h_q)$ has a smooth integral kernel given by*

$$K_{\check{R}_\Gamma(h_q)}(g,k) = \sum_{\sigma\in\hat{M}} \frac{\dim(V_\sigma)}{4\pi\omega_X} \int_{i\mathfrak{a}^*} \operatorname{Tr}\pi^{\sigma,\lambda}(g)h_q(\pi^{\sigma,\lambda})\pi^{\sigma,\lambda}(k^{-1})\mathrm{ext}\circ\pi_* \, p_\sigma(\lambda)d\lambda.$$

Proof. First of all note that the integral is well defined at $\lambda = 0$. If $\mathrm{ext}\circ\pi_*$ is singular at this point, then it has a pole of at most second order. But then the Plancherel density vanishes at least of second order, too.

Let $\phi \in C_c^\infty(\Gamma\backslash G)$. In the Plancherel decomposition, it is represented by the function $\pi \mapsto U_\Gamma(\phi)(\pi) \in M_\pi\hat{\otimes}V_\pi$. We fix λ for a moment. Let $\{v_i\}$ be an orthonormal basis of $L^2(B,V_B(\sigma_\lambda))$. Furthermore, let $\{v^i\}$ be the dual basis of $L^2(B,V_B(\tilde{\sigma}_{-\lambda}))$. Then we have

$$U_\Gamma(\phi)(\pi^{\sigma,\lambda}) = \sum_i \int_{\Gamma\backslash G} v^i \otimes \phi(k)\pi^{\sigma,\lambda}(k^{-1})\mathrm{ext}(v_i)\,\mu_G(dk).$$

We have

$$\begin{aligned} &\check{R}_\Gamma(h_q)(\phi)(g) \\ &\quad = \frac{\dim(V_\sigma)}{4\pi\omega_X} \sum_{\sigma\in\hat{M}} \int_{i\mathfrak{a}^*} \langle\langle(1\otimes\pi^{\sigma,\lambda}(g)h_q(\pi^{\sigma,\lambda}))U_\Gamma(\phi)(\pi^{\sigma,\lambda})\rangle\rangle p_\sigma(\lambda)d\lambda. \end{aligned} \tag{10}$$

We now choose $\{v_i\}$ such that $v_1,\ldots,v_r$ is a basis of the finite dimensional range of the operator $\pi_*\circ\pi^{\sigma,\lambda}(g)\circ h_q(\pi^{\sigma,\lambda})$ and compute

$$\begin{aligned} &\langle\langle(1\otimes\pi^{\sigma,\lambda}(g)h_q(\pi^{\sigma,\lambda}))U_\Gamma(\phi)(\pi^{\sigma,\lambda})\rangle\rangle \\ &\quad = \sum_{i=1}^\infty \int_{\Gamma\backslash G} \langle\mathrm{ext}(v^i), \pi^{\sigma,\lambda}(g)h_q(\pi^{\sigma,\lambda})\phi(k)\pi^{\sigma,\lambda}(k^{-1})\mathrm{ext}(v_i)\rangle\,\mu_G(dk) \\ &\quad = \sum_{i=1}^\infty \int_{\Gamma\backslash G} \langle v^i, \pi_*\circ\pi^{\sigma,\lambda}(g)h_q(\pi^{\sigma,\lambda})\phi(k)\pi^{\sigma,\lambda}(k^{-1})\mathrm{ext}(v_i)\rangle\,\mu_G(dk) \\ &\quad = \sum_{i=1}^r \int_{\Gamma\backslash G} \langle v^i, \pi_*\circ\pi^{\sigma,\lambda}(g)h_q(\pi^{\sigma,\lambda})\phi(k)\pi^{\sigma,\lambda}(k^{-1})\mathrm{ext}(v_i)\rangle\,\mu_G(dk) \\ &\quad = \int_{\Gamma\backslash G} \sum_{i=1}^r \langle v^i, \pi_*\circ\pi^{\sigma,\lambda}(g)h_q(\pi^{\sigma,\lambda})\phi(k)\pi^{\sigma,\lambda}(k^{-1})\mathrm{ext}(v_i)\rangle\,\mu_G(dk) \\ &\quad = \int_{\Gamma\backslash G} \phi(k)\operatorname{Tr}\left(\pi_*\circ\pi^{\sigma,\lambda}(g)h_q(\pi^{\sigma,\lambda})\pi^{\sigma,\lambda}(k^{-1})\mathrm{ext}\right)\,\mu_G(dk) \\ &\quad = \int_{\Gamma\backslash G} \phi(k)\operatorname{Tr}\left(\pi^{\sigma,\lambda}(g)h_q(\pi^{\sigma,\lambda})\pi^{\sigma,\lambda}(k^{-1})\mathrm{ext}\circ\pi_*\right)\,\mu_G(dk). \end{aligned}$$

Inserting this computation into (10), we obtain the desired formula for the integral kernel of $\check{R}_\Gamma(h_q)$. □

4 Poisson Transforms and Asymptotic Computations

4.1 Motivation

Let q be symmetric and define and h_q as in Sect. 3.3. We want to show that the function $g \mapsto [K_{\check{R}_\Gamma(h_q)}(g,g) - K_{\check{R}(h_q)}(g,g)]$ belongs to $L^1(\Gamma\backslash G)$. It follows that

$$\Psi'(\tilde{\check{h}}_q) = \int_{\Gamma\backslash G} [K_{\check{R}_\Gamma(h_q)}(\Gamma g, \Gamma g) - K_{\check{R}(h_q)}(g,g)]\mu_G(\mathrm{d}g)$$

is well defined, and we are asking for an expression of $\Psi'(\tilde{\check{h}}_q)$ (see (7) for $\check{h}_q$) in terms of q, respectively h_q. In this section, we show preparatory results using the language of Poisson transformations. The final result will be obtained in Sect. 5.1.

4.2 Poisson Transformation, c-Functions, and Asymptotics

We fix a K-type τ and an M-type σ. Let $T \in \mathrm{Hom}_M(V_\sigma, V_\tau)$ and $\lambda \in \mathfrak{a}^*_{\mathbb{C}}$. If $w \in V_{\tilde{\tau}}$, then by Frobenius reciprocity we consider $w \otimes T^*$ as an element of $H^{\tilde{\sigma},-\lambda}(\tilde{\tau})$, which is given by the function $k \mapsto T^*(\tilde{\tau}(k^{-1})w)$ under the canonical identification $\phi_{-\lambda} : H^{\tilde{\sigma},-\lambda}_\infty \xrightarrow{\sim} C^\infty(K \times_M V_{\tilde{\sigma}})$. We further put $\Phi_{\lambda,\mu} := \phi_\lambda^{-1} \circ \phi_\mu$. We will also use the notation $\Phi_{0,\lambda}$ for ϕ_λ.

The Poisson transformation

$$P^T_\lambda : H^{\sigma,\lambda}_{-\infty} \to C^\infty(G \times_K V_\tau)$$

is a G-equivariant map, which is defined by the relation

$$\langle w, P^T_\lambda(\psi)(g)\rangle = \langle w \otimes T^*, \pi^{\sigma,\lambda}(g^{-1})\psi\rangle,$$

for all $\psi \in H^{\sigma,\lambda}_{-\infty}$, $w \in V_{\tilde{\tau}}$.

For the definition of the function c_σ we refer to [4, Sect. 5]. We have the relation

$$c_\sigma(\lambda)c_{\tilde{\sigma}}(-\lambda) = p_\sigma(\lambda)^{-1}.$$

It turns out to be useful to introduce the normalized Poisson transformation ${}^0P^T_\lambda := c_\sigma(-\lambda)^{-1}P^T_\lambda$.

We introduce the family of operators

$$\mathscr{P}^T_{\lambda,a} : H^{\sigma,\lambda}_{-\infty} \to C^\infty(K \times_M V_\tau), \quad a \in A_+$$

by

$$\mathscr{P}^T_{\lambda,a}(f)(k) := {}^0P^T_\lambda(f)(ka).$$

In order to discuss the asymptotic behaviour of $\mathscr{P}^T_{\lambda,a}$ as $a \to \infty$, we need the normalized Knapp–Stein intertwining operators

$$J^w_{\sigma,\lambda} : H^{\sigma,\lambda}_{-\infty} \to H^{\sigma^w,-\lambda}_{-\infty}.$$

Note that $J^w_{\sigma,\lambda} \circ J^w_{\sigma^w,-\lambda} = \mathrm{id}$. We again refer to [4, Sect. 5], for more details. The following is a consequence of [4], Lemma 6.2. Let $\alpha \in \mathfrak{a}^*$ denote the short root of $(\mathfrak{a},\mathfrak{n})$. For $\lambda \in i\mathfrak{a}^*$, we have

$$\mathscr{P}^T_{\lambda,a} = a^{\lambda-\rho}\frac{c_\tau(\lambda)}{c_\sigma(-\lambda)} T \circ \Phi_{0,\lambda} + a^{-\lambda-\rho}\tau(w)T \circ \Phi_{0,-\lambda} \circ J^w_{\sigma,\lambda} + a^{-\rho-\alpha}\mathscr{R}_{-\infty}(\lambda,a), \tag{11}$$

where $\mathscr{R}_{-\infty}(\lambda,a) \circ \Phi_{\lambda,0}$ remains bounded in $C^\infty(i\mathfrak{a}^*, \mathrm{Hom}(H^{\sigma,0}_{-\infty}, C^{-\infty}(K \times_M V_\tau)))$ as $a \to \infty$. Multiplication by T (resp. $\tau(w)T$) is here considered as a map from $H^{\sigma,0}_{-\infty}$ (resp. $H^{\sigma^w,0}_{-\infty}$) to $C^{-\infty}(K \times_M V_\tau)$ in the natural way. If $\chi, \tilde{\chi}$ are smooth functions on K/M with disjoint support, then $\chi \mathscr{R}_{-\infty}(\lambda,a)\tilde{\chi} \circ \Phi_{\lambda,0}$ remains bounded in $C^\infty(i\mathfrak{a}^*, \mathrm{Hom}(H^{\sigma,0}_{-\infty}, C^\infty(K \times_M V_\tau)))$ as $a \to \infty$.

4.3 An Estimate

In order to formulate the result appropriately, we introduce the following space $C_\Gamma(G)$ of functions on G. For each compact $V \subset \Omega$ and integer N, we consider the seminorm

$$|\phi|_{V,N} := \sup_{kah \in VA_+K} (1 + |\log(a)|)^N a^{2\rho} |\phi(kah)|, \phi \in C(G).$$

Here we identify V (using $\partial X \cong K/M$) with the subset $\{k \in K \mid kM \in V\}$ of K. We define $C_\Gamma(G)$ as the space of all continuous functions ϕ on G such that $|\phi|_{V,N} < \infty$ for all compact $V \subset \Omega$ and $N \in \mathbb{N}$. If ϕ is Γ-invariant and belongs to $C_\Gamma(G)$, then clearly $\phi \in L^1(\Gamma \backslash G)$.

Now let $T \in \mathrm{Hom}_M(V_\sigma, V_\tau)$, $R \in \mathrm{Hom}_M(V_{\tilde\sigma}, V_{\tilde\tau})$, and $q \in C_c^\infty(i\mathfrak{a}^*)$. Then we can define the operator

$$\begin{aligned} A_q = A_q(T,R) &:= \int_{i\mathfrak{a}^*} {}^0P^T_\lambda \circ (\mathrm{ext} \circ \pi_* - 1) \circ ({}^0P^R_{-\lambda})^* q(\lambda) d\lambda \\ &\in \mathrm{Hom}(C_c^{-\infty}(G \times_K V_\tau), C^\infty(G \times_K V_\tau)). \end{aligned}$$

This operator has a smooth integral kernel $(g,h) \mapsto A_q(g,h) \in \mathrm{End}(V_\tau)$. The main result of this subsection is the following estimate.

Proposition 1. *The function $g \mapsto A_q(G,G)$ belongs to $C_\Gamma(G)$.*

Proof. Note that we only have to show finiteness of the norms $|.|_{V,N}$, where $V \subset \Omega$ is compact and has the additional property that V is contained in the interior of a compact subset $V_1 \subset \Omega$ satisfying $\gamma V_1 \cap V_1 = \emptyset$ for all $1 \neq \gamma \in \Gamma$. Indeed, any seminorm defining $C_\Gamma(G)$ can be majorized by the maximum of a finite number of these special ones.

We choose a smooth cut-off function $\tilde{\chi}$ on Ω such that $\mathrm{supp}(\tilde{\chi}) \subset V_1$ and $\mathrm{supp}(1-\tilde{\chi}) \cap V = \emptyset$. We further choose a cut-off function χ on Ω such that $\tilde{\chi}\chi = \chi$ and $\mathrm{supp}(1-\chi) \cap V = \emptyset$. Then we can write for $k \in V$

$$\begin{aligned} &{}^0P^T_\lambda \circ (\mathrm{ext} \circ \pi_* - 1) \circ \left({}^0P^R_{-\lambda}\right)^* (kah, kah) \\ &\quad = \tau(h)^{-1}\chi(k) \circ \left[\mathscr{P}^T_{\lambda,a} \circ (\mathrm{ext} \circ \pi_* - 1) \circ \left(\mathscr{P}^R_{-\lambda,a}\right)^*\right](k,k) \circ \chi(k)\tau(h). \quad (12) \end{aligned}$$

In order to employ the off-diagonal localization of the Poisson transformation, we write

$$\begin{aligned} &\chi \circ \mathscr{P}^T_{\lambda,a} \circ (\mathrm{ext} \circ \pi_* - 1) \circ \left(\mathscr{P}^R_{-\lambda,a}\right)^* \circ \chi \\ &\qquad = \chi \circ \mathscr{P}^T_{\lambda,a} \circ (1-\tilde{\chi}) \circ (\mathrm{ext} \circ \pi_* - 1) \circ \tilde{\chi} \circ (\mathscr{P}^R_{-\lambda,a})^* \circ \chi \\ &\qquad\quad + \chi \circ \mathscr{P}^T_{\lambda,a} \circ (\mathrm{ext} \circ \pi_* - 1) \circ (1-\tilde{\chi}) \circ (\mathscr{P}^R_{-\lambda,a})^* \circ \chi. \quad (13) \end{aligned}$$

In (13), we could insert the factor $(1-\tilde{\chi})$ since $\tilde{\chi} \circ (\mathrm{ext} \circ \pi_* - 1) \circ \tilde{\chi} = 0$. Using that $\mathrm{supp}(\chi) \cap \mathrm{supp}(1-\tilde{\chi}) = \emptyset$, we have

$$\chi \circ \mathscr{P}^T_{\lambda,a} \circ (1-\tilde{\chi}) = a^{-\lambda-\rho} \chi \circ \tau(w) T \circ \Phi_{0,-\lambda} \circ J^w_{\sigma,\lambda} \circ (1-\tilde{\chi}) + a^{-\rho-\alpha} \mathscr{R}_\infty(\lambda, a),$$

where $\mathscr{R}_\infty(\lambda,a) \circ \Phi_{\lambda,0}$ remains bounded in $C^\infty(i\mathfrak{a}^*, \mathrm{Hom}(H^{\sigma,0}_{-\infty}, C^\infty(K \times_M V_\tau)))$ as $a \to \infty$.

We obtain

$$\begin{aligned} &\chi \circ \mathscr{P}^T_{\lambda,a} \circ (\mathrm{ext} \circ \pi_* - 1) \circ (\mathscr{P}^R_{-\lambda,a})^* \circ \chi \\ &\quad = a^{-2\rho} a^{-2\lambda} \chi \circ \tau(w) T \circ \Phi_{0,-\lambda} \circ J^w_{\sigma,\lambda} \circ (1-\tilde{\chi}) \circ (\mathrm{ext} \circ \pi_* - 1) \circ \tilde{\chi} \\ &\qquad\qquad \circ \Phi_{\lambda,0} \circ R^* \frac{c_\tau(-\lambda)^*}{c_\sigma(\lambda)} \circ \chi \quad (14) \\ &\quad + a^{-2\rho} \chi \circ \tau(w) T \circ \Phi_{0,-\lambda} \circ J^w_{\sigma,\lambda} \circ (1-\tilde{\chi}) \circ (\mathrm{ext} \circ \pi_* - 1) \circ \tilde{\chi} \\ &\qquad\qquad \circ (J^w_{\tilde{\sigma},-\lambda})^* \circ \Phi_{-\lambda,0} \circ R^* \tilde{\tau}(w)^* \circ \chi \quad (15) \\ &\quad + a^{-2\rho} a^{2\lambda} \chi \circ \frac{c_\tau(\lambda)}{c_\sigma(-\lambda)} T \circ \Phi_{0,\lambda} \circ (\mathrm{ext} \circ \pi_* - 1) \circ (1-\tilde{\chi}) \\ &\qquad\qquad \circ (J^w_{\tilde{\sigma},-\lambda})^* \circ \Phi_{-\lambda,0} \circ R^* \tilde{\tau}(w)^* \circ \chi \quad (16) \end{aligned}$$

$$+a^{-2\rho}\chi\circ\tau(w)T\circ\Phi_{0,-\lambda}\circ J^{w}_{\sigma,\lambda}\circ(\mathrm{ext}\circ\pi_*-1)\circ(1-\tilde\chi)$$
$$\circ(J^{w}_{\tilde\sigma,-\lambda})^*\circ\Phi_{-\lambda,0}\circ R^*\tilde\tau(w)^*\circ\chi \qquad (17)$$
$$+Q(\lambda,a),$$

where $a^{2\rho+\alpha}Q(\lambda,a)$ remains bounded in

$$C^\infty(i\mathfrak{a}^*,\mathrm{Hom}(C^{-\infty}(K\times_M V_\tau),C^\infty(K\times_M V_\tau)))\text{ as }a\to\infty.$$

We further compute, using that the intertwining operators commute with $\mathrm{ext}\circ\pi_*$ (compare the proof of Lemma 3 for a similar argument), the functional equation of the intertwining operators, and $\chi\circ(\mathrm{ext}\circ\pi_*-1)\circ\chi=0$

$$\begin{aligned}
(15)+(17)&\\
&=a^{-2\rho}\chi\circ\tau(w)T\circ\Phi_{0,-\lambda}\circ J^{w}_{\sigma,\lambda}\circ(\mathrm{ext}\circ\pi_*-1)\circ\tilde\chi\\
&\qquad\circ(J^{w}_{\tilde\sigma,-\lambda})^*\circ\Phi_{-\lambda,0}\circ R^*\tilde\tau(w)^*\circ\chi\\
&+a^{-2\rho}\chi\circ\tau(w)T\circ\Phi_{0,-\lambda}\circ J^{w}_{\sigma,\lambda}\circ(\mathrm{ext}\circ\pi_*-1)\circ(1-\tilde\chi)\\
&\qquad\circ(J^{w}_{\tilde\sigma,-\lambda})^*\circ\Phi_{-\lambda,0}\circ R^*\tilde\tau(w)^*\circ\chi\\
&=a^{-2\rho}\chi\circ\tau(w)T\circ\Phi_{0,-\lambda}\circ(\mathrm{ext}\circ\pi_*-1)\circ J^{w}_{\sigma,\lambda}\circ(J^{w}_{\tilde\sigma,-\lambda})^*\circ\Phi_{-\lambda,0}\circ R^*\tilde\tau(w)^*\circ\chi\\
&=a^{-2\rho}\chi\circ\tau(w)T\circ\Phi_{0,-\lambda}\circ(\mathrm{ext}\circ\pi_*-1)\circ\Phi_{-\lambda,0}\circ R^*\tilde\tau(w)^*\circ\chi\\
&=a^{-2\rho}\tau(w)T\circ\Phi_{0,-\lambda}\circ\chi\circ(\mathrm{ext}\circ\pi_*-1)\circ\chi\circ\Phi_{-\lambda,0}\circ R^*\tilde\tau(w)^*\\
&=0.
\end{aligned}$$

Note that in (14) and (16) one of the intertwining operators is localized off-diagonally. We conclude that the following families of operators

$$\chi\circ\tau(w)T\circ\Phi_{0,-\lambda}\circ J^{w}_{\sigma,\lambda}\circ(1-\tilde\chi)\circ(\mathrm{ext}\circ\pi_*-1)\circ\tilde\chi\circ\Phi_{\lambda,0}\circ R^*\frac{c_\tau(-\lambda)}{c_\sigma(\lambda)}\circ\chi,$$

$$\chi\circ\frac{c_\tau(\lambda)}{c_\sigma(-\lambda)}T\circ\Phi_{0,\lambda}\circ(\mathrm{ext}\circ\pi_*-1)\circ(1-\tilde\chi)\circ(J^{w}_{\tilde\sigma,-\lambda})^*\circ\Phi_{-\lambda,0}\circ R^*\tilde\tau(w)^*\circ\chi$$

belong to $C^\infty(i\mathfrak{a}^*,\mathrm{Hom}(C^{-\infty}(K\times_M V_\tau),C^\infty(K\times_M V_\tau)))$. Restricting the smooth distribution kernel to the diagonal, multiplying by the smooth compactly supported function q, and integrating over $i\mathfrak{a}^*$ we obtain the following estimates using the standard theory of the Euclidean Fourier transform. For any $N\in\mathbb{N}$, we have

$$\sup_{k\in K}\sup_{a\in A_+}\left|\int_{i\mathfrak{a}^*}Q(\lambda,a)(k,k)q(\lambda)\mathrm{d}\lambda\right|a^{2\rho}(1+|\log(a)|)^N<\infty,$$

$$\sup_{k\in K}\sup_{a\in A_+}\left|\int_{i\mathfrak{a}^*}\chi(k)^2[\tau(w)T\circ\Phi_{0,-\lambda}\circ J^w_{\sigma,\lambda}\circ(1-\tilde{\chi})\circ(\mathrm{ext}\circ\pi_*-1)\right.$$

$$\left.\circ\Phi_{\lambda,0}\circ R^*\frac{c_\tau(-\lambda)}{c_\sigma(\lambda)}](k,k)q(\lambda)a^{-2\lambda}d\lambda\right|(1+|\log(a)|)^N<\infty,$$

$$\sup_{k\in K}\sup_{a\in A_+}\left|\int_{i\mathfrak{a}^*}\chi(k)^2[\frac{c_\tau(\lambda)}{c_\sigma(-\lambda)}T\circ\Phi_{0,\lambda}\circ(\mathrm{ext}\circ\pi_*-1)\circ(1-\tilde{\chi})\circ(J^w_{\tilde{\sigma},-\lambda})^*\right.$$

$$\left.\circ\Phi_{-\lambda,0}\circ R^*\tilde{\tau}(w)^*](k,k)q(\lambda)a^{2\lambda}d\lambda\right|(1+|\log(a)|)^N<\infty.$$

This implies the proposition. □

Remark 1. We have shown in fact that $q\mapsto|(A_q)_{|\mathrm{diag}}|$ is a continuous map from $C_c^\infty(i\mathfrak{a}^*)$ to $L^1(\Gamma\backslash G)$. It would be desirable to extend this map from $C_c^\infty(i\mathfrak{a}^*)$ to the Schwartz space $\mathscr{S}(i\mathfrak{a}^*)$. It is this technical problem that prevents us to prove that the Fourier transform Φ of Ψ restricted to the unitary principal series representations is a tempered distribution. If this would be true, then it is in fact a measure and given by our computations below.

If we would like to show that the map $q\mapsto|(A_q)_{|\mathrm{diag}}|$ extends to a map from the Schwartz space to $L^1(\Gamma\backslash G)$ along the lines above, we need estimates on the growth of ext as the parameter λ tends to infinity along the imaginary axis. If the imaginary axis is in the domain of convergence of ext, i.e. the critical exponent δ_Γ of Γ is negative, then such an estimate is easy to obtain. In the general case, one has to estimate the meromorphic continuation of ext, and this is an open problem.

4.4 A Computation

In this subsection, we want to express $\int_{\Gamma\backslash G}\mathrm{tr}\,A_q(g,g)\mu_G(dg)$ in terms of q.

Recall that the symmetric space $X=G/K$ can be compactified by adjoining the boundary $\partial X=G/P$. Since $\Gamma\subset G$ is convex-cocompact, it acts freely and properly on $X\cup\Omega$ with compact quotient. Therefore, we can choose a smooth function $\chi^\Gamma\in C_c^\infty(X\cup\Omega)$ such that $\sum_{\gamma\in\Gamma}\gamma^*\chi^\Gamma\equiv 1$ on $X\cup\Omega$. The restriction of χ^Γ to X can be lifted to G as a right-K-invariant function, which we still denote by χ^Γ. We denote by χ^Γ_∞ the right-M-invariant lift to K of the restriction of χ^Γ to $\partial X=K/M$. We write

$$\int_{\Gamma\backslash G}\mathrm{tr}\,A_q(g,g)\mu_G(dg)=\int_G\chi^\Gamma(g)\mathrm{tr}\,A_q(g,g)\mu_G(dg).$$

Let χ_U be the characteristic function of the ball B_U in X of radius U centered at the origin $[K]$. Again, we denote its right-K-invariant lift to G by the same symbol. Then we can write

$$\int_G \chi^\Gamma(g)\mathrm{tr}A_q(g,g)\mu_G(\mathrm{d}g) = \lim_{U\to\infty}\int_G \chi^\Gamma(g)\chi_U(g)\mathrm{tr}A_q(g,g)\mu_G(\mathrm{d}g).$$

Given U we fix a function $\chi_1 \in C_c^\infty(G/K)$ such that

$$\chi_1\chi_U\chi^\Gamma = \chi_U\chi^\Gamma. \tag{18}$$

The operator $\chi_1\chi^\Gamma\,{}^0P^T_\lambda \circ (\mathrm{ext}\circ\pi_* - 1)\circ({}^0P^R_{-\lambda})^*\chi_1$ has a compactly supported smooth integral kernel. It is therefore of trace class. Composing it with the multiplication operator by χ_U we see that $\chi_U\chi^\Gamma\,{}^0P^T_\lambda \circ (\mathrm{ext}\circ\pi_* - 1)\circ({}^0P^R_{-\lambda})^*\chi_1$ is of trace class, too. We can write

$$\int_G \chi^\Gamma(g)\chi_U(g)\mathrm{tr}A_q(g,g)\mu_G(\mathrm{d}g)$$

$$= \int_{i\mathfrak{a}^*} \mathrm{Tr}\,[\chi_U\chi^\Gamma\,{}^0P^T_\lambda \circ (\mathrm{ext}\circ\pi_* - 1)\circ({}^0P^R_{-\lambda})^*\chi_1]q(\lambda)\mathrm{d}\lambda.$$

Note that

$$i\mathfrak{a}^* \ni \lambda \mapsto \mathrm{Tr}\,[\chi_U\chi^\Gamma\,{}^0P^T_\lambda \circ (\mathrm{ext}\circ\pi_* - 1)\circ({}^0P^R_{-\lambda})^*\chi_1]$$

is a smooth function. We want to compute its limit in the sense of distributions as $U\to\infty$ using Green's formula.

Note that $V_B(1_{\rho+\alpha})$ is a complex bundle with a real structure, which is trivializable together with this structure. Indeed, B is orientable (note that we assume that G is connected), and $V_B(1_{\rho+\alpha})$ is a real power of $\Lambda^{\max}T^*B$. We choose any non-vanishing real section $\phi \in C^\infty(B, V_B(1_{\rho+\alpha}))$. Let $p: G\to\partial X$ be the natural projection. By changing the sign of ϕ if necessary, we will assume that ϕ is given by a positive smooth function $f: p^{-1}(\Omega)\to\mathbb{R}$ satisfying $f(\gamma gman) = a^\alpha f(g)$ for all $\gamma\in\Gamma$, $g\in p^{-1}(\Omega)$, $man\in P$. For any $z\in\mathbb{C}$, the complex power $f^z: p^{-1}(\Omega)\to\mathbb{C}$ defines a non-vanishing section $\phi^z \in C^\infty(B, V_B(1_{\rho+z\alpha}))$. In particular, if we take z such that $z\alpha = \lambda - \mu$, then multiplication by ϕ^z gives an isomorphism $\bar{\Phi}_{\lambda,\mu}$: $C^\infty(B, V_B(\sigma_\mu)) \to C^\infty(B, V_B(\sigma_\lambda))$, and similar isomorphisms of the spaces of L^2- and distribution sections. If $\mathrm{Re}(z) = 0$, then $\mathrm{ext}: C^\infty(B, V_B(1_{\rho+z\alpha})) \to {}^\Gamma H^{1,\rho+z\alpha}_{-\infty}$ is well defined (indeed $\rho + z\alpha$ belongs to the domain of convergence of the meromorphic family ext). Multiplication by $\mathrm{ext}(\phi^z)$ gives a continuous map

$$\bar{\Phi}_{\lambda,\mu} : H^{\sigma,\mu}_\infty \to H^{\sigma,\lambda}_{-\infty}.$$

This map is Γ-equivariant and extends in fact to larger subspaces of $H^{\sigma,\mu}_{-\infty}$ of distributions, which are smooth on neighbourhoods of the limit set Λ.

The usual trick to bring in Green's formula is to write

$$\begin{aligned}&\mathrm{Tr}\left[\chi_U\chi^{\Gamma 0}P^T_\lambda\circ(\mathrm{ext}\circ\pi_*-1)\circ({}^0P^R_{-\lambda})^*\chi_1\right]\\&=\lim_{\mu\to\lambda}\mathrm{Tr}\left[\chi_U\chi^{\Gamma 0}P^T_\lambda\circ(\mathrm{ext}\circ\bar{\Phi}_{\lambda,\mu}\circ\pi_*-\bar{\Phi}_{\lambda,\mu})\circ\left({}^0P^R_{-\mu}\right)^*\chi_1\right].\end{aligned}$$

Let ∇^τ denote the invariant connection of the bundle $V(\tau)=G\times_K V_\tau$ over X and $\Delta_\tau:=(\nabla^\tau)^*\nabla^\tau$ be the Laplace operator. Then there exists a constant $c\in\mathbb{R}$ such that $(\Delta_\tau+c+\lambda^2)\circ P^T_\lambda=0$. Let n denote the outer unit-normal vector field at ∂B_U. We set

$$\Psi_{\lambda,\mu}:=\mathrm{ext}\circ\bar{\Phi}_{\lambda,\mu}\circ\pi_*-\bar{\Phi}_{\lambda,\mu}.$$

By Green's formula, we have for $\lambda\neq\pm\mu$

$$\mathrm{Tr}\left[\chi_U\chi^{\Gamma 0}P^T_\lambda\circ\Psi_{\lambda,\mu}\circ\left({}^0P^R_{-\mu}\right)^*\chi_1\right]$$

$$=\frac{1}{\lambda^2-\mu^2}\mathrm{Tr}\left[\Delta_\tau,\chi^\Gamma\right]\chi_U\circ{}^0P^T_\lambda\circ\Psi_{\lambda,\mu}\circ\left({}^0P^R_{-\mu}\right)^*\chi_1 \tag{19}$$

$$-\frac{1}{\lambda^2-\mu^2}\mathrm{Tr}\,\chi^\Gamma_{|\partial B_U}\circ\left({}^0P^T_\lambda\right)_{|\partial B_U}\circ\Psi_{\lambda,\mu}\circ\left(\nabla^{\tau 0}_n P^R_{-\mu}\right)^*_{|\partial B_U} \tag{20}$$

$$+\frac{1}{\lambda^2-\mu^2}\mathrm{Tr}\,\chi^\Gamma_{|\partial B_U}\circ\left(\nabla^{\tau 0}_n P^T_\lambda\right)_{|\partial B_U}\circ\Psi_{\lambda,\mu}\circ\left({}^0P^R_{-\mu}\right)^*_{|\partial B_U}. \tag{21}$$

Note that the derivatives of χ_1 drop out because of (18). Moreover, $({}^0P^T_\lambda)_{|\partial B_U}$: $H^{\sigma,\lambda}_{-\infty}\to C^\infty(\partial B_U,V(\tau)_{|\partial B_U})$ denotes the composition of the Poisson transform and restriction to the boundary of B_U, and this operator can be expressed in terms of $\mathscr{P}^T_{\lambda,a}$.

We introduce the following notation. Let $a_U\in A_+$ be such that $\mathrm{dist}_X(a_UK,K)=U$. We define $\omega(U):=a_U^{-2\rho}\mathrm{vol}(\partial B_U)$. Note that $\omega_X:=\lim_{U\to\infty}\omega(U)$ exists. Let $\chi^\Gamma_a\in C^\infty(K)$ denote the function $k\mapsto\chi^\Gamma(ka)$. Note that $\lim_{a\to\infty}\chi_a=\chi_\infty$. Then we can write

$$\begin{aligned}&(20)+(21)\\&=-\frac{\omega_U a_U^{2\rho}}{\lambda^2-\mu^2}\mathrm{Tr}\,\chi^\Gamma_{a_U}\circ\mathscr{P}^T_{\lambda,a_U}\circ\Psi_{\lambda,\mu}\circ(\partial\mathscr{P}^R_{-\mu,a_U})^*\\&\quad+\frac{\omega_U a_U^{2\rho}}{\lambda^2-\mu^2}\mathrm{Tr}\,\chi^\Gamma_{a_U}\circ\partial\mathscr{P}^T_{\lambda,a_U}\circ\Psi_{\lambda,\mu}\circ(\mathscr{P}^R_{-\mu,a_U})^*.\end{aligned}$$

Here, $\partial\mathscr{P}^T_{\lambda,a}$ stands for the derivative of the function $a\mapsto\mathscr{P}^T_{\lambda,a}$ with respect to the positive fundamental unit vector field on A.

Let χ be a smooth cut-off function on $X\cup\Omega$ of compact support such that

$$\gamma(\mathrm{supp}(\chi))\cap\mathrm{supp}(\chi)=\emptyset,\quad\forall\gamma\in\Gamma,\gamma\neq 1. \tag{22}$$

Note that χ^Γ can be decomposed into a finite sum $\chi^\Gamma = \sum_i \chi^i$ such that each χ^i satisfies (22). We fix a cut-off function $\tilde{\chi}$ on Ω satisfying (22) and $\tilde{\chi} \equiv 1$ on a neighbourhood of $\mathrm{supp}(\chi) \cap \Omega$. We further define $\chi_a(k) = \chi(kaK)$, $\chi_\infty(k) = \chi_{|\partial X}(kM)$ and observe that $|\chi_a - \chi_\infty| = O(a^{-\alpha})$ for any seminorm $|.|$ of $C^\infty(K)$. Using that $\tilde{\chi}\Psi_{\lambda,\mu}\tilde{\chi} = 0$, we can write

$$
\begin{aligned}
&(20) + (21) \qquad (\chi^\Gamma \text{ replaced by } \chi)\\
&= -\frac{\omega_U a_U^{2\rho}}{\lambda^2 - \mu^2} \mathrm{Tr}\, \chi_{a_U} \circ \mathscr{P}^T_{\lambda,a_U} \circ (1-\tilde{\chi}) \circ \Psi_{\lambda,\mu} \circ (\partial \mathscr{P}^R_{-\mu,a_U})^*\\
&\quad -\frac{\omega_U a_U^{2\rho}}{\lambda^2 - \mu^2} \mathrm{Tr}\, \chi_{a_U} \circ \mathscr{P}^T_{\lambda,a_U} \circ \tilde{\chi} \circ \Psi_{\lambda,\mu} \circ (1-\tilde{\chi}) \circ (\partial \mathscr{P}^R_{-\mu,a_U})^*\\
&\quad +\frac{\omega_U a_U^{2\rho}}{\lambda^2 - \mu^2} \mathrm{Tr}\, \chi_{a_U} \circ \partial \mathscr{P}^T_{\lambda,a_U} \circ (1-\tilde{\chi}) \circ \Psi_{\lambda,\mu} \circ (\mathscr{P}^R_{-\mu,a_U})^*\\
&\quad +\frac{\omega_U a_U^{2\rho}}{\lambda^2 - \mu^2} \mathrm{Tr}\, \chi_{a_U} \circ \partial \mathscr{P}^T_{\lambda,a_U} \circ \tilde{\chi} \circ \Psi_{\lambda,\mu} \circ (1-\tilde{\chi}) \circ (\mathscr{P}^R_{-\mu,a_U})^*. \qquad (23)
\end{aligned}
$$

We now insert the asymptotic decomposition (11) of the operators $\mathscr{P}^T_{\lambda,a}$ as $a \to \infty$ noting that in each line one of these operators is localized off-diagonally.
In order to stay in trace class operators, we choose a function $\chi_1 \in C^\infty(K)$ such that $\mathrm{supp}(1-\chi_1) \cap \mathrm{supp}(\chi) = \emptyset$ and $\mathrm{supp}(1-\tilde{\chi}) \cap \mathrm{supp}(\chi_1) = \emptyset$. We obtain

$$
\begin{aligned}
&(20) + (21) \qquad (\chi^\Gamma \text{ replaced by } \chi)\\
&= -\frac{\omega_U a_U^{-\mu-\lambda}(-\mu-\rho)}{\lambda^2-\mu^2} \mathrm{Tr}\, R^* \frac{c_{\tilde{\tau}}(-\mu)^*}{c_\sigma(\mu)} \tau(w) T \chi_\infty \Phi_{0,-\lambda} J^w_{\sigma,\lambda} (1-\tilde{\chi}) \Psi_{\lambda,\mu} \Phi_{\mu,0} \chi_1\\
&\quad -\frac{\omega_U a_U^{\mu-\lambda}(\mu-\rho)}{\lambda^2-\mu^2} \mathrm{Tr}\, R^* T \chi_\infty \Phi_{0,-\lambda} J^w_{\sigma,\lambda} (1-\tilde{\chi}) \Psi_{\lambda,\mu} (J^w_{\tilde{\sigma},-\mu})^* \Phi_{-\mu,0} \chi_1\\
&\quad -\frac{\omega_U a_U^{\mu+\lambda}(\mu-\rho)}{\lambda^2-\mu^2} \mathrm{Tr}\, R^* \tau(w)^{-1} \frac{c_\tau(\lambda)}{c_\sigma(-\lambda)} T \chi_\infty \Phi_{0,\lambda} \tilde{\chi} \Psi_{\lambda,\mu} (1-\tilde{\chi}) (J^w_{\tilde{\sigma},-\mu})^* \Phi_{-\mu,0} \chi_1\\
&\quad -\frac{\omega_U a_U^{\mu-\lambda}(\mu-\rho)}{\lambda^2-\mu^2} \mathrm{Tr}\, R^* T \chi_\infty \Phi_{0,\lambda} J^w_{\sigma,\lambda} \tilde{\chi} \Psi_{\lambda,\mu} (1-\tilde{\chi}) (J^w_{\tilde{\sigma},-\mu})^* \Phi_{-\mu,0} \chi_1\\
&\quad +\frac{\omega_U a_U^{-\mu-\lambda}(-\lambda-\rho)}{\lambda^2-\mu^2} \mathrm{Tr}\, R^* \frac{c_{\tilde{\tau}}(-\mu)^*}{c_\sigma(\mu)} \tau(w) T \chi_\infty \Phi_{0,-\lambda} J^w_{\sigma,\lambda} (1-\tilde{\chi}) \Psi_{\lambda,\mu} \Phi_{\mu,0} \chi_1\\
&\quad +\frac{\omega_U a_U^{\mu-\lambda}(-\lambda-\rho)}{\lambda^2-\mu^2} \mathrm{Tr}\, R^* T \chi_\infty \Phi_{0,-\lambda} J^w_{\sigma,\lambda} (1-\tilde{\chi}) \Psi_{\lambda,\mu} (J^w_{\tilde{\sigma},-\mu})^* \Phi_{-\mu,0} \chi_1
\end{aligned}
$$

$$+\frac{\omega_U a_U^{\mu+\lambda}(\lambda-\rho)}{\lambda^2-\mu^2}\mathrm{Tr}\,R^*\tau(w)^{-1}\frac{c_\tau(\lambda)}{c_\sigma(-\lambda)}T\chi_\infty\Phi_{0,\lambda}\Psi_{\lambda,\mu}(1-\tilde{\chi})(J^w_{\tilde{\sigma},-\mu})^*\Phi_{-\mu,0}\chi_1$$

$$+\frac{\omega_U a_U^{\mu-\lambda}(-\lambda-\rho)}{\lambda^2-\mu^2}\mathrm{Tr}\,R^*T\chi_\infty\Phi_{0,-\lambda}J^w_{\sigma,\lambda}\tilde{\chi}\Psi_{\lambda,\mu}(1-\tilde{\chi})(J^w_{\tilde{\sigma},-\mu})^*\Phi_{-\mu,0}\chi_1$$

$$+a_U^{-\alpha}\frac{1}{\lambda^2-\mu^2}R_\chi(\lambda,\mu,a_U)$$

$$=-\frac{\omega_U a_U^{-\mu-\lambda}}{\lambda+\mu}\mathrm{Tr}\,R^*\frac{c_{\tilde{\tau}}(-\mu)^*}{c_\sigma(-\mu)}\tau(w)T\chi_\infty\Phi_{0,-\lambda}J^w_{\sigma,\lambda}(1-\tilde{\chi})\Psi_{\lambda,\mu}\Phi_{\mu,0}\chi_1$$

$$-\frac{\omega_U a_U^{\mu-\lambda}}{\lambda-\mu}\mathrm{Tr}\,R^*T\chi_\infty\Phi_{0,-\lambda}J^w_{\sigma,\lambda}(1-\tilde{\chi})\Psi_{\lambda,\mu}J^w_{\sigma,-\mu}\Phi_{-\mu,0}\chi_1$$

$$+\frac{\omega_U a_U^{\mu+\lambda}}{\lambda+\mu}\mathrm{Tr}\,R^*\tau(w)^{-1}\frac{c_\tau(\lambda)}{c_\sigma(-\lambda)}T\chi_\infty\Phi_{0,\lambda}\Psi_{\lambda,\mu}(1-\tilde{\chi})(J^w_{\tilde{\sigma},-\mu})^*\Phi_{-\mu,0}\chi_1$$

$$-\frac{\omega_U a_U^{\mu-\lambda}}{\lambda-\mu}\mathrm{Tr}\,R^*T\chi_\infty\Phi_{0,-\lambda}J^w_{\sigma,\lambda}\tilde{\chi}\Psi_{\lambda,\mu}(1-\tilde{\chi})(J^w_{\tilde{\sigma},-\mu})^*\Phi_{-\mu,0}\chi_1$$

$$+a_U^{-\alpha}\frac{1}{\lambda^2-\mu^2}R_\chi(\lambda,\mu,a_U).$$

The remainder $R_\chi(\lambda,\mu,a)$ is holomorphic and remains bounded in $C^\infty(i\mathfrak{a}^*\times i\mathfrak{a}^*)$ as $a\to\infty$.

We define $\langle R,T\rangle\in\mathbb{C}$ such that $R^*\circ T=\langle R,T\rangle\mathrm{id}_{V_\sigma}$. If σ is not Weyl-invariant, i.e. $\sigma^w\not\cong\sigma$, then the compositions $R^*c_{\tilde{\tau}}(-\mu)^*\tau(w)T$, $R^*\tau(w)^{-1}c_\tau(\lambda)T$ vanish and we define $\langle R,T\rangle(\lambda):=0$. If the representation σ is Weyl-invariant, then it can be extended to the normalizer $N_K(M)$ of M. In particular, we can define $\sigma(w)$. In this case, we define $\langle R,T\rangle(\lambda)\in\mathbb{C}$ such that

$$\sigma(w)R^*\tau(w)^{-1}\frac{c_\tau(\lambda)}{c_\sigma(-\lambda)}T=\langle R,T\rangle(\lambda)\mathrm{id}_{V_\sigma}.$$

Note that $R^*\frac{c_{\tilde{\tau}}(-\mu)^*}{c_\sigma(\mu)}\tau(w)T\sigma(w)^{-1}=\langle R,T\rangle(-\mu)\mathrm{id}_{V_\sigma}$. Further, we put

$$J_{\sigma,\lambda}=\sigma(w)J^w_{\sigma,\lambda}:H^{\sigma,\lambda}_{-\infty}\to H^{\sigma,-\lambda}_{-\infty}.$$

Then we can write

$$\mathrm{Tr}\left[\chi\chi_U\circ{}^0P^T_\lambda\circ(\mathrm{ext}\circ\bar{\Phi}_{\lambda,\mu}\circ\pi_*-\bar{\Phi}_{\lambda,\mu})\circ\left({}^0P^R_{-\mu}\right)^*\chi_1\right]$$

$$=-\frac{\omega_U a_U^{-\mu-\lambda}\langle R,T\rangle(-\mu)}{\lambda+\mu}\mathrm{Tr}\,\Phi_{\mu,-\lambda}\circ\chi_\infty\circ J_{\sigma,\lambda}\circ(1-\tilde{\chi})\circ\Psi_{\lambda,\mu}\circ\chi_1\quad(24)$$

$$-\frac{\omega_U a_U^{\mu-\lambda}\langle R,T\rangle}{\lambda-\mu}\mathrm{Tr}\,\chi_\infty\circ J^w_{\sigma,\lambda}\circ\Psi_{\lambda,\mu}\circ(J^w_{\tilde\sigma,-\mu})^*\circ\Phi_{-\mu,-\lambda}\circ\chi_1$$

$$+\frac{\omega_U a_U^{\mu+\lambda}\langle R,T\rangle(\lambda)}{\lambda+\mu}\mathrm{Tr}\,\chi_\infty\circ\Psi_{\lambda,\mu}\circ(1-\tilde\chi)\circ(J_{\tilde\sigma,-\mu})^*\circ\Phi_{-\mu,\lambda}\circ\chi_1 \qquad (25)$$

$$+\frac{1}{\lambda^2-\mu^2}\mathrm{Tr}'\,[\Delta_\tau,\chi]\circ{}^0P^T_\lambda\circ\Psi_{\lambda,\mu}\circ({}^0P^R_{-\mu})^*$$

$$+\frac{a_U^{-\alpha}}{\lambda^2-\mu^2}Q_\chi(\lambda,\mu,a_U), \qquad (26)$$

where

$$Q_\chi(\lambda,\mu,a_U):=R_\chi(\lambda,\mu,a_U)-a_U^\alpha\mathrm{Tr}'\,[\Delta_\tau,\chi](1-\chi_U)\circ{}^0P^T_\lambda\circ\Psi_{\lambda,\mu}\circ({}^0P^{R^*}_{-\mu})^*. \quad (27)$$

We defer the justification of the terms (26), (27) to Lemma 4 below. The functional Tr' here is applied to operators with distribution kernels which are continuous on the diagonal, and it takes the integral of its local trace over the diagonal. Note that the remainder Q_χ is independent of the choice of χ_1.

The left-hand side of this formula is holomorphic on $\mathfrak{a}^*_\mathbb{C}\times\mathfrak{a}^*_\mathbb{C}$. The individual terms on the right-hand side may have poles. To aim of the following discussion is to understand these singularities properly.

Lemma 3. *The meromorphic family*

$$(\mu,\lambda)\mapsto\frac{1}{\lambda-\mu}\mathrm{Tr}\,\chi_\infty\circ J^w_{\sigma,\lambda}\circ\Psi_{\lambda,\mu}\circ(J^w_{\tilde\sigma,-\mu})^*\circ\Phi_{-\mu,-\lambda}\circ\chi_1$$

is regular for $\mu=\lambda$.

Proof. We must show that

$$\mathrm{Tr}\,\chi_\infty\circ J^w_{\sigma,\lambda}\circ(\mathrm{ext}\circ\pi_*-1)\circ(J^w_{\tilde\sigma,-\lambda})^*\circ\chi_1=0.$$

Recall the definition of the scattering matrix $S^w_{\sigma,\lambda}$ from [4], Def. 5.6. We are going to employ the relations

$$\mathrm{ext}\circ S^w_{\sigma,\lambda}=J^w_{\sigma,\lambda}\circ\mathrm{ext},\ \pi_*\circ(J^w_{\tilde\sigma,-\lambda})^*=(S^w_{\tilde\sigma,-\lambda})^*\circ\pi_*,\ S^w_{\sigma,\lambda}\circ(S^w_{\tilde\sigma,-\lambda})^*=\mathrm{id}.$$

We now compute

$$\begin{aligned}&\mathrm{Tr}\,\chi_\infty\circ J^w_{\sigma,\lambda}\circ(\mathrm{ext}\circ\pi_*-1)\circ(J^w_{\tilde\sigma,-\lambda})^*\circ\chi_1\\&=\mathrm{Tr}\,\chi_\infty\circ\mathrm{ext}\circ\hat S^w_{\sigma,\lambda}\circ(S^w_{\tilde\sigma,-\lambda})^*\circ\pi_*\circ\chi_1-\chi_\infty\circ J^w_{\sigma,\lambda}\circ(J^w_{\tilde\sigma,-\lambda})^*\circ\chi_1\\&=\mathrm{Tr}\,\chi_\infty\circ(\mathrm{ext}\circ\pi_*-1)\circ\chi_1\\&=0.\end{aligned}$$

□

In particular, we have

$$\lim_{U\to\infty}\lim_{\mu\to\lambda} -\frac{\omega(U)a_U^{\mu-\lambda}\langle R,T\rangle}{\lambda-\mu}\operatorname{Tr}\chi_\infty\circ J^w_{\sigma,\lambda}\circ\Psi_{\lambda,\mu}\circ(J^w_{\tilde\sigma,-\mu})^*\circ\Phi_{-\mu,-\lambda}\circ\chi_1$$
$$=\omega_X\langle R,T\rangle\frac{\mathrm{d}}{\mathrm{d}\mu}_{|\mu=\lambda}\operatorname{Tr}\chi\circ J^w_{\sigma,\lambda}\circ\Psi_{\lambda,\mu}\circ(J^w_{\tilde\sigma,-\mu})^*\circ\Phi_{-\mu,-\lambda}\circ\chi_1. \tag{28}$$

If the distribution kernel of an operator A admits a continuous restriction to the diagonal, then let $\mathscr{D}A$ denote this restriction.

Lemma 4. *1. For any compact subset $Q\subset i\mathfrak{a}^*$ there is a constant C such that for all $k\in K$ and $\mu,\lambda\in Q$ we have*

$$|\mathscr{D}[\Delta_\tau,\chi]\circ{}^0P^T_\lambda\circ\Psi_{\lambda,\mu}\circ({}^0P^R_{-\mu})^*(ka)|<Ca^{-2\rho-\alpha}.$$

2. We have

$$\operatorname{Tr}'[\Delta_\tau,\chi^\Gamma]\circ{}^0P^T_\lambda\circ\Psi_{\lambda,\mu}\circ({}^0P^{R^*}_{-\mu})^*=0$$

(note that we consider the cut-off function χ^Γ here).

Proof. The reason that 1. holds true is that $|d\chi(ka)|\le Ca^{-\alpha}$ and $|\Delta\chi(ka)|\le Ca^{-\alpha}$ uniformly in $k\in K$ and $a\in A$. We use the decomposition

$$\mathscr{D}[\Delta_\tau,\chi]\circ{}^0P^T_\lambda\circ\Psi_{\lambda,\mu}\circ({}^0P^R_{-\mu})^*(ka)$$
$$=\mathscr{D}[\Delta_\tau,\chi]\circ{}^0P^T_\lambda\circ\tilde\chi\circ\Psi_{\lambda,\mu}\circ(1-\tilde\chi)\circ({}^0P^R_{-\mu})^*(ka) \tag{29}$$
$$+\mathscr{D}[\Delta_\tau,\chi]\circ{}^0P^T_\lambda\circ(1-\tilde\chi)\circ\Psi_{\lambda,\mu}\circ({}^0P^R_{-\mu})^*(ka). \tag{30}$$

The asymptotic behaviour (11) of the operators $\mathscr{P}^T_{\lambda,a}$ is uniform for λ in compact subsets of $i\mathfrak{a}^*$ and can be differentiated with respect to a. We conclude that for any compact subset $Q\subset i\mathfrak{a}^*$ there is a constant $C\in\mathbb{R}$ such for all $\lambda,\mu\in Q$ we have

$$\sup_k|\mathscr{D}[\Delta_\tau,\chi]_a\circ{}^0P^T_\lambda\circ(1-\tilde\chi)\circ\Psi_{\lambda,\mu}\circ({}^0P^R_{-\mu})^*(ka)|\le Ca^{-2\rho-\alpha},$$
$$\sup_k|\mathscr{D}[\Delta_\tau,\chi]_a\circ{}^0P^T_\lambda\circ\tilde\chi\circ\Psi_{\lambda,\mu}\circ(1-\tilde\chi)\circ({}^0P^R_{-\mu})^*(ka)|\le Ca^{-2\rho-\alpha}.$$

We can write χ^Γ as a finite sum $\chi^\Gamma=\sum_i\chi^i$, where the cut-off functions χ^i obey (22). For each index i, we choose an appropriate cut-off function χ^i_1 as above. It follows from 1. that $\operatorname{tr}\mathscr{D}[\Delta_\tau,\chi^\Gamma]\circ{}^0P^T_\lambda\circ\Psi_{\lambda,\mu}\circ({}^0P^R_{-\mu})^*$ is integrable over G. We observe that

$$\pi^{\sigma,\lambda}(\gamma)\circ\Psi_{\lambda,\mu}=\operatorname{ext}\circ\bar\Phi_{\lambda,\mu}\circ\pi_*-\bar\Phi_{\lambda,\mu}\circ\pi^{\sigma,\mu}(\gamma)=\Psi_{\lambda,\mu}\circ\pi^{\sigma,\mu}(\gamma)$$

and compute

$$\begin{aligned}
&\mathrm{Tr}'\,[\Delta_\tau,\chi^\Gamma]\circ{}^0P^T_\lambda\circ\Psi_{\lambda,\mu}\circ({}^0P^R_{-\mu})^*\\
&=\sum_{\gamma\in\Gamma}\mathrm{Tr}'\,\pi(\gamma)^{-1}\chi^\Gamma\pi(\gamma)\circ[\Delta_\tau,\chi^\Gamma]\circ{}^0P^T_\lambda\circ\Psi_{\lambda,\mu}\circ({}^0P^R_{-\mu})^*\\
&=\sum_{\gamma\in\Gamma}\mathrm{Tr}'\,\chi^\Gamma\circ[\Delta_\tau,(\gamma^{-1})^*\chi^\Gamma]\circ\pi(\gamma)\circ{}^0P^T_\lambda\circ\Psi_{\lambda,\mu}\circ\pi^{\sigma,\mu}(\gamma)^{-1}\circ({}^0P^R_{-\mu})^*\\
&=\sum_{\gamma\in\Gamma}\mathrm{Tr}'\,\chi^\Gamma\circ[\Delta_\tau,(\gamma^{-1})^*\chi^\Gamma]\circ{}^0P^T_\lambda\circ\pi^{\sigma,\lambda}(\gamma)\circ\Psi_{\lambda,\mu}\circ\pi^{\sigma,\mu}(\gamma)^{-1}\circ({}^0P^{R^*}_{-\mu})^*\\
&=\sum_{\gamma\in\Gamma}\mathrm{Tr}'\,\chi^\Gamma\circ[\Delta_\tau,(\gamma^{-1})^*\chi^\Gamma]\circ{}^0P^T_\lambda\circ\Psi_{\lambda,\mu}\circ({}^0P^R_{-\mu})^*\\
&=0,
\end{aligned}$$

since $\sum_{\gamma\in\Gamma}(\gamma^{-1})^*\chi^\Gamma\equiv 1$. □

Note that the second assertion of the lemma implies

$$\sum_i \mathrm{Tr}\,[\Delta_\tau,\chi^i]\circ{}^0P^T_\lambda\circ\Psi_{\lambda,\mu}\circ({}^0P^{R^*}_{-\mu})^*\circ\chi^i_1=0.$$

We now combine (24) and (25) and write

$$\begin{aligned}
&(24)+(25)\\
&=\frac{\omega(U)(a_U^{\mu+\lambda}-a_U^{-\mu-\lambda})}{\lambda+\mu}\langle R,T\rangle(-\mu)\\
&\quad\mathrm{Tr}\,\Phi_{\mu,-\lambda}\circ\chi_\infty\circ J_{\sigma,\lambda}\circ(1-\tilde\chi)\circ\Psi_{\lambda,\mu}\circ\chi_1 \qquad (31)\\
&\quad+\frac{\omega(U)a_U^{\mu+\lambda}}{\lambda+\mu}\mathrm{Tr}\,\big[\langle R,T\rangle(\lambda)\chi_\infty\circ\Psi_{\lambda,\mu}\circ(1-\tilde\chi)\circ(J_{\tilde\sigma,-\mu})^*\circ\Phi_{-\mu,\lambda}\circ\chi_1\\
&\quad-\langle R,T\rangle(-\mu)\Phi_{\mu,-\lambda}\circ\chi_\infty\circ J_{\sigma,\lambda}\circ(1-\tilde\chi)\circ\Psi_{\lambda,\mu}\circ\chi_1\big] \qquad (32)
\end{aligned}$$

Note that (31) is regular at $\lambda=\mu=0$. In fact, if ext has a pole at $\lambda=0$, then it is of first order and $J_{\sigma,0}=\mathrm{id}$ (see [4], Prop. 7.4). Therefore, $\chi_\infty\circ J_{\sigma,0}\circ(1-\tilde\chi)$ vanishes.

In order to see that (32) is regular at $\lambda=\mu=0$, too, observe in addition that

$$\begin{aligned}
&\mathrm{Tr}\,\big[\chi\circ(\mathrm{ext}\circ\pi_*-1)\circ(1-\tilde\chi)\circ(J_{\tilde\sigma,0})^*\circ\chi_1\\
&\quad-\chi\circ J_{\sigma,0}\circ(1-\tilde\chi)\,(\mathrm{ext}\circ\pi_*-1)\circ\chi_1\big]=0. \qquad (33)
\end{aligned}$$

Combining Lemma 3 and 4, (2), and (33) we conclude that $\frac{1}{\lambda^2-\mu^2}Q_{\chi^\Gamma}(\lambda,\mu,a_U)$ is regular at $\mu=\lambda$, where we set $Q_{\chi^\Gamma} := \sum_i Q_{\chi^i}$. Furthermore, locally uniformly in λ one has

$$\lim_{U\to\infty} \frac{a_U^{-\alpha}}{\lambda^2-\mu^2} Q_{\chi^\Gamma}(\lambda,\mu,a_U) = 0.$$

By the Lemma of Riemann–Lebesgue, we have

$$\begin{aligned}\lim_{U\to\infty} \frac{\omega(U)a_U^{2\lambda}}{2\lambda} \mathrm{Tr}\, &\big[\langle R,T\rangle(\lambda)\chi_\infty \circ (\mathrm{ext}\circ\pi_* - 1)\circ(1-\tilde{\chi})\circ(J_{\tilde{\sigma},-\lambda})^* \circ \Phi_{-\lambda,\lambda}\circ\chi_1 \\ &-\langle R,T\rangle(-\lambda)\Phi_{\lambda,-\lambda}\circ\chi_\infty\circ J_{\sigma,\lambda}\circ(1-\tilde{\chi})\circ(\mathrm{ext}\circ\pi_*-1)\circ\chi_1\big] \\ &= 0\end{aligned}$$

as distributions on $i\mathfrak{a}^*$. Moreover,

$$\begin{aligned}\lim_{U\to\infty} &\frac{\omega(U)(a_U^{2\lambda}-a_U^{-2\lambda})}{2\lambda} \langle R,T\rangle(-\lambda) \\ &\mathrm{Tr}\,\Phi_{\lambda,-\lambda}\circ\chi_\infty\circ J_{\sigma,\lambda}\circ(1-\tilde{\chi})\circ(\mathrm{ext}\circ\pi_*-1)\circ\chi_1 \qquad (34)\\ = \pi\omega_X &\langle R,T\rangle(0)\delta_0(\lambda)\lim_{\lambda\to 0}\big[\mathrm{Tr}\,\Phi_{\lambda,-\lambda}\circ\chi_\infty\circ J_{\sigma,\lambda}\circ(1-\tilde{\chi})\circ(\mathrm{ext}\circ\pi_*-1)\circ\chi_1\big] \qquad (35)\end{aligned}$$

in the sense of distributions on $i\mathfrak{a}^*$. Combining (28) and (35) we now have shown the following proposition.

Proposition 2. *We have*

$$\begin{aligned}\int_{\Gamma\backslash G} &\mathrm{tr}\, A_q(g,g)\mu_G(dg) \\ &= \omega_X\langle R,T\rangle \int_{i\mathfrak{a}^*} \sum_i \frac{d}{d\mu}_{|\mu=\lambda} \mathrm{Tr}\,\Big[\chi^i_\infty\circ J^w_{\sigma,\lambda}\circ\big(\mathrm{ext}\circ\bar{\Phi}_{\lambda,\mu}\circ\pi_* - \bar{\Phi}_{\lambda,\mu}\big) \\ &\qquad\circ(J^w_{\tilde{\sigma},-\mu})^*\circ\Phi_{-\mu,-\lambda}\circ\chi^i_1\Big]\, q(\lambda)d\lambda \\ &\quad +\pi\omega_X\langle R,T\rangle(0)\sum_i \lim_{\lambda\to 0}\big(\mathrm{Tr}\,\big[\Phi_{\lambda,-\lambda}\circ\chi^i_\infty\circ J_{\sigma,\lambda}\circ(\mathrm{ext}\circ\pi_*-1)\circ\chi^i_1\big]\big)\, q(0). \qquad (36)\end{aligned}$$

Observe that we can rewrite this in the more invariant form

$$\begin{aligned}\omega_X\langle R,T\rangle \int_{i\mathfrak{a}^*} \frac{d}{d\mu}_{|\mu=\lambda} &\mathrm{Tr}'\,\Big[\chi^\Gamma_\infty\circ J^w_{\sigma,\lambda}\circ\big(\mathrm{ext}\circ\bar{\Phi}_{\lambda,\mu}\circ\pi_*-\bar{\Phi}_{\lambda,\mu}\big) \\ &\qquad\circ(J^w_{\tilde{\sigma},-\mu})^*\circ\Phi_{-\mu,-\lambda}\Big]\, q(\lambda)d\lambda \\ +\pi\omega_X\langle R,T\rangle(0) &\lim_{\lambda\to 0}\big(\mathrm{Tr}'\,\big[\Phi_{\lambda,-\lambda}\circ\chi^\Gamma_\infty\circ J_{\sigma,\lambda}\circ(\mathrm{ext}\circ\pi_*-1)\big]\big)\, q(0).\end{aligned}$$

5 The Fourier Transform of Ψ

5.1 The Contribution of the Scattering Matrix

We consider a symmetric function $q \in C_c^\infty(i\mathfrak{a}^*, \mathrm{End}(V_\tau \otimes \mathrm{Hom}_M(V_\tau, V_\sigma)))$. For $\pi \in \hat{G}$, we form $h_q(\pi) \in \mathrm{End}(V_\pi)$ as in Sect. 3.3. There we have seen that $\check{R}(h_q)$ and $\check{R}_\Gamma(h_q)$ have smooth integral kernels.

We choose a basis $\{v_i\}_{i=1,\dots,\dim(\tau)}$ of V_τ and a basis $\{T_j\}_{j=1,\dots,\dim(\mathrm{Hom}_M(V_\tau,V_\sigma))}$ of $\mathrm{Hom}_M(V_\tau, V_\sigma)$. Let $\{v^i\}$ be a dual basis of $V_{\tilde{\tau}}$ and $\{T^j\}$ be a dual basis of $\mathrm{Hom}_M(V_\sigma, V_\tau) = \mathrm{Hom}_M(V_\tau, V_\sigma)^*$ (with respect to the pairing $\langle T', T\rangle = \mathrm{tr}_{V_\tau} T' \circ T$, $T \in \mathrm{Hom}_M(V_\tau, V_\sigma)$, $T' \in \mathrm{Hom}_M(V_\sigma, V_\tau)$). Then $\{\phi_{ij} := v_i \otimes T_j\}$ can be considered as a basis of $H^{\sigma,\lambda}(\tau)$. Furthermore, $\{\phi^{ij} := v^i \otimes (T^j)^*\}$ can be considered as a basis of $H^{\tilde{\sigma},-\lambda}(\tilde{\tau}) = H^{\sigma,\lambda}(\tau)^*$, and we have

$$\begin{aligned}
\langle \phi_{ij}, \phi^{hl}\rangle &= \int_K \langle T_j \circ \tau(k)^{-1}(v_i), (T^l)^* \circ \tilde{\tau}(k)^{-1}(v^h)\rangle \mathrm{d}k \\
&= \int_K \langle \tau(k) \circ T^l \circ T_j \circ \tau(k)^{-1}(v_i), v^h\rangle \mathrm{d}k \\
&= \frac{\langle T^l, T_j\rangle}{\dim(V_\tau)} \langle v_i, v^h\rangle \\
&= \frac{1}{\dim(V_\tau)} \delta_j^l \delta_i^h. \qquad (37)
\end{aligned}$$

For $g \in G$ and $w \in V_{\tilde{\tau}}$, we define $p^T_{\lambda,w}(g) \in H_\infty^{\tilde{\sigma},-\lambda}$ such that $\langle P^T_\lambda(\phi)(g), w\rangle = \langle p^T_{\lambda,w}(g), \phi\rangle$ for all $\phi \in H_{-\infty}^{\sigma,\lambda}$. Using (9), we can write $p^T_{\lambda,w}(g) = \pi^{\tilde{\sigma},-\lambda}(g)(w \otimes T^*)$. We can write

$$q = \sum_{i,j,k,l} q_{ijkl} v_i \otimes T_j \otimes v^k \otimes T^l,$$

where the functions $q_{ijkl} := \langle v^i \otimes T^j, q(v_k \otimes T_l)\rangle$ belong to $C_c^\infty(i\mathfrak{a}^*)$. Now we can compute

$$\begin{aligned}
\mathrm{Tr}\, \pi^{\sigma,\lambda}(g) h_q(\pi^{\sigma,\lambda}) &= \dim(V_\tau) \sum_{i,j} \left\langle \phi^{ij}, \pi^{\sigma,\lambda}(g) h_q\left(\pi^{\sigma,\lambda}\right) \phi_{ij}\right\rangle \\
&= \dim(V_\tau) \sum_{i,j} \left\langle P_\lambda^{T^j}\left(h_q(\pi^{\sigma,\lambda})\phi_{ij}\right)(g^{-1}), v^i\right\rangle \\
&= \dim(V_\tau) \sum_{i,j} \left\langle h_q\left(\pi^{\sigma,\lambda}\right) \phi_{ij}, p^{T^j}_{\lambda,v^i}(g^{-1})\right\rangle \\
&= \dim(V_\tau) \sum_{i,j,k,l} q_{klij}(\lambda) \left\langle v_k \otimes T_l, p^{T^j}_{\lambda,v^i}(g^{-1})\right\rangle
\end{aligned}$$

$$= \dim(V_\tau) \sum_{i,j,k,l} q_{klij}(\lambda) \left\langle P^{T_l^*}_{-\lambda} \left(p^{T^j}_{\lambda,v^i}(g^{-1}) \right)(1), v_k \right\rangle$$

$$= \dim(V_\tau) \sum_{i,j,k,l} q_{klij}(\lambda) \left\langle v^i, P^{T^j}_{\lambda} \circ \left(P^{T_l^*}_{-\lambda} \right)^* (g^{-1},1)(v_k) \right\rangle .$$

In the last line of this computation $P^{T^j}_{\lambda} \circ (P^{T_l^*}_{-\lambda})^*(g,g')$ is the integral kernel of the G-equivariant operator

$$P^{T^j}_{\lambda} \circ \left(P^{T_j^*}_{-\lambda} \right)^* : C_c^{-\infty}(G \times_K V_\tau) \to C^\infty(G \times_K V_\tau).$$

We can express the integral kernel of $\check{R}(h_q)$ via Poisson transforms as follows:

$$\begin{aligned}
K_{\check{R}(h_q)}(g,g_1) &= \int_{\hat{G}} \operatorname{Tr} \pi(g) h_q(\pi) \pi(g_1^{-1}) p(\mathrm{d}\pi) \\
&= \frac{\dim(V_\sigma)}{4\pi\omega_X} \int_{i\mathfrak{a}^*} \operatorname{Tr} \pi(g_1^{-1}g) h_q(\pi^{\sigma,\lambda}) p_\sigma(\lambda) \mathrm{d}\lambda \\
&= \sum_{i,j,k,l} \frac{\dim(V_\sigma)\dim(V_\tau)}{4\pi\omega_X} \int_{i\mathfrak{a}^*} q_{klij}(\lambda) \left\langle v^i, P^{T^j}_{\lambda} \circ \left(P^{T_l^*}_{-\lambda} \right)^* (g^{-1}g_1, 1)(v_k) \right\rangle p_\sigma(\lambda) \mathrm{d}\lambda \\
&= \sum_{i,j,k,l} \frac{\dim(V_\sigma)\dim(V_\tau)}{4\pi\omega_X} \int_{i\mathfrak{a}^*} q_{klij}(\lambda) \left\langle v^i, P^{T^j}_{\lambda} \circ \left(P^{T_l^*}_{-\lambda} \right)^* (g_1, g)(v_k) \right\rangle p_\sigma(\lambda) \mathrm{d}\lambda \\
&= \sum_{i,j,k,l} \frac{\dim(V_\sigma)\dim(V_\tau)}{4\pi\omega_X} \int_{i\mathfrak{a}^*} q_{klij}(\lambda) \left\langle v^i, {}^0P^{T^j}_{\lambda} \circ \left({}^0P^{T_l^*}_{-\lambda} \right)^* (g_1, g)(v_k) \right\rangle \mathrm{d}\lambda. \qquad (38)
\end{aligned}$$

In a similar manner, we obtain

$$\begin{aligned}
&\operatorname{Tr} \pi^{\sigma,\lambda}(g) h_q(\pi^{\sigma,\lambda}) \pi^{\sigma,\lambda}(g_1^{-1}) \operatorname{ext} \pi_* \\
&\quad = \operatorname{Tr} h_q(\pi^{\sigma,\lambda}) \pi^{\sigma,\lambda}(g_1^{-1}) \operatorname{ext} \pi_* \, \pi^{\sigma,\lambda}(g) \\
&\quad = \dim(V_\tau) \sum_{k,l} \left\langle \phi^{kl}, h_q(\pi^{\sigma,\lambda}) \pi^{\sigma,\lambda}(g_1^{-1}) \operatorname{ext} \pi_* \, \pi^{\sigma,\lambda}(g) \phi_{kl} \right\rangle \\
&\quad = \dim(V_\tau) \sum_{k,l} \left\langle h_q(\pi^{\sigma,\lambda})^* \phi^{kl}, \pi^{\sigma,\lambda}(g_1^{-1}) \operatorname{ext} \pi_* \, \pi^{\sigma,\lambda}(g) \phi_{kl} \right\rangle \\
&\quad = \dim(V_\tau) \sum_{i,j,k,l} q_{klij}(\lambda) \left\langle v^i \otimes T^j, \pi^{\sigma,\lambda}(g_1^{-1}) \operatorname{ext} \pi_* \, \pi^{\sigma,\lambda}(g) \phi_{kl} \right\rangle \\
&\quad = \dim(V_\tau) \sum_{i,j,k,l} q_{klij}(\lambda) \left\langle v^i, P^{T^j}_{\lambda} (\operatorname{ext} \pi_* \, \pi^{\sigma,\lambda}(g) \phi_{kl})(g_1) \right\rangle \\
&\quad = \dim(V_\tau) \sum_{i,j,k,l} q_{klij}(\lambda) \left\langle \operatorname{ext} \pi_* \, \pi^{\sigma,\lambda}(g) \phi_{kl}, p^{T^j}_{\lambda,v^i}(g_1) \right\rangle
\end{aligned}$$

$$= \dim(V_\tau) \sum_{i,j,k,l} q_{klij}(\lambda) \left\langle \phi_{kl}, \pi^{\tilde{\sigma},-\lambda}(g^{-1}) \mathrm{ext}\, \pi_* \left(p^{T^j}_{\lambda,v^i}(g_1) \right) \right\rangle$$

$$= \dim(V_\tau) \sum_{i,j,k,l} q_{klij}(\lambda) \left\langle v_k, P^{T_l^*}_{-\lambda} \left(\mathrm{ext}\, \pi_* \left(p^{T^j}_{\lambda,v^i}(g_1) \right) \right)(g) \right\rangle$$

$$= \dim(V_\tau) \sum_{i,j,k,l} q_{klij}(\lambda) \left\langle v^i, P^{T^j}_{\lambda} \circ \mathrm{ext} \circ \pi_* \circ \left(P^{T_l^*}_{-\lambda} \right)^* (g_1, g)(v_k) \right\rangle$$

and thus

$$K_{\check{R}_\Gamma(h_q)}(g, g_1) = \frac{\dim(V_\sigma)}{4\pi\omega_X} \int_{i\mathfrak{a}^*} \mathrm{Tr}\, \pi^{\sigma,\lambda}(g) h_q \left(\pi^{\sigma,\lambda} \right) \pi^{\sigma,\lambda} \left(g_1^{-1} \right) p_\sigma(\lambda) \mathrm{d}\lambda$$

$$= \sum_{i,j,k,l} \frac{\dim(V_\sigma)\dim(V_\tau)}{4\pi\omega_X} \int_{i\mathfrak{a}^*} q_{klij}(\lambda) \left\langle v^i, P^{T^j}_{\lambda} \circ \mathrm{ext} \circ \pi_* \circ \left(P^{T_l^*}_{-\lambda} \right)^* (g_1, g)(v_k) \right\rangle p_\sigma(\lambda) \mathrm{d}\lambda$$

$$= \sum_{i,j,k,l} \frac{\dim(V_\sigma)\dim(V_\tau)}{4\pi\omega_X} \int_{i\mathfrak{a}^*} q_{klij}(\lambda) \left\langle v^i, {}^0P^{T^j}_{\lambda} \circ \mathrm{ext} \circ \pi_* \circ \left({}^0P^{T_l^*}_{-\lambda} \right)^* (g_1, g)(v_k) \right\rangle \mathrm{d}\lambda. \quad (39)$$

We conclude that

$$K_{\check{R}_\Gamma(h_q)}(g, g_1) - K_{\check{R}(h_q)}(g, g_1)$$

$$= \sum_{i,j,k,l} \frac{\dim(V_\sigma)\dim(V_\tau)}{4\pi\omega_X} \int_{i\mathfrak{a}^*} q_{klij}(\lambda) \langle v^i, {}^0P^{T^j}_{\lambda} \circ (\mathrm{ext} \circ \pi_* - 1) \circ ({}^0P^{T_l^*}_{-\lambda})^* (g_1, g)(v_k) \rangle \mathrm{d}\lambda$$

$$= \frac{\dim(V_\sigma)\dim(V_\tau)}{4\pi\omega_X} \sum_{i,j,k,l} \langle v^i, A_{q_{klij}}(T^j, T_l^*)(g_1, g) v_k \rangle.$$

By Proposition 1, the difference

$$g \mapsto (K_{\check{R}_\Gamma(h_q)}(g, g) - K_{\check{R}(h_q)}(g, g))$$

is integrable over $\Gamma\backslash G$. Using the fact that τ is irreducible, we compute

$$\sum_{i,k} \int_{\Gamma\backslash G} \langle v^i, A_{q_{klij}}(T^j, T_l^*)(g, g) v_k \rangle \mu_G(\mathrm{d}g)$$

$$= \sum_{i,k} \int_{\Gamma\backslash G} \int_K \langle v^i, A_{q_{klij}}(T^j, T_l^*)(gh, gh) v_k \rangle \mu_K(\mathrm{d}h) \mu_G(\mathrm{d}g)$$

$$= \sum_{i,k} \int_{\Gamma\backslash G} \int_K \langle v^i, \tau(h)^{-1} A_{q_{klij}}(T^j, T_l^*)(g, g) \tau(h) v_k \rangle \mu_K(\mathrm{d}h) \mu_G(\mathrm{d}g)$$

$$= \sum_{k} \int_{\Gamma\backslash G} \mathrm{tr}\, A_{q_{klkj}}(T^j, T_l^*)(g, g) \mu_G(\mathrm{d}g).$$

The following formula is now an immediate consequence of Proposition 2.

$$\int_{\Gamma\backslash G}\left(K_{\check{R}_\Gamma(h_q)}(g,g)-K_{\check{R}(h_q)}(g,g)\right)\mu_G(\mathrm{d}g) \tag{40}$$

$$= \frac{\dim(V_\sigma)\dim(V_\tau)}{4\pi}\int_{i\mathfrak{a}^*}\sum_i\sum_{k,j,l}\langle T_l^*,T^j\rangle\frac{\mathrm{d}}{\mathrm{d}\mu}_{|\mu=\lambda}\mathrm{Tr}\left[\chi_\infty^i\circ J^w_{\sigma,\lambda}\circ(\mathrm{ext}\circ\bar{\Phi}_{\lambda,\mu}\circ\pi_*-\bar{\Phi}_{\lambda,\mu})\right.$$

$$\left.\circ(J^w_{\tilde{\sigma},-\mu})^*\circ\Phi_{-\mu,-\lambda}\circ\chi_1^i\right]q_{klkj}(\lambda)\mathrm{d}\lambda$$

$$+\frac{\dim(V_\sigma)\dim(V_\tau)}{4}\sum_i\sum_{k,j,l}\langle T_l^*,T^j\rangle(0)$$

$$\lim_{\lambda\to 0}\left(\mathrm{Tr}\left[\Phi_{\lambda,-\lambda}\circ\chi_\infty^i\circ J_{\sigma,\lambda}\circ(\mathrm{ext}\circ\pi_*-1)\circ\chi_1^i\right]\right)q_{klkj}(0).$$

Now we will rewrite this formula in a more invariant fashion. Using (37), we first compute

$$\sum_{k,j,l}\langle T_l^*,T^j\rangle q_{klkj}(\lambda) \tag{41}$$

$$= \frac{1}{\dim(V_\sigma)}\sum_{j,l,k}T_l(T^j)\langle v^k\otimes T^l,q(\lambda)v_k\otimes T_j\rangle$$

$$= \frac{1}{\dim(V_\sigma)}\sum_{l,k}\langle v^k\otimes T^l,q(\lambda)v_k\otimes T_l\rangle$$

$$= \frac{1}{\dim(V_\tau)\dim(V_\sigma)}\mathrm{Tr}\,h_q(\pi^{\sigma,\lambda}). \tag{42}$$

We assume for a moment that σ is Weyl-invariant. For $T\in\mathrm{Hom}_M(V_\sigma,V_\tau)$, we define $T^\sharp\in\mathrm{Hom}(H^{\sigma,\lambda}_{-\infty},V_\tau)$ by $T^\sharp(f)=P^T_\lambda(f)(1)$. Recall the relation ([4], Lemma 5.5, 1.)

$$T^\sharp\circ\sigma(w)\circ J^w_{\sigma,\lambda}=c_\sigma(-\lambda)[\tau(w)\circ c_\tau(\lambda)\circ T\circ\sigma(w)^{-1}]^\sharp.$$

We compute

$$\langle R,T\rangle(0)$$

$$= \frac{1}{\dim(V_\sigma)c_\sigma(0)}\mathrm{Tr}\,\sigma(w)\circ R^*\circ\tau(w)^{-1}\circ c_\tau(0)\circ T$$

$$= \frac{1}{\dim(V_\sigma)c_\sigma(0)}\mathrm{Tr}\,\tau(w)\circ c_\tau(0)\circ T\circ\sigma(w)^{-1}\circ R^*$$

$$= \frac{1}{\dim(V_\sigma)c_\sigma(0)}\int_K\mathrm{Tr}\,\tau(k)^{-1}\circ\tau(w)\circ c_\tau(0)\circ T\circ\sigma(w)^{-1}\circ R^*\circ\tau(k)\mu_K(\mathrm{d}k)$$

$$= \frac{1}{\dim(V_\sigma)c_\sigma(0)}\sum_i\int_K\langle\tau(k)^{-1}\circ\tau(w)\circ c_\tau(0)\circ T\circ\sigma(w)^{-1}\circ R^*\circ\tau(k)(v_i),v^i\rangle\mu_K(\mathrm{d}k)$$

$$= \frac{1}{\dim(V_\sigma)c_\sigma(0)} \sum_i \langle P_0^{\tau(w)\circ c_\tau(0)\circ T\circ\sigma(w)^{-1}}(v_i \otimes R^*)(1), v^i\rangle$$

$$= \frac{1}{\dim(V_\sigma)c_\sigma(0)} \sum_i \langle [\tau(w)\circ c_\tau(0)\circ T\circ \sigma(w)^{-1}]^\sharp (v_i\otimes R^*), v^i\rangle$$

$$= \frac{1}{\dim(V_\sigma)} \sum_i \langle T^\sharp \circ \sigma(w)\circ J^w_{\sigma,0}(v_i\otimes R^*), v^i\rangle$$

$$= \frac{1}{\dim(V_\sigma)} \sum_i P_0^T \circ J_{\sigma,0}(v_i\otimes R)(1), v^i\rangle$$

$$= \frac{1}{\dim(V_\sigma)} \sum_i \langle J_{\sigma,0}(v_i\otimes R^*), v^i\otimes T^*\rangle.$$

Since $J_{\sigma,0}$ is K-equivariant, we can write

$$J_{\sigma,0}(v_i\otimes R) = v_i \otimes j_{\sigma,0}(R)$$

for some $j_{\sigma,0}\in \mathrm{End}(\mathrm{Hom}_M(V_\sigma,V_\tau))$. We now have

$$\begin{aligned}
&\sum_{k,j,l} \langle T_l^*, T^j\rangle(0) q_{klkj}(\lambda)\\
&\quad = \frac{1}{\dim(V_\sigma)} \sum_{k,j,l}\sum_i \langle v_i\otimes j_{\sigma,0}(T_l^*), v^i\otimes T^j\rangle\langle v^k\otimes T^l, q(\lambda)v_k\otimes T_j\rangle\\
&\quad = \frac{1}{\dim(V_\sigma)} \sum_{k,l}\langle v^k\otimes T^l, q(\lambda)(v_k\otimes j_{\sigma,0}(T_l^*))\rangle \qquad (43)\\
&\quad = \frac{1}{\dim(V_\sigma)} \sum_{k,l}\langle v^k\otimes T^l, q(\lambda)\circ J_{\sigma,0}(v_k\otimes T_l^*)\rangle\\
&\quad = \frac{1}{\dim(V_\sigma)\dim(V_\tau)} \mathrm{Tr}\, h_q(\pi^{\sigma,0})\circ J_{\sigma,0}. \qquad (44)
\end{aligned}$$

Inserting (42) and (44) into (40), we obtain the following theorem.

Theorem 1. *If q is a smooth compactly supported symmetric function on $\hat{M}\times i\mathfrak{a}^*$ such that $q(\sigma,\lambda)\in \mathrm{End}(V_\tau\otimes \mathrm{Hom}_M(V_\tau,V_\sigma))$, then the difference*

$$g\mapsto K_{\check{R}_\Gamma(h_q)}(g,g) - K_{\check{R}(h_q)}(g,g)$$

is integrable over $\Gamma\backslash G$, and we have

$$\int_{\Gamma\backslash G}\left(K_{\check{R}_\Gamma(h_q)}(g,g) - K_{\check{R}(h_q)}(g,g)\right)\mu_G(dg)$$

$$= \sum_{\sigma\in\hat{M}} \frac{1}{4\pi} \sum_i \int_{i\mathfrak{a}^*} \frac{\mathrm{d}}{\mathrm{d}\mu}_{|\mu=\lambda} \mathrm{Tr}\, \Big[\chi^i_\infty \circ J^w_{\sigma,\lambda} \circ (\mathrm{ext}\circ \bar{\Phi}_{\lambda,\mu} \circ \pi_* - \bar{\Phi}_{\lambda,\mu})$$

$$\circ (J^w_{\tilde{\sigma},-\mu})^* \circ \Phi_{-\mu,-\lambda} \circ \chi^i_1\Big] \,\mathrm{Tr}\, h_q(\pi^{\sigma,\lambda})\mathrm{d}\lambda$$

$$+\frac{1}{4}\sum_i \lim_{\lambda\to 0} \big(\mathrm{Tr}\,\big[\Phi_{\lambda,-\lambda} \circ \chi^i_\infty \circ J_{\sigma,\lambda} \circ (\mathrm{ext}\circ \pi_* - 1) \circ \chi^i_1\big]\big)\,\mathrm{Tr}\, h_q(\pi^{\sigma,0}) \circ J_{\sigma,0}.$$

We can again rewrite this formula as follows

$$\int_{\Gamma\backslash G} \left(K_{\check{R}_\Gamma(h_q)}(g,g) - K_{\check{R}(h_q)}(g,g)\right) \mu_G(\mathrm{d}g)$$

$$= \sum_{\sigma\in\hat{M}} \frac{1}{4\pi} \int_{i\mathfrak{a}^*} \frac{\mathrm{d}}{\mathrm{d}\mu}_{|\mu=\lambda} \mathrm{Tr}'\, \Big[\chi^\Gamma_\infty \circ J^w_{\sigma,\lambda} \circ (\mathrm{ext}\circ \bar{\Phi}_{\lambda,\mu} \circ \pi_* - \bar{\Phi}_{\lambda,\mu})$$

$$\circ (J^w_{\tilde{\sigma},-\mu})^* \circ \Phi_{-\mu,-\lambda}\Big] \,\mathrm{Tr}\, h_q(\pi^{\sigma,\lambda})\mathrm{d}\lambda$$

$$+\frac{1}{4} \lim_{\lambda\to 0} \big(\mathrm{Tr}'\,\big[\Phi_{\lambda,-\lambda} \circ \chi^\Gamma_\infty \circ J_{\sigma,0} \circ (\mathrm{ext}\circ \pi_* - 1)\big]\big)\,\mathrm{Tr}\, h_q(\pi^{\sigma,0}) \circ J_{\sigma,0}.$$

5.2 *The Fourier Transform of Ψ: A Conjecture*

Observe that Theorem 1 does not solve our initial problem of computing the Fourier transform of the distribution Ψ. The point is that there is no function $f \in C^\infty_c(G)$ such that its Fourier transform $\hat{f}$ has compact support, too.[3] In order to extend Theorem 1 to $\hat{f}$, we must extend Proposition 1 to Schwartz functions. As explained in Remark 1, the main obstacle to do this is an estimate of the growth of the extension map $\mathrm{ext} = \mathrm{ext}_\lambda : C^{-\infty}(B, V_B(\sigma,\lambda)) \to H^{\sigma,\lambda}_{-\infty}$ as λ tends to infinity along the imaginary axis.

The goal of this subsection is to rewrite the result of the computation of Ψ' in terms of characters thus obtaining the candidate of the measure Φ. We will also take the discrete spectrum of $L^2(\Gamma\backslash G)$ into account.

Recall that

$$\mathrm{Tr}\, \hat{f}(\pi^{\sigma,\lambda}) = \theta_{\pi^{\sigma,\lambda}}(f).$$

If $\pi^{\sigma,0}$ is reducible, then it decomposes into a sum of $\pi^{\sigma,+} \oplus \pi^{\sigma,-}$ of limits of discrete series representations which are just the ± 1 eigenspaces of $J_{\sigma,0}$. In this case, ext is regular at $\lambda = 0$ ([4], Prop. 7.4.). We can write

$$\mathrm{Tr}\, \hat{f}(\pi^{\sigma,0}) \circ J_{\sigma,0} = \theta_{\pi^{\sigma,+}}(f) - \theta_{\pi^{\sigma,-}}(f).$$

[3]Compare the footnote on page 99.

If we replace h_q by $\hat{f}$, then formulas (38) and (39) just give the contributions of the continuous spectrum $K_{R^{\mathrm{ac}}(f)}$ and $K_{R^{\mathrm{ac}}_\Gamma(f)}$ to the integral kernels $K_{R(f)}$ and $K_{R_\Gamma(f)}$. We have $K_{R(f)} = K_{R^{\mathrm{ac}}(f)} + K_{R^{\mathrm{d}}(f)}$, where

$$K_{R^{\mathrm{d}}(f)}(g,g_1) = \sum_{\pi\in\hat{G}_{\mathrm{d}}} \operatorname{Tr}\pi(g)\hat{f}(\pi)\pi(g_1^{-1}).$$

Since we assume that f is K-finite, and there are only finitely many discrete series representations containing a given K-type, this sum is finite.

Furthermore, $K_{R_\Gamma(f)} = K_{R^{\mathrm{ac}}_\Gamma(f)} + K_{R^{\mathrm{d}}_\Gamma(f)} + K_{R^{\mathrm{c}}_\Gamma(f)}$. Here $K_{R^{\mathrm{d}}_\Gamma(f)} = \sum_{\pi\in\hat{G}_{\mathrm{d}}} K_{R^{\mathrm{d},\pi}_\Gamma(f)}$ is the contribution of discrete series representations (again a finite sum),

$$K_{R^{\mathrm{d},\pi}_\Gamma(f)}(g,g_1) = \sum_{i,j} \overline{\langle\psi_i,\phi_j\rangle}\langle\psi_i,\pi(g)\hat{f}(\pi)\pi(g_1^{-1})\phi_j\rangle,$$

where $\{\phi_j\}$ and $\{\psi_i\}$ are orthonormal bases of the infinite-dimensional Hilbert spaces V_π and M_π, respectively. The finite sum $K_{R^{\mathrm{c}}_\Gamma(f)} = \sum_{\pi\in\hat{G}_{\mathrm{c}}} K_{R^{\mathrm{c},\pi}_\Gamma(f)}$ is the discrete contribution of representations belonging to $\hat{G}_{\mathrm{c}}$. If we choose orthogonal bases $\{\phi_j\}$ and $\{\psi_i\}$ of the Hilbert spaces V_π and M_π (note that $\dim(M_\pi) < \infty$), then we can write

$$K_{R^{\mathrm{c},\pi}_\Gamma(f)}(g,g_1) = \sum_{i,j} \overline{\langle\psi_i,\phi_j\rangle}\langle\psi_i,\pi(g)\hat{f}(\pi)\pi(g_1^{-1})\phi_j\rangle.$$

We define the multiplicity of π by $N_\Gamma(\pi) := \dim(M_\pi)$. It is clear that

$$\int_{\Gamma\backslash G} K_{R^{\mathrm{c}}_\Gamma(f)}(g,g)\mu_G(\mathrm{d}g) = \sum_{\pi\in\hat{G}_{\mathrm{c}}} N_\Gamma(\pi)\theta_\pi(f).$$

Now we discuss discrete series contributions. In Lemma 14, we will show that if f is K-finite and invariant under conjugation by K, then for each $\pi\in\hat{G}_{\mathrm{d}}$ we have

$$K_{R^{\mathrm{d},\pi}_\Gamma(f)}(g,g) - K_{R^{\mathrm{d},\pi}(f)}(g,g) \in L^1(\Gamma\backslash G). \tag{45}$$

Given any $A\in V_{\tilde{\pi},K}\otimes V_{\pi,K}$ we define the function $\hat{G}\ni\pi'\mapsto h_A(\pi')$ to be zero for all $\pi'\neq\pi$ and $h_A(\pi) := \bar{A}$, where $\bar{A} := \int_K \tilde{\pi}(k)\otimes\pi(k)A\,dk$. The map

$$V_{\tilde{\pi},K}\otimes V_{\pi,K}\ni A\mapsto T(A) := \int_{\Gamma\backslash G}\left[K_{\check{R}^{\mathrm{d}}_\Gamma(h_{\bar{A}})}(g,g) - K_{\check{R}^{\mathrm{d}}(h_{\bar{A}})}(g,g)\right]\mu_G(\mathrm{d}g)$$

is well defined and a $(\mathfrak{g},K)$-invariant functional on $V_{\tilde{\pi},K}\otimes V_{\pi,K}$. Since $V_{\pi,K}$ is irreducible it follows that

$$T(A) = N_\Gamma(\pi)\operatorname{Tr}(A) \tag{46}$$

for some number $N_\Gamma(\pi)\in\mathbb{C}$, which plays the role of the multiplicity of π. Here, we consider A as a finite-dimensional operator on V_π.

A discrete series representation is called integrable if its bi-K-finite matrix coefficients belong to $L^1(G)$. We split $\hat{G}_{\mathrm{d}}$ into integrable and non-integrable representations

$$\hat{G}_{\mathrm{d}} = \hat{G}_{\mathrm{di}} \cup \hat{G}_{\mathrm{dn}}.$$

Note that discrete series representations with sufficiently regular parameter are integrable, but if $\hat{G}_{\mathrm{d}} \neq \emptyset$, then also the set G_{dn} is infinite (except for $G = SL(2,\mathbb{R})$). In contrast to the general case, for integrable discrete series representations π, the assertion (45) is easy to obtain. In fact, more is true.

Proposition 3. *Let $\pi \in \hat{G}_{\mathrm{di}}$, $v, w \in V_{\pi,K}$, and let $f = f_{v,w}$ be the matrix coefficient $f_{v,w}(g) := \langle v, \pi(g)w \rangle$. Then $\Psi(f)$, $\Psi'(f)$ are well-defined, and we have*

$$\Psi(f) = \Psi'(\hat{f}) = 0.$$

Moreover, $N_\Gamma(\pi) = 0$.

Proof. The function f belongs to an appropriate L^1-Schwartz space for G. Using Weyl's integral formula and that Γ is convex cocompact, one observes that for any such Schwartz function h the function $(\gamma, g) \mapsto \mathrm{vol}(\Gamma_\gamma \backslash G_\gamma) h(g^{-1}\gamma g)$ belongs to $L^1(\bigcup_{\gamma \in \widetilde{C\Gamma}} G_\gamma \backslash G)$. Hence, by Fubini the computation in Lemma 1 remains valid. This proves the first equation. Hyperbolic orbital integrals of matrix coefficients of discrete series vanish ([21], 7.5.4). Hence $\Psi(f) = 0$. For any unitary representation (ρ, Y) of G, any $y \in Y$, $w \in V_{\pi,K}$ the map $V_{\pi,K} \ni v \mapsto \pi(f_{v,w})y \in Y_{\infty,K}$ is an intertwining operator of $(\mathfrak{g}, K)$-modules. It follows that $R(f) = R^{\mathrm{d}}(f) = R^{\mathrm{d},\pi}(f)$ and $R_\Gamma(f) = R_\Gamma^{\mathrm{d}}(f) = R_\Gamma^{\mathrm{d},\pi}(f)$. Hence, $T(\pi(f)) = \Psi'(\hat{f}) = 0$. Since $\mathrm{Tr}(\pi(f_{v,v})) \neq 0$ for $v \neq 0$, we eventually find (see (46)) that $N_\Gamma(\pi) = 0$. □

Proposition 3 can be extended to a larger subset of $\hat{G}_{\mathrm{d}}$ that depends on the critical exponent of Γ. We will not discuss this issue here. It follows from Lemma 14 that $K_{R_\Gamma^{\mathrm{ac}}(f)}(g,g) - K_{R^{\mathrm{ac}}(f)}(g,g)$ belongs to $L^1(\Gamma \backslash G)$ not only if $\hat{f}$ is smooth of compact support, K-finite and K-invariant (Proposition 1), but also in the case that $f \in C_c^\infty(G)$ is K-finite and K-conjugation invariant.

The following conjecture provides the candidate for the measure Φ. Its discrete part is expressed in terms of multiplicities $N_\Gamma(\pi)$. If $\pi \in \hat{G}_{\mathrm{c}}$, then $N_\Gamma(\pi)$ is just the dimension of the space of multiplicities M_π and thus a non-negative integer. If $\pi \in \hat{G}_{\mathrm{dn}}$, then $N_\Gamma(\pi)$ is a sort of regularized dimension of M_π. We will show in Proposition 5 that $N_\Gamma(\pi)$ is an integer in this case, too. The continuous part of the spectrum will contribute a point measure supported on the irreducible constitutents of the representations $\pi^{\sigma,0}$, and the corresponding weight will be denoted by $\tilde{N}_\Gamma(\pi)$. The remaining contribution of the continuous spectrum is absolute continuous to the Lebesgue measure $d\lambda$ on $\hat{G}_{\mathrm{ac}} = \hat{M} \times i\mathfrak{a}_+^*$ and will be described by the density $L_\Gamma(\pi^{\sigma,\lambda})$. By Theorem 3, this density appears in the functional equation of the Selberg zeta function. In particular, as it can be already seen from the definition below, $L_\Gamma(\pi^{\sigma,\lambda})$ admits a meromorphic continuation to all of $\mathfrak{a}_\mathbb{C}^*$ as a function of λ. Its residues are closely related to the multiplicities of resonances.

Conjecture 1. If $f \in C_c^\infty(G)$ is bi-K-finite, then we have

$$\begin{aligned}\Psi(f) = & \sum_{\sigma\in\hat{M}} \frac{1}{4\pi}\int_{i\mathfrak{a}^*} L_\Gamma(\pi^{\sigma,\lambda})\theta_{\pi^{\sigma,\lambda}}(f)\mathrm{d}\lambda \\ & + \sum_{\sigma\in\hat{M},\sigma\cong\sigma^w,\pi^{\sigma,0}\text{red.}} \sum_{\varepsilon\in\{+,-\}} \tilde{N}_\Gamma(\pi^{\sigma,\varepsilon})\theta_{\pi^{\sigma,\varepsilon}}(f) \\ & + \sum_{\sigma\in\hat{M},\sigma\cong\sigma^w,\pi^{\sigma,0}\text{ irred.}} \tilde{N}_\Gamma(\pi^{\sigma,0})\theta_{\pi^{\sigma,0}}(f) \\ & + \sum_{\pi\in\hat{G}_c\cup\hat{G}_{dn}} N_\Gamma(\pi)\theta_\pi(f),\end{aligned}$$

where

$$L_\Gamma(\pi^{\sigma,\lambda}) := \frac{\mathrm{d}}{\mathrm{d}\mu}_{|\mu=\lambda} \mathrm{Tr}' \left[\chi_\infty^\Gamma \circ J_{\sigma,\lambda}^w \circ \left(\mathrm{ext}\circ \bar{\Phi}_{\lambda,\mu}\circ\pi_* - \bar{\Phi}_{\lambda,\mu}\right)\circ (J_{\tilde{\sigma},-\mu}^w)^*\circ \Phi_{-\mu,-\lambda}\right] \tag{47}$$

and

$$\begin{aligned}\tilde{N}_\Gamma(\pi^{\sigma,\pm}) &:= \pm\frac{1}{4}\mathrm{Tr}' \lim_{\lambda\to 0}\left(\left[\Phi_{\lambda,-\lambda}\circ\chi_\infty^\Gamma\circ J_{\sigma,\lambda}\circ(\mathrm{ext}\circ\pi_* - 1)\right]\right),\\ \tilde{N}_\Gamma(\pi^{\sigma,0}) &:= \frac{1}{4}\mathrm{Tr}' \lim_{\lambda\to 0}\left(\left[\Phi_{\lambda,-\lambda}\circ\chi_\infty^\Gamma\circ J_{\sigma,\lambda}\circ(\mathrm{ext}\circ\pi_* - 1)\right]\right).\end{aligned}$$

Note that the discussion above does not prove this conjecture. What it does prove is the following theorem. Here, we call a measurable operator valued function h on $\hat{G}$ smooth if it depends smoothly on the inducing parameter of unitary principal series representations. More precisely, for $\sigma\in\hat{M}$ we consider the natural map $i_\sigma : i\mathfrak{a}^* \to \hat{G}$ given by $i_\sigma(\lambda) := [\pi^{\sigma,\lambda}] \in \hat{G}$. We require that $h\circ i_\sigma : i\mathfrak{a}^* \to \mathrm{End}(L^2(K\times_M V_\sigma))$ is smooth. We do not impose any condition at the remaining representations $\pi\in\hat{G}$.

Theorem 2. *If $\hat{G}\ni\pi\mapsto h(\pi)\in\mathrm{End}_K(V_\pi)$ is smooth, of compact support, and factorizes over finitely many K-types, then we have*

$$\begin{aligned}\Psi'(\tilde{\tilde{h}}) = & \sum_{\sigma\in\hat{M}} \frac{1}{4\pi}\int_{i\mathfrak{a}^*} L_\Gamma(\pi^{\sigma,\lambda})\mathrm{Tr}\, h(\pi^{\sigma,\lambda})\mathrm{d}\lambda \\ & + \sum_{\sigma\in\hat{M},\sigma\cong\sigma^w,\pi^{\sigma,0}\textit{red.}} \sum_{\varepsilon\in\{+,-\}} \tilde{N}_\Gamma(\pi^{\sigma,\varepsilon})\mathrm{Tr}\, h(\pi^{\sigma,\varepsilon}) \\ & + \sum_{\sigma\in\hat{M},\sigma\cong\sigma^w,\pi^{\sigma,0}\textit{ irred.}} \tilde{N}_\Gamma(\pi^{\sigma,0})\mathrm{Tr}\, h(\pi^{\sigma,0}) \\ & + \sum_{\pi\in\hat{G}_c\cup\hat{G}_{dn}} N_\Gamma(\pi)\mathrm{Tr}\, h(\pi).\end{aligned}$$

6 Resolvent Kernels and Selberg Zeta Functions

6.1 Meromorphic Continuation of Resovent Kernels

We fix some K-type τ and introduce the vector bundles

$$V(\tau) := G \times_K V_\tau \to X := G/K, \quad V_Y(\tau) := \Gamma\backslash V(\tau) \to Y := \Gamma\backslash X.$$

Fixing the (scale of the) invariant Riemannian metric on X defines an Ad-invariant bilinear form b on $\mathfrak{g}$. The Casimir operator C of G corresponding to $-b$ gives rise to an unbounded self-adjoint operator on the Hilbert space of sections $L^2(X,V(\tau))$. To be precise, it is the unique self-adjoint extension of the restriction of C to the space of smooth sections with compact support. It is bounded from below. For each complex number z which is not contained in the spectrum $\sigma(C)$, we let $R(z)$ be the operator $(z-C)^{-1}$. By $r(z)$, we denote its distribution kernel.

The Casimir operator descends to an operator acting on sections of $V_Y(\tau)$ and induces a unique unbounded self-adjoint operator C_Y on $L^2(Y,V_Y(\tau))$. For $z \notin \sigma(C_Y)$, we define the resolvent $R_Y(z)$ of the operator $(z-C_Y)^{-1}$ and denote by $r_Y(z)$ its distribution kernel. We consider both, $r(z)$ and $r_Y(z)$, as distribution sections of the bundle $V(\tau) \boxtimes V(\tilde{\tau})$ over $X \times X$. We will in particular be interested in the difference $d(z) := r_Y(z) - r(z)$.

Let $\hat{M}(\tau) \subset \hat{M}$ denote the set of irreducible representations of M which appear in the restriction of τ to M. For each $\sigma \in \hat{M}$ and $\lambda \in \mathfrak{a}^*_\mathbb{C}$ let $z_\sigma(\lambda)$ denote the value of C on the principal series representation $H^{\sigma,\lambda}$. We will fix the unique order-preserving isometric identification $\mathfrak{a}^* \cong \mathbb{R}$. Then $z_\sigma(\lambda)$ is of the form $z_\sigma(\lambda) = c_\sigma - \lambda^2$ for some $c_\sigma \in \mathbb{R}$. We define $\mathbb{C}_\tau$ to be the branched cover of $\mathbb{C}$ to which the inverse functions

$$\lambda_\sigma(z) = \sqrt{c_\sigma - z}$$

extend holomorphically for all $\sigma \in \hat{M}(\tau)$. We fix one sheet $\mathbb{C}^{\mathrm{phys}}$ of $\mathbb{C}_\tau$ over the set $\mathbb{C} \setminus [b,\infty)$, $b := \min_{\sigma \in \hat{M}(\tau)} c_\sigma$, which we call physical. We will often consider $\mathbb{C}^{\mathrm{phys}}$ as a subset of $\mathbb{C}$. It follows from the Plancherel theorem for $L^2(X,V(\tau))$ and $L^2(Y,V_Y(\tau))$ (see [4]) that $[b,\infty)$ is the continuous spectrum of C and C_Y. Thus, $d(z)$ is defined on the complement of finitely many points of $\mathbb{C}^{\mathrm{phys}}$, which belong to the discrete spectrum of C and C_Y.

Let Δ denote the diagonal in $X \times X$ and define $S := \bigcup_{1 \neq \gamma \in \Gamma} (1 \times \gamma)\Delta \subset X \times X$. Let C_i, $i = 1,2$, denote the Casimir operators of G acting on the first and the second variable of the product $X \times X$. The distribution $d(z)$ satisfies the elliptic differential equation

$$(2z - C_1 - C_2)d(z) = 0 \tag{48}$$

on $X \times X \setminus S$ and is therefore smooth on this set.

Lemma 5. *The kernel $d(z)$ extends to $\mathbb{C}_\tau$ as a meromorphic family of smooth sections of $V(\tau)\boxtimes V(\tilde{\tau})$ on $X\times X\setminus S$.*

Proof. We first show

Lemma 6. *$r(z)$ and $r_Y(z)$ extend to $\mathbb{C}_\tau$ as meromorphic families of distributions.*

Proof. We give the argument for $r_Y(z)$ since $r(z)$ can be considered as a special case, where Γ is trivial. Let $V_i\subset X$, $i=1,2$ be open subsets such that the restriction to V_i of the projection $X\to Y$ is a diffeomorphism. We consider $\phi\in C_c^\infty(V_1,V(\tilde{\tau}))$, $\psi\in C_c^\infty(V_2,V(\tau))$ as compactly supported sections over Y.

We now employ the Plancherel theorem [4] for $L^2(Y,V_Y(\tau))$ in order to show that

$$r_{\phi,\psi}(z):=r_Y(z)(\phi\otimes\psi)=\langle\phi,(z-C_Y)^{-1}\psi\rangle$$

extends meromorphically to $\mathbb{C}_\tau$. We decompose

$$\phi=\sum_{s\in\sigma_p(C_Y)}\phi_s+\phi_{\mathrm{ac}},\quad \psi=\sum_{s\in\sigma_p(C_Y)}\psi_s+\psi_{\mathrm{ac}}$$

according to the discrete and continuous spectrum of C_Y. We have

$$r_{\phi,\psi}(z)=\sum_{s\in\sigma_p(C_Y)}(z-s)^{-1}\langle\phi_s,\psi_s\rangle+\langle\phi_{\mathrm{ac}},(z-C)^{-1}\psi_{\mathrm{ac}}\rangle.$$

We now employ the Eisenstein Fourier transformation in order to rewrite the last term of this equation.

For each $\sigma\in\hat{M}(\tau)$, we consider the normalized Eisenstein series as a meromorphic family of maps

$$C^{-\infty}(B,V_B(\sigma_\lambda))\otimes\mathrm{Hom}_M(V_\sigma,V_\tau)\ni f\otimes T\mapsto {}^0E_\lambda^T(f):={}^0P_\lambda^T\circ\mathrm{ext}(f)\in C^\infty(Y,V_Y(\tau))$$

parametrized by $\lambda\in\mathfrak{a}_{\mathbb{C}}^*$. For each σ let $\{T_i(\sigma)\}_i$, $T_i(\sigma)\in\mathrm{Hom}_M(V_\sigma,V_\tau)$, be a basis, and let $T^j(\sigma)\in\mathrm{Hom}_M(V_{\tilde{\sigma}},V_{\tilde{\tau}})$ be the dual basis in the following sense: $T_i(\sigma)^*T^j(\sigma)=\delta_i^j\mathrm{id}_{V_{\tilde{\sigma}}}$. If $\phi\in C_c^\infty(Y,V_Y(\tilde{\tau}))$, then its Eisenstein Fourier transform $\mathrm{EFT}_{\tilde{\sigma}}(\phi)(\lambda)\in C^\infty(B,V_B(\tilde{\sigma}_\lambda))\otimes\mathrm{Hom}_M(V_{\tilde{\sigma}},V_{\tilde{\tau}})$ is given by

$$\langle\mathrm{EFT}_{\tilde{\sigma}}(\phi)(\lambda),f\otimes T_i(\sigma)\rangle:=\langle\phi,{}^0E_{-\lambda}^{T_i(\sigma)}(f)\rangle=\langle({}^0E_{-\lambda}^{T_i(\sigma)})^*(\phi),f\rangle.$$

As a consequence of the Plancherel theorem, we obtain for $\phi\in C_c^\infty(V_1,V(\tilde{\tau}))$, $\psi\in C_c^\infty(V_2,V(\tau))$ that

$$\begin{aligned}\langle\phi_{\mathrm{ac}},\psi_{\mathrm{ac}}\rangle&=\sum_{\sigma\in\hat{M}(\tau)}\frac{1}{4\pi\omega_X}\int_{i\mathfrak{a}^*}\langle\mathrm{EFT}_{\tilde{\sigma}}(\phi)(-\lambda),\mathrm{EFT}_\sigma(\psi)(\lambda)\rangle d\lambda\\&=\sum_{\sigma\in\hat{M}(\tau)}\frac{1}{4\pi\omega_X}\sum_j\int_{i\mathfrak{a}^*}\langle({}^0E_\lambda^{T_j(\sigma)})^*(\phi),({}^0E_{-\lambda}^{T^j(\sigma)})^*(\psi)\rangle d\lambda\end{aligned}$$

$$= \sum_{\sigma \in \hat{M}(\tau)} \frac{1}{4\pi\omega_X} \sum_j \int_{i\mathfrak{a}^*} \langle \phi, {}^0E_\lambda^{T_j(\sigma)} \circ ({}^0E_{-\lambda}^{T^j(\sigma)})^*(\psi) \rangle d\lambda.$$

In a similar manner, we obtain

$$\langle \phi_{\mathrm{ac}}, (z-C)^{-1}\psi_{\mathrm{ac}} \rangle$$
$$= \sum_{\sigma \in \hat{M}(\tau)} \frac{1}{4\pi\omega_X} \sum_j \int_{i\mathfrak{a}^*} (z - z_\sigma(\lambda))^{-1} \left\langle \phi, {}^0E_\lambda^{T_j(\sigma)} \circ {}^0 \left(E_{-\lambda}^{T^j(\sigma)} \right)^* (\psi) \right\rangle d\lambda \tag{49}$$

We further investigate the summands in (49) for each σ seperately. So for $z \in \mathbb{C}^{\mathrm{phys}}$ we put

$$u(z) := \int_{i\mathfrak{a}^*} (z - z_\sigma(\lambda))^{-1} \sum_j \left\langle \phi, {}^0E_\lambda^{T_j(\sigma)} \circ {}^0 \left(E_{-\lambda}^{T^j(\sigma)} \right)^* (\psi) \right\rangle d\lambda. \tag{50}$$

If we set $F(\mu) := u(z_\sigma(\mu))$, then it is defined and holomorphic for $\mathrm{Re}(\mu) > 0$. Thus

$$F(\mu) = \int_{-i\infty}^{i\infty} \frac{f(z)}{\mu^2 - z^2} dz, \qquad \mathrm{Re}(\mu) > 0, \tag{51}$$

for some meromorphic function f on $\mathbb{C}$ that is regular and integrable on the imaginary axis and satisfies $f(z) = f(-z)$. The last property is a consequence of the functional equation of the Eisenstein series and the unitarity of the scattering matrix on the imaginary axis. Here, we interpret the integral as a complex contour integral. Elementary residue calculus shows that any function of the form (51) extends meromorphically to all of $\mathbb{C}$ obeying the functional equation

$$F(\mu) - F(-\mu) = 2\pi i \frac{f(\mu)}{\mu}. \tag{52}$$

Indeed, let us denote the right half-plane by U. For $\nu \in i\mathbb{R} \setminus \{0\}$, we consider a compact change of the contour of integration in (51), symmetric with respect to $z = 0$, without crossing singularities of f and leaving ν to the right (and hence $-\nu$ to the left). We obtain holomorphic functions F_ν, $\nu \in i\mathbb{R} \setminus \{0\}$ on open neighborhoods U_ν of ν such that $F_{\nu|U \cap U_\nu} = F_{|U \cap U_\nu}$ and $F_{\nu|U_\nu \cap U_{\nu'}} = F_{\nu'|U_\nu \cap U_{\nu'}}$ and thus a holomorphic extension of F to an open set U_+ containing $U \cup i\mathbb{R} \setminus \{0\}$. For $\mu \in U_+ \cap (-U_+)$ we get by construction of the extension of F

$$F(\mu) - F(-\mu) = 2\pi i \left(-\mathrm{res}_{z=\mu} \frac{f(z)}{\mu^2 - z^2} + \mathrm{res}_{z=-\mu} \frac{f(z)}{\mu^2 - z^2} \right) = 2\pi i \frac{f(\mu)}{\mu}.$$

We can now define F on the left half-plane by this functional equation and obtain a meromorphic continuation of F to $\mathbb{C} \setminus \{0\}$, where 0 is an isolated singularity of F. This singularity is removable if $f(0) = 0$. In general, it is a pole of at most first order since $\int_{-i\infty}^{i\infty} \frac{1}{\mu^2 - z^2} dz = \frac{\pi i}{\mu}$. Hence, F is also meromorphic at $\mu = 0$.

Thus, we have shown that $r_Y(z)$ is a meromorphic family of distributions if we put the topology induced by the evaluations against sections of the form $\phi \otimes \psi$. We now argue that it is indeed meromorphic with respect to the strong topology. It is clear that $r_Y(z)$ is so on the sheet $\mathbb{C}^{\mathrm{phys}}$. If $z_1, z_2 \in \mathbb{C}_\tau$ vary in different sheets of $\mathbb{C}_\tau$ but project to the same $z \in \mathbb{C}^{\mathrm{phys}}$, then by the construction above the difference $r_Y(z_1) - r_Y(z_2)$ is a meromorphic family of smooth sections. We conclude that $r_Y(z)$ is a meromorphic family of distributions on $\mathbb{C}_\tau$. □

We see that $\mathrm{d}(z)$ extends to a meromorphic family of distributions on $\mathbb{C}_\tau$. Since it fulfills the differential equation (48) on $\Delta \setminus S$, we conclude that its restriction to this set is in fact a meromorphic family of smooth sections. This finishes the proof of Lemma 5. □

6.2 Finite Propagation Speed Estimates

In Sect. 6.1, we have fixed the scaling of the Riemannian metric, which we use in order to define the distance function $d_1 : X \times X \to \mathbb{R}$. On $X \times X$ we furthermore define the function $d_0(x,y) := \inf_{1 \neq \gamma \in \Gamma} d_1(x, \gamma y)$. Note that $S = \{d_0 = 0\}$. Given $\varepsilon > 0$ we define the neighbourhood $S_\varepsilon := \{d_0 \leq \varepsilon\}$ of S. Let $b := \inf \sigma(C_Y)$.

Lemma 7. *Given $\varepsilon > 0$ and a compact subset $W \subset \{\mathrm{Re}(z) < b\}$ there is a constant $C > 0$ such that $|d(z)(x,y)| < C\mathrm{e}^{-(d_0(x,y)-\varepsilon)\sqrt{b-\mathrm{Re}(z)}}$ for all $(x,y) \notin S_{2\varepsilon}$ and $z \in W$.*

Proof. We are going to use the finite propagation speed method, which has been introduced in [6]. In this case, we employ the finite propagation speed of the wave operators $\cos(tA)$ and $\cos(tA_Y)$, where $A := \sqrt{C-b}$, $A_Y := \sqrt{C_Y - b}$. We write

$$R(z) = (b-z)^{-\frac{1}{2}} \int_0^\infty \mathrm{e}^{-t\sqrt{b-z}} \cos(tA)\mathrm{d}t,$$

$$R_Y(z) = (b-z)^{-\frac{1}{2}} \int_0^\infty \mathrm{e}^{-t\sqrt{b-z}} \cos(tA_Y)\mathrm{d}t.$$

Finite propagation speed gives

$$d(z)(x,y) = (b-z)^{-\frac{1}{2}} \int_{d_0(x,y)-\varepsilon}^\infty \mathrm{e}^{-t\sqrt{b-z}} [\cos(tA_Y) - \cos(tA)]\mathrm{d}t\,(x,y)$$

on the level of distribution kernels. By partial integration

$$\begin{aligned}
&(C_1 - b)^N (C_2 - b)^N d(z)(x,y) \\
&\quad = (b-z)^{2N-\frac{1}{2}} \int_{d_0(x,y)-\varepsilon}^\infty \mathrm{e}^{-t\sqrt{b-z}} [\cos(tA_Y) - \cos(tA)]\mathrm{d}t\,(x,y).
\end{aligned}$$

We now employ the fact that $\int_{d_0(x,y)-\varepsilon}^{\infty} \mathrm{e}^{-t\sqrt{b-z}}\cos(tA_Y)\mathrm{d}t$ is a bounded operator on $L^2(Y,V_Y(\tau))$ with norm bounded by $C_1\mathrm{e}^{-(d_0(x,y)-\varepsilon)\mathrm{Re}(\sqrt{b-z})}$. A similar estimate holds for the other term. If we choose $N > \dim(X)/4$, then we can conclude that

$$|d(z)(x,y)| < C\mathrm{e}^{-(d_0(x,y)-\varepsilon)\mathrm{Re}(\sqrt{b-z})},$$

where C depends on W, C_1, and a uniform estimate of norms of delta distributions as functionals on the Sobolev spaces $W^{2N,2}(X,V(\tau))$ and $W^{2N,2}(Y,V_Y(\tau))$. This estimate holds because X,Y as well as the bundles $V(\tau),V_Y(\tau)$ have bounded geometry. We further have employed the fact (which is again a consequence of bounded geometry) that we can use powers of the operator C,C_Y in order to define the norm of the Sobolev spaces. The assertion of the lemma now follows from $\mathrm{Re}(\sqrt{b-z}) \geq \sqrt{b-\mathrm{Re}(z)}$. □

The distribution $r(z)$ is smooth outside the diagonal Δ because it satisfies a differential equation similar to (48). For $\varepsilon > 0$, we define the neighbourhood $\Delta_\varepsilon := \{d \leq \varepsilon\}$ of Δ.

Lemma 8. *For $\varepsilon > 0$ and a compact subset $W \subset \{\mathrm{Re}(z) < b\}$, there is a constant and $C > 0$ such that $|r(z)(x,y)| < C\mathrm{e}^{-(d_1(x,y)-\varepsilon)\sqrt{b-\mathrm{Re}(z)}}$ for all $(x,y) \notin \Delta_{2\varepsilon}$ and $z \in W$.*

Proof. The proof is similar to that of Lemma 7. Using finite propagation speed, we can write

$$r(z)(x,y) = (b-z)^{-\frac{1}{2}} \int_{d(x,y)-\varepsilon}^{\infty} \mathrm{e}^{-t\sqrt{b-z}}\cos(tA)\mathrm{d}t(x,y),$$

$$(C_1-b)^N(C_2-b)^N r(z)(x,y) = (b-z)^{2N-\frac{1}{2}} \int_{d(x,y)-\varepsilon}^{\infty} \mathrm{e}^{-t\sqrt{b-z}}\cos(tA)\mathrm{d}t(x,y).$$

We now argue as in the proof of Lemma 7 in order to conclude the estimate. □

Let L_γ denote the action of $\gamma \in \Gamma$ on sections of $V(\tau)$. By $\delta_\Gamma \in \mathfrak{a}^*$ we denote the critical exponent of Γ (see [4]).

Lemma 9. *If $\sqrt{b-\mathrm{Re}(z)} > \delta_\Gamma + \rho$, then on $X \times X \setminus S$ we have*

$$d(z) = \sum_{1\neq\gamma\in\Gamma} (1 \otimes L_\gamma)r(z).$$

Proof. It follows from Lemma 8 that $|(1\otimes L_\gamma)r(z)(x,y)| < C\mathrm{e}^{-d_1(x,\gamma^{-1}y)\sqrt{b-\mathrm{Re}(z)}}$. In view of the definition of the critical exponent δ_Γ, this sum converges locally uniformly on $X \times X \setminus S$. The distribution $u(z) := r_Y(z) - \sum_{\gamma\in\Gamma}(1\otimes L_\gamma)r(z)$ on $X \times X$ satisfies the differential equations

$$(z-C_1)u(z) = 0, \quad (z-C_2)u(z) = 0,$$

and is therefore a smooth section depending meromorphically on $z \in \mathbb{C}^{\mathrm{phys}}$. We further have the estimate

$$\sum_{1 \neq \gamma \in \Gamma} |(1 \otimes L_\gamma) r(z)(x,y)| < C e^{-s d_0(x,y)}, \quad s < \sqrt{b - \mathrm{Re}(z)} - \delta_\Gamma - \rho .$$

For $\mathrm{Re}(z) \ll 0$, we see that $u(z)$ defines a bounded operator on $L^2(Y, V_Y(\tau))$ with image contained the z-eigenspace of C. Since z is outside the spectrum it vanishes. Since u is meromorphic in z, it vanishes for all z with $\sqrt{b - \mathrm{Re}(z)} > \delta_\Gamma + \rho$. This proves the lemma. □

6.3 Boundary Values

We define $\partial S := \bigcup_{1 \neq \gamma \Gamma} (\gamma \times 1)(\partial\Delta \cap \Omega \times \Omega)$, where $\partial\Delta \subset \partial X \times \partial X$ is the diagonal in the boundary. The meromorphic family of eigensections $d(z)$ on $X \times X \setminus S$ has meromorphic families of hyperfunction boundary values. Since we consider a product of rank one spaces it is easy to determine the leading exponents of a joint eigensection of C_1, C_2 with eigenvalue z. These exponents are pairs of elements of $\mathfrak{a}^*_{\mathbb{C}}$.

Lemma 10. *The set of leading exponents of a joint eigensection of C_1, C_2 in the bundle $V(\tau) \otimes V(\tilde{\tau})$ with generic eigenvalue z is*

$$\{\mu_{(\sigma,\varepsilon),(\sigma',\varepsilon')}(z) \mid \sigma \in \hat{M}(\tau), \sigma' \in \hat{M}(\tilde{\tau}), \varepsilon, \varepsilon' \in \{+,-\}\},$$

where $\mu_{(\sigma,\varepsilon),(\sigma',\varepsilon')}(z) = (-\rho + \varepsilon\lambda_\sigma(z), -\rho + \varepsilon'\lambda_{\sigma'}(z))$. The corresponding boundary value is a section of the bundle $V(\tau(\sigma)_{\varepsilon\lambda_\sigma(z)}) \otimes V(\tilde{\tau}(\sigma')_{\varepsilon'\lambda_{\sigma'}(z)}) \to \partial X \times \partial X$, where $\tau(\sigma), \tilde{\tau}(\sigma')$ denote the isotypic components.

Proof. An eigensection of C in $V(\tau) \to X$ has leading exponents $-\rho + \varepsilon\lambda_\sigma(z)$, $\varepsilon \in \{+,-\}$, and the corresponding boundary value is a section of $V(\tau(\sigma)_{\varepsilon\lambda_\sigma(z)})$. This implies the lemma. □

Note that $d(z)$ is a joint eigensection in a neighbourhood in $X \times X$ of $\Omega \times \Omega \setminus \partial S$. Therefore, for generic z it has hyperfunction boundary values along this set [15]. We denote the boundary value associated with the leading exponent $\nu := \mu_{(\sigma,\varepsilon),(\sigma',\varepsilon')}(z)$ by $\beta_\nu(f)$.

Lemma 11. *We have $\beta_\nu(d(z)) = 0$ (the meromorphic family of hyperfunctions vanishes) except for $\nu = \mu_{(\sigma,-),(\tilde{\sigma}^w,-)}(z)$, $\sigma \in \hat{M}(\tau)$, in which case $\beta_\nu(d(z))$ is a meromorphic family of real analytic sections.*

Proof. We employ the fact that $\beta_\nu(d(z))$ depends meromorphically on z. Let $U \subset \Omega$ be such that the restriction of the projection $\Omega \to B$ to U is a diffeomorphism. The identification $\mathfrak{a}^* \cong \mathbb{R}$ fixed above induces an identification $A \cong \mathbb{R}^+$, which we use in the following estimates. There is a constant $c > 0$ such that for all $(k_1,k_2) \in U \times U$ we have $d_0(k_1a_1,k_2a_2) > c|\max(a_1,a_2)|$ (see [4], Cor. 2.4). Using Lemma 7, we see that for $\mathrm{Re}(z) < b$ we have $|d(z)(k_1a_1,k_2a_2)| < C|\max(a_1,a_2)|^{-\sqrt{b-\mathrm{Re}(z)}}$, where C depends on z. If one of the signs ε,ε' is positive, for $z \ll 0$ we have $\sqrt{b-z} - 2\rho + \varepsilon\lambda_\sigma(z) + \varepsilon'\lambda_{\sigma'}(z) > 0$. For those z, we have $\lim_{\min(a_1,a_2)\to\infty} d(z)(k_1a_1,k_2a_2)(a_1,a_2)^{-\mu_{(\sigma,\varepsilon),(\sigma',\varepsilon')}(z)} = 0$ uniformly in (k_1,k_2), where $(a,b)^{(\mu,\nu)} := a^\mu b^\nu$. This shows that $\beta_\nu(d(z)) = 0$ if one of ε,ε' is positive.

We now consider the kernel $r(z)$ on $X \times X \setminus \Delta$. It is a joint eigenfunction of C_1, C_2 to the eigenvalue z on a neighbourhood of $\partial X \times \partial X \setminus \partial\Delta$ and therefore has hyperfunction boundary values along this set. A similar argument as above but using Lemma 8 instead of 7 shows that $\beta_\nu(r(z)) = 0$ except for $\varepsilon = \varepsilon' = -$.

Note that $r(z)$ is G-invariant in the sense that for $g \in G$ we have $L_g \otimes L_g r(z) = r(z)$. If $\nu = \mu_{(\sigma,-),(\sigma',-)}(z)$, then $\beta_\nu(r(z))$ is a G-invariant hyperfunction section of $V(\tau(\sigma)_{-\lambda_\sigma(z)}) \otimes V(\tilde{\tau}(\sigma')_{-\lambda_{\sigma'(z)}})$ over $\partial X \times \partial X \setminus \partial\Delta$. Since this set is an orbit of G, an invariant hyperfunction on this set is smooth, and the evaluation at the point $(w,1) \in \partial X \times \partial X$ provides an injection of the space of invariant sections into $V_\tau(\sigma) \otimes V_{\tilde{\tau}}(\sigma')$. If b is such a G-invariant section, then we have for $ma \in MA$

$$\begin{aligned} b(w,1) &= b(maw, ma) \\ &= b(wm^w a^{-1}, ma) \\ &= \tau(wm^{-1}w) \otimes \tilde{\tau}(m^{-1}) a^{\lambda_{\sigma'}(z) - \lambda_\sigma(z)} b(w,1). \end{aligned}$$

Thus, $b(w,1) \in [V_{\tau^w}(\sigma^w) \otimes V_{\tilde{\tau}}(\sigma')]^M$. We conclude that $\sigma' \cong \tilde{\sigma}^w$, and in this case $\lambda_{\sigma'}(z) = \lambda_\sigma(z)$ holds automatically. Thus, $\beta_\nu(r(z)) = 0$ if $\sigma' \ncong \tilde{\sigma}^w$.

We write

$$\begin{aligned} & V(\tau(\sigma)_{-\lambda}) \otimes V(\tilde{\tau}(\tilde{\sigma}^w)_{-\lambda}) \\ & \qquad = V(\sigma_{-\lambda}) \otimes V(\tilde{\sigma}^w_{-\lambda}) \otimes \mathrm{Hom}_M(V_\sigma, V_\tau) \otimes \mathrm{Hom}_M(V_{\tilde{\sigma}^w}, V_{\tilde{\tau}}). \end{aligned}$$

The space of invariant sections of $V(\sigma_{-\lambda}) \otimes V(\tilde{\sigma}^w_{-\lambda})$ over $\partial X \times \partial X \setminus \partial\Delta$ is spanned by the distribution kernel $\hat{j}^w_{\sigma^w,\lambda}$ of the Knapp–Stein intertwining operator $\hat{J}^w_{\sigma^w,\lambda}$. We conclude that for each $\sigma \in \hat{M}(\tau)$ there is a meromorphic family $A_\sigma(z) \in \mathrm{Hom}_M(V_\sigma, V_\tau) \otimes \mathrm{Hom}_M(V_{\tilde{\sigma}^w}, V_{\tilde{\tau}})$ such that for $\nu = \mu_{(\sigma,-),(\tilde{\sigma}^w,-)}(z)$ we have under the identifications above $\beta_\nu(r(z)) = \hat{j}^w_{\sigma^w,\lambda_\sigma(z)} \otimes A_\sigma(z)$.

Let $\nu = \mu_{(\sigma,-),(\tilde{\sigma}^w,-)}(z)$. We now employ Lemma 9 which states that for $\mathrm{Re}(z) \ll 0$ we have $d(z) = \sum_{1 \neq \gamma \in \Gamma} (L_\gamma \otimes 1) r(z)$. The sum converges locally uniformly and thus in the space of smooth section over $X \times X \setminus S$. We further see that convergence holds locally uniformly in a neighbourhood of $\Omega \times \Omega \setminus \partial S$. Thus by [18], we can consider distribution boundary values, and by continuity of the boundary value map we have on $\Omega \times \Omega \setminus \partial S$

$$\begin{aligned}\beta_\nu(d(z)) &= \sum_{1\neq\gamma\in\Gamma} (\pi^{\sigma,-\lambda_\sigma(z)}(\gamma)\otimes 1)\beta_\nu(d(z)) \\ &= \sum_{1\neq\gamma\in\Gamma} (\pi^{\sigma,-\lambda_\sigma(z)}(\gamma)\otimes 1)\hat{j}^w_{\sigma^w,\lambda_\sigma(z)}\otimes A_\sigma(z) \\ &= (\hat{s}^w_{\sigma^w,\lambda_\sigma(z)} - \hat{j}^w_{\sigma^w,\lambda_\sigma(z)})\otimes A_\sigma(z),\end{aligned}$$

where $\hat{s}^w_{\sigma^w,\lambda_\sigma(z)}$ is the distribution kernel of the scattering matrix $\hat{S}^w_{\sigma^w,\lambda_\sigma(z)}$. Here, we use the identity $\pi_* \circ \hat{J}^w_{\sigma^w,\lambda_\sigma(z)} = \hat{S}^w_{\sigma^w,\lambda_\sigma(z)} \circ \pi_*$, which implies that the distribution kernel of the scattering matrix $\hat{S}^w_{\sigma^w,\lambda}$ can be obtained by averaging the distribution kernel of the Knapp–Stein intertwining operator $\hat{J}^w_{\sigma^w,\lambda}$ for $\mathrm{Re}(\lambda) \gg 0$.

It follows from the results of [4] that $\hat{s}^w_{\sigma^w,\lambda_\sigma(z)} - \hat{j}^w_{\sigma^w,\lambda_\sigma(z)}$ extends to a meromorphic family of smooth sections on all of $\mathfrak{a}^*_\mathbb{C}$. By [5], Lemma 2.19, 2.20, it is indeed a meromorphic family of real analytic sections. Strictly speaking, in [5] we only considered the spherical M-type for $G = SO(1,n)$, the straightforward extension of these results to the general case can be found in [17, Chap. 3]. We conclude that $\beta_\nu(d(z))$ is real analytic as required.

A similar reasoning shows that $\beta_\nu(d(z)) = 0$ for all ν which are not of the form $\mu_{(\sigma,-),(\tilde{\sigma}^w,-)}(z)$ for some $\sigma \in \hat{M}(\tau)$. □

Lemma 12. *We have an asymptotic expansion (for generic z)*

$$d(z)(k_1a,k_2a) \overset{a\to\infty}{\sim} \sum_{\sigma\in\hat{M}(\tau)}\sum_{n=0}^{\infty} a^{-2\rho-2\lambda_\sigma(z)-n\alpha} p_{z,\sigma,n}(k_1,k_2), \tag{53}$$

which holds locally uniformly for $k \in \Omega\times\Omega\setminus\partial S$, and where the real analytic sections $p_{z,\sigma,n}(k_1,k_2)$ of $V(\tau)\otimes V(\tilde{\tau})$ depend meromorphically on z.

Proof. Since the boundary value of $d(z)$ along $\Omega\times\Omega\setminus\partial S$ is real analytic we can employ [18], Prop. 2.16, in order to conclude that $d(z)$ has an asymptotic expansion with coefficients, which depend meromorphically on z. The formula follows from an inspection of the list of leading exponents Lemma 11. □

Lemma 12 has the following consequence. For generic z, we have

$$\mathrm{tr}\, d(z)(ka,ka) \overset{a\to\infty}{\sim} \sum_{\sigma\in\hat{M}(\tau)}\sum_{n=0}^{\infty} a^{-2\rho-2\lambda_\sigma(z)-n\alpha} p_{z,\sigma,n}(k),$$

which holds locally uniformly for $k\in\Omega$, and where the real analytic functions $p_{z,\sigma,n}$ depend meromorphically on z.

6.4 The Regularized Trace of the Resolvent

Recall that $\chi^\Gamma \in C_c^\infty(X \cup \Omega)$ is a cut-off function such that $\sum_{\gamma\in\Gamma} \gamma^*\chi^\Gamma = 1$ on $X \cup \Omega$.

Lemma 13. *The integral*

$$Q_\tau(z) := \int_X \chi^\Gamma(x)\, \mathrm{tr}\, d(z)(x,x)\mathrm{d}x$$

converges for $\mathrm{Re}(z) \ll 0$ *and admits a meromorphic continuation to all of* $\mathbb{C}_\tau$.

Proof. Convergence for $\mathrm{Re}(z) \ll 0$ follows from Lemma 7. Fix $R \in A$. We write $Q_\tau(z) = Q_1(z,R) + Q_2(z,R)$, where

$$Q_1(z,R) := \int_1^R \int_K \chi^\Gamma(ka)\mathrm{tr}\, d(z)(ka,ka)\mathrm{d}k v(a)\mathrm{d}a,$$

where v is such that $\mathrm{d}k\, v(a)\mathrm{d}a$ is the volume measure on X. Note that $v(a) \sim a^{2\rho}(\omega_X + a^{-\alpha}c_1 + a^{-2\alpha}c_2 + \cdots)$ as $a \to \infty$.

It is clear that $Q_1(z,R)$ admits a meromorphic continuation. We have an asymptotic expansion as $a \to \infty$.

$$u(z,a) := \int_K \chi^\Gamma(ka)\mathrm{tr}\, d(z)(ka,ka)\mathrm{d}k v(a) \sim \sum_{\sigma\in\hat{M}(\tau)} \sum_{n=0}^{\infty} a^{-2\lambda_\sigma(z)-n\alpha} q_{z,\sigma,n},$$

where q depends meromorphically on z. For $m \in \mathbb{N}$, let

$$u_m(z,a) := u(z,a) - \sum_{\sigma\in\hat{M}(\tau)} \sum_{n=0}^{m} a^{-2\lambda_\sigma(z)-n\alpha} q_{z,\sigma,n}.$$

Given a compact subset W of $\mathbb{C}_\tau$ we can choose $m \in \mathbb{N}_0$ such that $\int_R^\infty u_m(z,a)\mathrm{d}a$ converges (for generic z) and depends meromorphically on z for all $z \in W$. We further have

$$\sum_{\sigma\in\hat{M}(\tau)} \sum_{n=0}^{m} \int_R^\infty a^{-2\lambda_\sigma(z)-n\alpha} q_{z,\sigma,n}\mathrm{d}a = \sum_{\sigma\in\hat{M}(\tau)} \sum_{n=0}^{m} \frac{R^{-2\lambda_\sigma(z)-n\alpha} q_{z,\sigma,n}}{2\lambda_\sigma(z)+n\alpha},$$

and this function extends meromorphically to C_τ. Since we can choose W arbitrary large, we conclude that $Q_\tau(z)$ admits a meromorphic continuation to all of $\mathbb{C}_\tau$. □

6.5 A Functional Equation

Let $L(\tau) := \{c_\sigma \mid \sigma \in \hat{M}(\tau)\}$ be the set of ramification points of $\mathbb{C}_\tau$, define $\mathbb{C}^\sharp := \mathbb{C} \setminus L(\tau)$, and let $\mathbb{C}_\tau^\sharp \subset \mathbb{C}_\tau$ be the preimage of $\mathbb{C}^\sharp$ under the projection $\mathbb{C}_\tau \to \mathbb{C}$.

Then $\mathbb{C}^\sharp_\tau \to \mathbb{C}^\sharp$ is a Galois covering with group of deck transformations $\Pi := \oplus_{L(\tau)}\mathbb{Z}_2$. The action of Π extends to $\mathbb{C}_\tau$ such $\mathbb{C}_\tau \setminus \mathbb{C}^\sharp_\tau$ consists of fixed points. For $l \in L(\tau)$ let $q_l \in \Pi$ be the corresponding generator. Then we have $\lambda_\sigma(q_l z) = -\lambda_\sigma(z)$ for all $\sigma \in \hat{M}(\tau)$ with $c_\sigma = l$ and $\lambda_{\sigma'}(q_l z) = \lambda_{\sigma'}(z)$ else.

Recall the definition (47) of the function $L_\Gamma(\pi^{\sigma,\lambda})$.

Proposition 4. *For $l \in L(\tau)$, we have*

$$Q_\tau(q_l z) - Q_\tau(z) = \sum_{\sigma\in\hat{M}(\tau), c_\sigma = l} \frac{-[\tau:\sigma]}{2\lambda_\sigma(z)} L_\Gamma(\pi^{\sigma,\lambda_\sigma(z)}).$$

Proof. The proof of Lemma 5, in particular the functional equation (52), gives that (in the notation introduced there)

$$\langle r_Y(q_l z) - r_Y(z), \phi\otimes\psi\rangle = \sum_{\sigma\in\hat{M}(\tau), c_\sigma = l} \frac{-1}{2\omega_X\lambda_\sigma(z)} \sum_j \left\langle \phi, {}^0E^{T_j(\sigma)}_{\lambda_\sigma(z)} \circ {}^0\left(E^{T^j(\sigma)}_{-\lambda_\sigma(z)}\right)^*(\psi)\right\rangle.$$

We conclude that

$$r_Y(q_l z) - r_Y(z) = \sum_{\sigma\in\hat{M}(\tau), c_\sigma = l} \frac{-1}{2\omega_X\lambda_\sigma(z)} \sum_j {}^0E^{T_j(\sigma)}_{\lambda_\sigma(z)} \circ {}^0\left(E^{T^j(\sigma)}_{-\lambda_\sigma(z)}\right)^*.$$

The same reasoning applies to the trivial group Γ, where the Eisenstein series get replaced by the Poisson transformations. Thus, we can write

$$d(q_l z) - d(z) = \sum_{\sigma\in\hat{M}(\tau), c_\sigma = l} \frac{-1}{2\omega_X\lambda_\sigma(z)} \sum_j {}^0P^{T_j(\sigma)}_{\lambda_\sigma(z)} \circ (\mathrm{ext}\circ\pi_* - 1) \circ \left({}^0P^{T^j(\sigma)}_{-\lambda_\sigma(z)}\right)^*.$$

The proof of Lemma 13 shows that $Q_1(z,R)$ has an asymptotic expansion

$$Q_1(z,R) \sim Q_\tau(z) + \sum_{\sigma\in\hat{M}(\tau)} \sum_{n=0}^\infty \frac{R^{-2\lambda_\sigma(z)-n\alpha} q_{z,\sigma,n}}{2\lambda_\sigma(z) + n\alpha}.$$

In particular, if $2\lambda_\sigma(z) \notin -\mathbb{N}_0\alpha$ for all σ, then $Q_\tau(z)$ is the constant term in the asymptotic expansion of $Q_1(z,R)$.

We can now apply the part of the proof of Proposition 2 in which we determined the constant term (as $R\to\infty$) of

$$\int_1^R \int_K \chi^\Gamma(ka)\mathrm{tr}\left[{}^0P^{T_j(\sigma)}_\lambda \circ (\mathrm{ext}\circ\pi_* - 1) \circ ({}^0P^{T^j(\sigma)}_{-\lambda})^*\right](ka,ka)dk\nu(a)da$$

as a distribution on $i\mathfrak{a}^* \setminus \{0\}$. This shows the desired equation first on $z_\sigma(i\mathfrak{a}^*)$, and then everywhere by meromorphic continuation. □

6.6 Selberg Zeta Functions

For a detailed investigation of Selberg zeta functions associated with bundles (for cocompact Γ), we refer to [3]. Here, we assume that σ is irreducible and Weyl invariant, or that it is of the form $\sigma' \oplus (\sigma')^w$ for some Weyl non-invariant irreducible M-type σ'. In the latter case, we define $L_\Gamma(\pi^{\sigma,\lambda}) := L_\Gamma(\pi^{\sigma',\lambda}) + L_\Gamma(\pi^{(\sigma')^w,\lambda})$.

Let $P = MAN$ be a parabolic subgroup of G. If $\gamma \in \Gamma \setminus 1$, then it is conjugate in G to an element $m_g a_g \in MA$ with $a_g > 1$. Let $\bar{\mathfrak{n}}$ be the negative root space of $(\mathfrak{g}, \mathfrak{a})$. For $\mathrm{Re}(\lambda) > \rho$, we can define the Selberg zeta function $Z_S(\lambda, \sigma)$ by the converging infinite product

$$Z_S(\lambda,\sigma) := \prod_{1 \neq [g] \in C\Gamma} \prod_{k=0}^{\infty} \det\left(1 - \sigma(m_g) \otimes S^k(\mathrm{Ad}(m_g a_g)_{|\bar{\mathfrak{n}}}) a_g^{-\lambda-\rho}\right).$$

In the case of cocompact Γ, it was shown by [7] that $Z_S(\lambda, \sigma)$ has a meromorphic continuation to all of $\mathfrak{a}^*_{\mathbb{C}}$. In [20], it was explained that the argument of [7] extends to the case of convex cocompact subgroups since it is the compactness of the non-wandering set of the geodesic flow of Y that matters and not the compactness of Y. Strictly speaking, [20] deals with the spherical case of $SO(1,2n)$, but the argument extends to the general case.

There is a virtual representation τ of K (i.e. an element of the integral representation ring of K) such that $\tau_{|M} = \sigma$ in the integral representation ring of M (see [16], [3]). We call τ a lift of σ. Note that τ is not unique. We can extend the material developed above to virtual K-types by extending the traces linearly. Because of the factor $[\tau : \sigma]$, Proposition 4 has the following corollary.

Corollary 1. *If τ is a lift of σ, then $Q_\tau(z)$ extends to a twofold branched cover of $\mathbb{C}$ associated with $\lambda_\sigma(z)$.*

Theorem 3. *The Selberg zeta function satisfies*

$$\frac{Z_S(\lambda,\sigma)}{Z_S(-\lambda,\sigma)} = \exp \int_0^\lambda L_\Gamma(\pi^{\sigma,z})dz.$$

In particular, the residues of $L_\Gamma(\pi^{\sigma,\lambda})$ are integral.

Proof. By [3], Prop. 3.8. we have

$$Q_\tau(z_\sigma(\lambda)) = \frac{1}{2\lambda} Z_S'(\lambda,\sigma)/Z_S(\lambda,\sigma) \tag{54}$$

for $\mathrm{Re}(\lambda) \gg 0$. Indeed, for $\mathrm{Re}(\lambda) \gg 0$ the function f, $f(g) := \mathrm{tr}\, d(z_\sigma(\lambda))(g,g)$, satisfies Lemma 1. Therefore, $Q_\tau(z_\sigma(\lambda))$ is just what is called in [3] the hyperbolic contribution associated to the resolvent.

So Corollary 1 and Proposition 4 yield the functional equation of the logarithmic derivative of the Selberg zeta function

$$\frac{Z_S'(-\lambda,\sigma)}{Z_S(-\lambda,\sigma)}+\frac{Z_S'(\lambda,\sigma)}{Z_S(\lambda,\sigma)}=L_\Gamma\left(\pi^{\sigma,\lambda}\right). \tag{55}$$

Integrating and employing the a priori information that $Z_S(\lambda,\sigma)$ is meromorphic, we obtain the desired functional equation. □

As explained in the introduction, it is known (from the approach to Z_S using symbolic dynamics and Ruelles thermodynamic formalism) that $Z_S(\lambda)$ is a meromorphic function of finite order. It follows that $L_\Gamma(\pi^{\sigma,\lambda})$, as a function of λ, grows at most polynomially.

In order to describe the singularities of $Z_S(\lambda,\sigma)$ we assume – if X has even dimension – that τ is an admissible lift of σ (see [3], Def. 1.17). Let $n_{\lambda,\sigma}$ denote the (virtual) dimension of the subspace of the L^2-kernel of $C_Y - z_\sigma(\lambda)$ on $V_Y(\tau)$, which is generated by non-discrete series representations of G. Using (54) and (55), one can derive the following corollary.

Corollary 2. *The orders of the singularities of the Selberg zeta function $Z_S(\lambda,\sigma)$ associated to Γ at $\mu \neq 0$ are given by*

$$\mathrm{ord}_{\lambda=\mu} Z_S(\lambda,\sigma)=\begin{cases} n_{\mu,\sigma} & \mathrm{Re}(\mu)>0,\\ \mathrm{res}_{\lambda=\mu}L_\Gamma(\pi^{\sigma,\lambda})+n_{-\mu,\sigma} & \mathrm{Re}(\mu)<0.\end{cases}$$

If μ is non-integral and if ext has a pole at μ of at most first order, then it is not difficult to see that

$$\mathrm{res}_{\lambda=\mu}L_\Gamma(\pi^{\sigma,\lambda})=\dim {}^\Gamma C^{-\infty}(\Lambda,V(\sigma_\mu))-\dim {}^\Gamma C^{-\infty}(\Lambda,V(\sigma_{-\mu})),$$

where ${}^\Gamma C^{-\infty}(\Lambda,V(\sigma_{\pm\mu}))$ is the space of Γ-invariant distribution sections on ∂X with support on the limit set Λ. If ext has higher order singularity, then the residue has a similar interpretation (see [5], Prop. 5.6) This provides an independent argument for the integrality of the residues of $L_\Gamma(\pi^{\sigma,\lambda})$ at non-integral μ. For $\mathrm{Re}(\mu)>0$, μ non-integral again, one has in addition

$$n_{\mu,\sigma}=\dim {}^\Gamma C^{-\infty}(\Lambda,V(\sigma_\mu)).$$

This gives the plain formula (provided μ is non-integral and ext has an at most first order pole at μ)

$$\mathrm{ord}_{\lambda=\mu} Z_S(\lambda,\sigma)=\dim {}^\Gamma C^{-\infty}(\Lambda,V(\sigma_\mu)),$$

which is in accordance with Patterson's general conjecture [5, 14, 17, 19] on a cohomological description of the singularities of $Z_S(\lambda,\sigma)$. The conjecture has been established in a number of cases. The difficulty in the general case is to achieve a good understanding of the residues of $L_\Gamma(\pi^{\sigma,\lambda})$ at integral μ.

6.7 Integrality of $N_\Gamma(\pi)$ for Discrete Series Representations

Let π be a discrete series representation of G containing the K-type $\tilde{\tau}$. There are embeddings $M_\pi \otimes V_\pi(\tilde{\tau}) \hookrightarrow L^2(\Gamma\backslash G)(\tilde{\tau})$ and $V_{\tilde{\pi}} \otimes V_\pi(\tilde{\tau}) \hookrightarrow L^2(G)(\tilde{\tau})$. Let $A \in \mathrm{End}_K(V_\pi(\tilde{\tau}))$ be given. We extend A by zero to the orthogonal complement of $V_\pi(\tilde{\tau})$, thus obtaining an operator in $\mathrm{End}_K(V_\pi)$, which we will still denote by A. The operator A induces operators $\check{R}_\Gamma(h_A)$ and $\check{R}(h_A)$ on $L^2(\Gamma\backslash G)$ and $L^2(G)$, where h_A is supported on $\{\pi\} \subset \hat{G}$ and $h_A(\pi) := A$.

Here, we are mainly interested in the case of a non-integrable discrete series representation π, compare the discussion in Sect. 5.2.

Lemma 14. *We have*

$$K_{\check{R}_\Gamma(h_A)}(g,g) - K_{\check{R}(h_A)}(g,g) \in L^1(\Gamma\backslash G).$$

Proof. Let $D(G,\tau)$ be the algebra of invariant differential operators on $V(\tau)$. It is isomorphic to $(\mathscr{U}(\mathfrak{g}) \otimes_{\mathscr{U}(\mathfrak{k})} \mathrm{End}(V_\tau))^K$. If π' is an admissible representation of G, then $D(G,\tau)$ acts in a natural way on $(V_{\pi'} \otimes V_\tau)^K$. If π' is irreducible, then $(V_{\pi'} \otimes V_\tau)^K$ is an irreducible representation of $D(G,\tau)$. The correspondence $\pi' \mapsto (V_{\pi'} \otimes V_\tau)$ provides a bijection between the sets of equivalence classes of irreducible representations of G containing the K-type $\tilde{\tau}$ and irreducible representations of $D(G,\tau)$, see e.g. [21, 3.5.4].

Note that $\mathrm{End}_K(V_\pi(\tilde{\tau})) \cong \mathrm{End}((V_\pi \otimes V_\tau)^K)$. We conclude that there is $D_A \in D(G,\tau)$ that induces the endomorphism A on $V_\pi(\tilde{\tau})$. Let z_0 be the eigenvalue of the Casimir operator on π and Z be the finite set of irreducible representations of G containing the K-type $\tilde{\tau}$ such that C acts with eigenvalue z_0. Then we can choose D_A such that it vanishes on all $(V_{\pi'} \otimes V_\tau)^K$ for $\pi' \in Z$, $\pi' \neq \pi$.

For simplicity, we assume that z_0 is not a branching point of $R(z)$. In the latter case, the following argument can easily be modified. The operators $D_A R_Y(z)$ and $D_A R(z)$ have poles at z_0 with residues $K_{\check{R}_\Gamma(h_A)}$ and $K_{\check{R}(h_A)}$. The difference $(D_A)_1 d(z) := (D_A)_1 r_Y(z) - (D_A)_1 r(z)$ of distribution kernels is still a meromorphic family of joint eigenfunctions with real analytic boundary values along $\Omega \times \Omega \setminus \partial S$. We have the asymptotic expansion

$$(D_A)_1 d(z)(k_1 a, k_2 a) \stackrel{a\to\infty}{\sim} \sum_{\sigma \in \hat{M}(\tau)} \sum_{n=0}^{\infty} a^{-2\rho - 2\lambda_\sigma(z) - n\alpha} p_{z,\sigma,n,A}(k_1,k_2). \tag{56}$$

The residue of $(D_A)_1 d(z)$ at z_0 can be computed by integrating $(D_A)_1 d(z)$ along a small circle counter-clockwise surrounding z_0. If we insert the asymptotic expansion (56) into this integral, then we obtain an asymptotic expansion

$$\mathrm{res}_{z=z_0}(D_A)_1 d(z)(k_1 a, k_2 A) \stackrel{a\to\infty}{\sim} \sum_{\sigma \in \hat{M}(\tau)} \sum_{n=0}^{\infty} \mathrm{res}_{z=z_0} a^{-2\rho - 2\lambda_\sigma(z) - n\alpha} p_{z,\sigma,n,A}(k_1,k_2)$$

$$\sim \sum_{\sigma\in\hat{M}(\tau)} \sum_{n=0}^{\infty} \sum_{m=0}^{\text{finite}} \log(a)^m a^{-2\rho-2\lambda_\sigma(z_0)-n\alpha} p_{z_0,\sigma,n,m,A}(k_1,k_2),$$

where $p_{z_0,\sigma,n,m,A}$ is a real analytic section on $\Omega\times\Omega\setminus\partial S$. Since $K_{\check{R}_\Gamma(h_A)}$ and $K_{\check{R}(h_A)}$ project onto eigenspaces of square integrable sections we have for $k_1\neq k_2$

$$K_{\check{R}(h_A)}(k_1a_1,k_2a_2) \overset{a_i\to\infty}{\sim} \sum_{n_1,n_2=0}^{\infty} \sum_{m_1,m_2}^{\text{finite}} \log(a_1)^{m_1}\log(a_2)^{m_2}$$
$$a_1^{-\rho-\lambda_\sigma(z_0)-n_1\alpha} a_2^{-\rho-\lambda_\sigma(z_0)-n_2\alpha} p_{z_0,\sigma,n_1,n_2,m_1,m_2,A}(k_1,k_2),$$

with $p_{z_0,\sigma,n_1,n_2,m_1,m_2,A}(k_1,k_2)=0$ as long as $-\lambda_\sigma(z_0)-n_1\alpha\geq 0$, $-\lambda_\sigma(z_0)-n_2\alpha\geq 0$, and similarly for $K_{\check{R}_\Gamma(h_A)}$. We conclude that $-2\lambda_\sigma(z_0)-n\alpha<0$ if $p_{z_0,\sigma,n,m,A}\neq 0$. The assertion of the lemma now follows. □

Recall the definition (46) of the regularized multiplicity $N_\Gamma(\pi)$ of a discrete series representation π. By Proposition 3, it vanishes for integrable π. For the general case, we have the following

Proposition 5. *If π be a representation of the discrete series of G, then $N_\Gamma(\pi)\in\mathbb{Z}$.*

Proof. There exists an invariant generalized Dirac operator D acting on a graded vector bundle $E\to X$, $E=E^+\oplus E^-$, such that $V_{\tilde{\pi}}\oplus\{0\}$ is the kernel of D [1]. If τ is the virtual K-representation associated with E, then

$$\tau_{|M}=0. \tag{57}$$

Let D_Y be the induced operator on Y. The distribution kernels of $r(z):=(z-D^2)^{-1}$ and $r_Y(z):=(z-D_Y^2)^{-1}$ have meromorphic continuations to a branched covering of $\mathbb{C}$. Their difference goes into the functional Ψ'. The function

$$Q(z):=\int_{\Gamma\backslash G}\operatorname{tr}\,(r_Y(z)-r(z))(g,g)dg$$

has a meromorphic continuation to all of $\mathbb{C}$ by (57) and Proposition 4. Its residue at $z=0$ is given by

$$\operatorname{res}_{z=0}Q(z)=N_\Gamma(\pi)+\sum_{\pi'\in\hat{G}_c} n(\tau,\pi')N_\Gamma(\pi'),$$

where the sum reflects the fact that a finite number of representations belonging to $\hat{G}_c$ may contribute to the kernel of D_Y. Here, $n(\tau,\pi')\in\mathbb{Z}$ is the virtual multiplicity of τ in π', and $N_\Gamma(\pi')\in\mathbb{N}_0$ is the multiplicity of the complementary series representation π' in $L^2(\Gamma\backslash G)$. We show that $\operatorname{res}_{z=0}Q(z)=0$ in order to conclude that $N_\Gamma(\pi)=-\sum_{\pi'\in G_c} n(\tau,\pi')N_\Gamma(\pi')\in\mathbb{Z}$.

It suffices to show that $Q(z) \equiv 0$ for $\mathrm{Re}(z) \ll 0$. This follows from (54), but we will give an independent argument. We can write

$$r_Y(z) - r(z) = \frac{1}{z}((D_Y^2)_1 r_Y(z) - (D^2)_1 r(z)).$$

Integrating the restriction of this difference to the diagonal over $\Gamma\backslash G$, we obtain

$$\begin{aligned}
Q(z) &= \frac{1}{z}\int_{\Gamma\backslash G} \mathrm{tr}\,((D_Y^2)_1 r_Y(z) - (D^2)_1 r(z))(g,g)\mathrm{d}g \\
&= \frac{1}{z}\int_G \chi^\Gamma(g)\mathrm{tr}\,((D_Y^2)_1 r_Y(z) - (D^2)_1 r(z))(g,g)\mathrm{d}g \\
&= -\frac{1}{z}\int_G \chi^\Gamma(g)\mathrm{tr}\,((D_Y)_1(D_Y)_2 r_Y(z) - D_1 D_2 r(z))(g,g)\mathrm{d}g \\
&\quad -\frac{1}{z}\int_G \mathrm{tr}\, c(d\chi^\Gamma)_1((D_Y)_1 r_Y(z) - D_1 r(z))(g,g)\mathrm{d}g \\
&= -\frac{1}{z}\int_G \chi^\Gamma(g)\mathrm{tr}\,((D_Y^2)_1 r_Y(z) - D_1^2 r(z))(g,g)\mathrm{d}g \\
&\quad -\frac{1}{z}\int_G \mathrm{tr}\, c(d\chi^\Gamma)_1((D_Y)_1 r_Y(z) - D_1 r(z))(g,g)\mathrm{d}g,
\end{aligned}$$

where $c(d\chi^\Gamma)$ denotes Clifford multiplication. We conclude that

$$Q(z) = -\frac{1}{2z}\int_G \mathrm{tr}\, c(d\chi^\Gamma)_1((D_Y)_1 r_Y(z) - D_1 r(z))(g,g)\mathrm{d}g.$$

The right-hand side of this equation vanishes as a consequence of $\sum_{\gamma\in\Gamma}\gamma^*\chi^\Gamma \equiv 1$ and the Γ-invariance of $((D_Y)_1 r_Y(z) - D_1 r(z))(g,g)$. This finishes the proof of the proposition. □

Acknowledgements This paper has been written[4] in the years 1999/2000. It is an outcome of the exciting research environment of the Göttingen Mathematical Institute which was heavily stimulated by the interests and mathematical influence of Samuel J. Patterson.

References

1. M. Atiyah and W. Schmid. A geometric construction of the discrete series for semisimple Lie groups. *Invent. Math.*, 42(1977), 1–62.
2. J. N. Bernstein. On the support of Plancherel measure. *J. of Geom. and Phys.*, 5(1988), 663–710.

[4]Except for some minor corrections and additions including more recent references.

3. U. Bunke and M. Olbrich. *Selberg Zeta and Theta Functions*. Akademie Verlag, 1995.
4. U. Bunke and M. Olbrich. The spectrum of Kleinian manifolds. *J. Funct. Anal.*, 172(2000), 76–164.
5. U. Bunke and M. Olbrich. Group cohomology and the singularities of the Selberg zeta function associated to a Kleinian group. *Annals. of. Math.*, 149 (1999), 627–689.
6. J. Cheeger, M. Gromov, and and M. Taylor. Finite propagation speed, kernel estimates for functions of the Laplace operator, and the geometry of complete Riemannian manifolds. *J. Differential Geom.*, 17, (1982), 15–53.
7. D. Fried. The zeta functions of Ruelle and Selberg I. *Ann. scient. éc. norm. sup.* 4e *Série*, 19 (1986), 491–517.
8. C. Guillarmou. Generalized Krein formula, determinants, and Selberg zeta function in even dimension. *Amer. J. Math.*, 131(2009), 1359–1417.
9. C. Guillarmou and F. Naud. Wave 0-trace and length spectrum on convex co-compact hyperbolic manifolds. *Comm. Anal. Geom.*, 14(2006), 945–967.
10. C. Guillarmou, S. Moroianu and J. Park. Eta invariant and Selberg Zeta function of odd type over convex co-compact hyperbolic manifolds *Adv. Math.* 225(2010), 2464–2516.
11. L. Guillopé. Fonctions zéta de Selberg et surface de géométrie finie. *Adv. Stud. Pure Math.*, 21(1992), 33–70.
12. L. Guillopé, K. K. Lin, and M. Zworski. The Selberg zeta function for convex co-compact Schottky groups. *Comm. Math. Phys.*, 245(2004), 149–176.
13. Harish-Chandra. Harmonic analysis on real reductive groups III. The Maass-Selberg relations and the Plancherel formula. *Ann. of Math.*, 104 (1976), 117–201.
14. A. Juhl, *Cohomological theory of dynamical Zeta functions*. Progress in Mathematics 194, Birkhäuser, Basel, 2001.
15. M. Kashiwara and T. Oshima. Systems of differential equations with regular singularities and their boundary value problems. *Ann. Math.*, 106(1977), 145–200.
16. R. J. Miatello and J. A. Vargas. On the distribution of the principal series in $L^2(\Gamma\backslash G)$. *Trans. AMS*, 279 (1983), 63–75.
17. M. Olbrich. Cohomology of convex cocompact subgroups and invariant distributions on limit sets. *arXiv:math/0207301*.
18. T. Oshima and J. Sekiguchi. Eigenspaces of invariant differential operators on an affine symmetric space. *Invent. math.*, 57(1980), 1–81.
19. S. J. Patterson. Two conjectures on Kleinian groups. Handwritten notes, Talk at Warwick 1993.
20. S. J. Patterson and P. A. Perry. The Divisor of Selberg's zeta function for Kleinian groups. *Duke Math. J.*, 106 (2001)
21. N. R. Wallach, *Real reductive groups. I.* Pure and Applied Mathematics 132, Academic Press Inc., Boston, MA, 1988.
22. N. R. Wallach, *Real reductive groups. II.* Pure and Applied Mathematics 132-II, Academic Press Inc., Boston, MA, 1992.

Double Dirichlet Series and Theta Functions

Gautam Chinta, Solomon Friedberg, and Jeffrey Hoffstein

Abstract Generalized theta functions are residues of metaplectic Eisenstein series. Even in the case of the n-fold cover of $GL(2)$, the Fourier coefficients of these mysterious functions have not been determined beyond $n = 3$. However, a conjecture of Patterson illuminates the case $n = 4$. In this paper, we make a new conjecture concerning the Fourier coefficients of the theta function on the sixfold cover of $GL(2)$, present some evidence for the conjecture, and prove it in the case that the base field is a rational function field. Although the conjecture involves a single complex variable, our approach makes critical use of double Dirichlet series.

Dedicated to Professor Samuel J. Patterson in honor of his 60th birthday

1 Introduction

The quadratic theta function

$$\theta(z) = \sum_{n\in\mathbb{Z}} e^{2\pi i n^2 z}$$

G. Chinta (✉)
Department of Mathematics, The City College of CUNY, New York, NY 10031, USA
e-mail: chinta@sci.ccny.cuny.edu

S. Friedberg
Department of Mathematics, Boston College, Chestnut Hill, MA 02467-3806, USA
e-mail: friedber@bc.edu

J. Hoffstein
Department of Mathematics, Brown University, Providence, RI 02912-9032, USA
e-mail: jhoff@math.brown.edu

V. Blomer and P. Mihăilescu (eds.), *Contributions in Analytic and Algebraic Number Theory*, Springer Proceedings in Mathematics 9, DOI 10.1007/978-1-4614-1219-9_6,

has been a familiar object since the nineteenth century and it has found many applications in number theory and other fields. Weil observed that $\theta(z)$ can be interpreted as an automorphic form on the twofold cover of $GL(2)$. An Eisenstein series $E^{(2)}(z,s)$ on this group can be constructed which has a pole at $s = 3/4$, and whose residue at this pole is a constant multiple of $\theta(z)$.

Kubota [9] investigated automorphic forms on the corresponding n-fold cover of $GL(2)$, $n \geq 3$. He defined a metaplectic Eisenstein series $E^{(n)}(z,s)$ on this group whose constant coefficient has a pole at $s = 1/2 + 1/(2n)$. It follows that $E^{(n)}(z,s)$ has a pole at this point, and Kubota defined the nth-order analog of the theta function as

$$\theta^{(n)}(z) = \mathrm{Res}_{s=1/2+1/(2n)} E^{(n)}(z,s).$$

The precise nature of this general nth order theta function seems to be far more mysterious than the familiar $n = 2$ case. Patterson [10, 11] determined (by means of a metaplectic converse theorem) that in the case $n = 3$ its Fourier coefficients are essentially cubic Gauss sums. Kazhdan and Patterson [8] then showed that on the n-cover of $GL(r)$ the Whittaker–Fourier coefficients of an analogously defined theta function satisfy certain periodicity properties. However, even for $GL(2)$, for $n \geq 4$ the Fourier coefficients of $\theta^{(n)}(z)$ have proved quite difficult to determine. Since they are naturally defined from an arithmetic situation (the n-fold cover is built using nth power local Hilbert symbols; in some sense, the existence of such a group is a reflection of nth order reciprocity), it would be of great interest to do this, and such a determination would be likely to have applications. See, for example, [2] for such an application which does not rely on a precise understanding of the coefficients.

Patterson [13] has made a beautiful conjecture about the Fourier coefficients of $\theta^{(n)}(z)$ in the case $n = 4$ (see also [5] for a refinement of the original conjecture). It was proved when the ground field is a rational function field in [6] (see [14] for a version of [6] from Patterson's point of view). In addition, [5, 16], extensive numerical investigations have been made in the cases $n = 4, 6$. But aside from some suggestions concerning the algebraic number field in which these coefficients ought to lie, the values of the coefficients are not in general understood, even heuristically.

The purpose of this paper is to formulate a conjecture about some of the Fourier coefficients of $\theta^{(n)}(z)$ in the case $n = 6$, and to prove this conjecture in the case of a rational function field. This conjecture seems to be "almost" true in a more general setting than $n = 6$ but some extra insight is still missing.

In the next section, we will set up some notation and explain in a rough, but hopefully informative way, what is known, what has been conjectured, and what is still inscrutable.

2 A Formulation of the Conjecture

Although a great deal of number theory is concerned with Euler products, constructions on the metaplectic group frequently give rise to Dirichlet series with analytic continuation and functional equation that are *not* Euler products. Remarkably, it is an observation of Patterson that the equality of two such series may nonetheless encode deep information about the Fourier coefficients of the higher order theta functions. In this section, we build on Patterson's insight to arrive at a new conjecture concerning $\theta^{(6)}$ that is formulated as such an equality, and we explain its consequences.

The series we require are Rankin–Selberg convolutions of metaplectic forms. Unfortunately, such convolutions require a great deal of care at bad places (indeed, even in the non-metaplectic case the treatment of such places is delicate). To avoid these difficulties, we will work heuristically at first, following the style in the early sections of [7]. We will then give a full, precise proof of the conjecture in the rational function field case in Sect. 6. One expects many aspects of the theory of automorphic forms over global fields to be uniform in terms of the base field, so the proof in this case is a likely indication of a more general phenomenon.

It is worth mentioning a caveat: there is an important difference between the function field and the number field cases. Namely, there are typically many elements (modulo units) of given norm in a function field, and there may be much cancellation in corresponding sums, a phenomenon that does not occur in the number field case. Hence, the function field analog can oversimplify the number field scenario, rather than producing a true likeness. The situations would be more comparable if twists by characters were introduced. We have, in fact, checked in unpublished computations that the heuristic arguments still hold if character twists are included, but we have not checked the corresponding function field calculations.

Let F be a global field containing the $2n$th roots of unity. Let $\mathfrak{o}$ denote the ring of integers of F. To give the heuristic treatment, we will imagine that the class number of $\mathfrak{o}$ is one and that all primes are unramified. These assumptions are never truly satisfied, but the S-integer formalism, introduced by Patterson in this context, allows one to make the heuristic definitions we give below precise. In addition to these simplifying assumptions, we will not keep track of powers of the numbers 2 and π in gamma factors, and we will neglect values of characters whose conductors consist of ramified primes (simplifying, for example, the statement of the Davenport–Hasse relation). A rational function field $\mathbb{F}_q(t)$ with q congruent to 1 modulo $4n$ comes close to satisfying these simplifying assumptions, and thus conjectures formulated via such simplifying assumptions can usually be stated, and occasionally proved, rigorously in this case. That is the situation with the conjecture we present below.

A fundamental object for us is the normalized Gauss sum with numerator m and denominator d formed with the jth power of the kth power residue symbol:

$$G_j^{(k)}(m,d) = \mathbf{N}d^{-1/2} \sum_{\alpha \pmod{d}} \left(\frac{\alpha}{d}\right)_k^j e\left(\frac{\alpha m}{d}\right),$$

where $e(x)$ is an additive character of conductor $\mathfrak{o}$ and $\mathbf{N}d$ denotes the absolute norm of d. With this normalization, $|G_j^{(k)}(m,d)| = 1$ when d is square free and $(m,d) = 1$.

Because we will later work with both the n and $2n$-fold covers, let us begin with a discussion of the k-fold cover, $k \geq 2$. In this context, the mth Fourier coefficient of Kubota's Eisenstein series consists of an arithmetic part times a Whittaker function (essentially a K-Bessel function with index $1/k$). The arithmetic part is a Dirichlet series

$$D_j^{(k)}(s,m) = \sum_d \frac{G_j^{(k)}(m,d)}{\mathbf{N}d^s}.$$

Here, j is prime to k, and arbitrary; it may be regarded as parametrizing the different embeddings of the abstract group of kth roots of unity into $\mathbb{C}^\times$. The sum is over d sufficiently congruent to 1. (More carefully, one would keep track of the dependence on the inducing data for the Eisenstein series and obtain a sum over non-zero ideal classes, see [1].) The product

$$\tilde{D}_j^{(k)}(s,m^2) = \Gamma_k(s)\zeta^*(ks - k/2 + 1)D_j^{(k)}(s,m^2) \tag{1}$$

has an analytic continuation and satisfies a functional equation

$$\mathbf{N}m^{s/2}\tilde{D}_j^{(k)}(s,m^2) = \tilde{D}_j^{(k)}(1-s,m^2)\mathbf{N}m^{(1-s)/2}. \tag{2}$$

Here,

$$\Gamma_k(s) = \Gamma\left(s - \frac{1}{2} + \frac{1}{k}\right)\Gamma\left(s - \frac{1}{2} + \frac{2}{k}\right)\cdots\Gamma\left(s - \frac{1}{2} + \frac{k-1}{k}\right) \tag{3}$$

and ζ^* denotes the completed zeta function of the field F. The normalized series (1) is analytic except for simple poles at $s = 1/2 + 1/k, 1/2 - 1/k$, and its residue at $s = 1/2 + 1/k$ is given by

$$\mathrm{Res}_{s=1/2+1/k}\tilde{D}_j^{(k)}(s,m) = c\,\frac{\tau_j^{(k)}(m)}{\mathbf{N}m^{1/2k}}, \tag{4}$$

where c is a nonzero constant. The numerator $\tau_j^{(k)}(m)$ is the object we are investigating: the mth Fourier coefficient of the theta function on the k-fold cover of $GL(2)$.

The Eisenstein series is an eigenfunction of the Hecke operators T_{p^k} for every prime p and consequently so is its residue, the theta function. This forces the $\tau_j^{(k)}(m)$ to obey certain Hecke relations (see [7, 8, 12]). These are:

$$\tau_1^{(k)}(mp^i) = G_{i+1}^{(k)}(m,p)\tau_1^{(k)}(mp^{k-2-i}), \tag{5}$$

valid for $k \geq 2$, p a prime, $0 \leq i \leq k-2$, and $(m,p) = 1$.

For the moment, we will restrict ourselves to the case $m = 1$. Our object is to understand the nature of the coefficients $\tau_1^{(k)}(p^i)$, that is the coefficients at prime power indices of the theta function formed from the first power of the kth order residue symbol. The periodicity relation proved by Kazhdan and Patterson reduces, in this case, to the relation

$$\tau_1^{(k)}(mp^k) = \mathbf{N}p^{1/2}\tau_1^{(k)}(m)$$

for any m. Thus when studying $\tau_1^{(k)}(p^i)$, we need go no higher than $i = n-1$.

Referring to (5), we see from taking $i = k-1$ that $\tau_1^{(k)}(p^{k-1}) = 0$. Also, from $i = k-2$ we see (normalizing so $\tau_1^{(k)}(1) = 1$), that

$$\tau_1^{(k)}(p^{k-2}) = G_{k-1}^{(k)}(1,p).$$

Thus, the Hecke relations *completely determine* the coefficients in the cases $k = 2$, the familiar quadratic theta function, and $k = 3$, the cubic theta function whose coefficients were found by Patterson. In particular, when $k = 3$

$$\tau_1^{(3)}(p) = G_2^{(3)}(1,p) = \overline{G_1^{(3)}(1,p)}$$

and $\tau_1^{(3)}(p^2) = 0$.

Unfortunately, for $k \geq 4$, the information provided by the Hecke operators is incomplete. The first undetermined case, $k = 4$, was studied by Patterson in [13]. The Hecke relations in this case give $\tau_1^{(4)}(p^3) = 0$ and

$$\tau_1^{(4)}(p^2) = G_3^{(4)}(1,p) = \overline{G_1^{(4)}(1,p)},$$

but $\tau_1^{(4)}(p)$ is just related to itself. When the m is reintroduced and we use periodicity we have the more refined information

$$\tau_1^{(4)}(mp) = G_2^{(4)}(m,p)\tau_1^{(4)}(mp).$$

The Gauss sum is

$$G_2^{(4)}(m,p) = \left(\frac{m}{p}\right)_4^2 G_2^{(4)}(1,p) = \left(\frac{m}{p}\right)_2 G_1^{(2)}(1,p) = \left(\frac{m}{p}\right)_2,$$

as the quadratic Gauss sum is trivial by our simplifying assumption. Thus, $\tau_1^{(4)}(mp)$ must vanish unless m is a quadratic residue modulo p.

Patterson observed that there are two natural Dirichlet series that can be formed:

$$D_1(w) = \zeta(4w-1)\sum \frac{G_3^{(4)}(1,m)}{\mathbf{N}m^w}$$

and

$$D_2(w) = \zeta(4w-1) \sum \frac{\tau_1^{(4)}(m)^2}{\mathbf{N}m^w}.$$

The first is the first Fourier coefficient of the Eisenstein series on the 4-cover of $GL(2)$, multiplied by its normalizing zeta function, and with the variable change $2s - 1/2 \to w$. The second is the Rankin–Selberg convolution of the theta function with itself (not its conjugate), also multiplied by its normalizing zeta factor. He conjectured that

$$D_2(w) = D_1(w)^2.$$

This conjecture was based on the fact that both sides had double poles in the same places, both had identical gamma factors, and when corresponding coefficients were matched, all provable properties of the coefficients of $D_2(w)$ were consistent with the completely known $D_1(w)$. If this conjecture were true it would follow that

$$\tau_1^{(4)}(m)^2 = G_3^{(4)}(1,m) \sum_{d_1 d_2 = m} \left(\frac{d_1}{d_2}\right)_2$$

and in particular that

$$\tau_1^{(4)}(p)^2 = 2G_3^{(4)}(1,p).$$

To date, Patterson's conjecture has remained unproved and even ungeneralized. A remarkable aspect of it is that it states that a naturally occurring Dirichlet series *without* an Euler product is equal to a square of another such Dirichlet series. In fact, one side (D_2) is the Rankin–Selberg convolution of a theta function on the 4-cover of $GL(2)$ with itself. The other side (D_1^2) is the square of a Rankin-Selberg convolution. The object being squared, D_1, is the first Fourier coefficient of the Eisenstein series on the 4-cover of $GL(2)$. Using [8], it may also be regarded as the analog of the standard L- series associated with the theta function on the 4-cover of $GL(3)$.

A weaker conjecture, that has been generalized, was made in [3]. It implies that $\tau_1^{(4)}(p)G_1^{(4)}(1,p) = \tau_3^{(4)}(p)$, i.e. that the argument of $\tau_1^{(4)}(p)$ is the square root of the conjugate Gauss sum. This was proved by Suzuki [15] in the case where the ground field is a function field.

We now make a new conjecture relating Rankin–Selberg convolutions involving coefficients of the higher-order theta functions. We specify the 6th order residue symbol by (11) below wth $n = 3$.

Conjecture 1.

$$\zeta(3u-1/2) \sum \frac{\tau_1^{(6)}(m^2)}{\mathbf{N}m^u} = \zeta(3u-1/2) \sum \frac{G_1^{(3)}(1,d)}{\mathbf{N}d^u} \cdot \sum \frac{\overline{\tau_1^{(3)}(m)}}{\mathbf{N}m^u}.$$

The left-hand side is the convolution of the theta function on the sixfold cover of $GL(2)$ with the theta function on the double cover of $GL(2)$. The right-hand side is the product of two terms: the first coefficient of the cubic Kubota Eisenstein series, multiplied by its normalizing zeta factor, and the Mellin transform of the theta function on the threefold cover of $GL(2)$. In this case $(n = 3)$, the two factors on the right are equal, but we write it this way with an eye toward potential future generalizations. We include the apparently extraneous zeta functions as they arise naturally in the normalizing factors.

Writing $m = m_1 m_2^2 m_3^3$, with m_1, m_2 square free and relatively prime, m_3 unrestricted we see by the periodicity properties of $\tau_1^{(6)}$ and the known valuation of $\tau_1^{(3)}$ that this conjectured equality translates to

$$\sum \frac{\tau_1^{(6)}(m_1^2 m_2^4)}{\mathbf{N}m_1^u \mathbf{N}m_2^{2u}} = \left(\sum \frac{G_1^{(3)}(1,d)}{\mathbf{N}d^u} \right)^2,$$

another striking identity involving the square of a series without an Euler product. Note that the Gauss sums $G_1^{(3)}(1,d)$ on the right-hand side vanish unless d is square free.

If we cancel the zeta factor, and equate corresponding coefficients, we have the following predicted behavior for the coefficients $\tau_1^{(6)}(m_1^2 m_2^4)$:

$$\tau_1^{(6)}(m_1^2 m_2^4) = G_1^{(3)}(1,m_2)^2 G_1^{(3)}(1,m_1) \left(\frac{m_2}{m_1} \right)_3^2 \sum_{m_1 = d_1 d_2} \left(\frac{d_2}{d_1} \right)_3 .$$

Bearing in mind that $G_1^{(3)}(1,m_2)^2 = \overline{G_1^{(6)}(1,m_2)}$ by the Davenport–Hasse relation [4], this relation is consistent with what is implied by setting $k = 6$ in the Hecke relations (5). Similarly, all aspects of the identity given above are consistent with the Hecke relations. Setting $m_2 = 1$ and $m_1 = p$, this reduces to

$$\tau_1^{(6)}(p^2) = 2G_1^{(3)}(1,p).$$

The Conjecture is made after verifying that the polar behaviour and gamma factors of the left and right-hand sides are identical. This verification is the content of Sects. 3–5. Indeed, as will be seen, a similar conjecture is almost true for the general case where 3 is replaced by n and 6 by $2n$. The difficulty is that the identity is partially, but not completely, compatible with the Hecke relations.

3 A Double Dirichlet Series Obtained from the $2n$-Cover of $GL(2)$

We will obtain the desired information about the poles and gamma factors of the series above by first performing the easier task of determining the analytic continuation, polar lines and functional equations of several related multiple

Dirichlet series. We begin by defining the following double Dirichlet series, initially for $\Re(s), \Re(w) > 1$. Let

$$Z_1(s,w) = \sum_{d,m} \frac{G_1^{(2n)}(m^2,d)}{\mathbf{N}d^s\mathbf{N}m^w}. \tag{6}$$

Also, for n odd, let

$$Z_2(s,w) = \sum_{d,m} \frac{G_1^{(n)}(m^2,d)}{\mathbf{N}d^s\mathbf{N}m^w}, \tag{7}$$

and for n even, let

$$Z_2(s,w) = \sum_{d,m} \frac{G_{n+1}^{(2n)}(m^2,d)}{\mathbf{N}d^s\mathbf{N}m^w}. \tag{8}$$

The corresponding normalized series are

$$\tilde{Z}_1(s,w) = \zeta^*(\delta n(s+w-1/2) - \delta n/2+1)\zeta^*(2ns-n+1)Z_1(s,w) \tag{9}$$

and

$$\tilde{Z}_2(s,w) = \zeta^*(\delta ns - \delta n/2+1)\zeta^*(2ns+2nw-2n+1)Z_2(s,w). \tag{10}$$

Here,

$$\delta = \begin{cases} 1 & \text{if } n \text{ is odd} \\ 2 & \text{if } n \text{ is even,} \end{cases}$$

and the $*$ in the zeta functions again means that the appropriate gamma factors have been included.

Let χ_d and ψ_d be multiplicative characters of conductor d, with $\psi_d^2 = 1$. Then if $\tau(\chi)$ refers to the usual Gauss sum corresponding to χ, normalized to have absolute value 1, the Davenport–Hasse relation [4] states (ignoring characters ramified at primes dividing $2n$) that

$$\tau(\chi_d)\,\tau(\chi_d\psi_d) = \tau(\chi_d^2).$$

We have also suppressed the quadratic Gauss sum as it is trivial with our simplifying assumptions. In the case, n is odd we choose a non-standard definition of the $2n$th order power residue symbol:

$$\left(\frac{\alpha}{d}\right)_{2n} = \left(\frac{\alpha}{d}\right)_n \left(\frac{\alpha}{d}\right)_2, \tag{11}$$

as it makes our formulas cleaner. (The right-hand side is actually the usual symbol raised to the $(n+2)$th power.) Thus, the Davenport–Hasse relation implies that

$$G_1^{(2n)}(1,d)\,G_1^{(n)}(1,d) = G_2^{(n)}(1,d). \tag{12}$$

In the case $n = 3$ this translates into the familiar

$$G_1^{(6)}(1,d) = G_2^{(3)}(1,d)\,\overline{G_1^{(3)}(1,d)} = \overline{G_1^{(3)}(1,d)^2}.$$

In the case n is even

$$\left(\frac{\alpha}{d}\right)_{2n}^{n+1} = \left(\frac{\alpha}{d}\right)_{2n}\left(\frac{\alpha}{d}\right)_{2}$$

and

$$\left(\frac{\alpha}{d}\right)_{2n}^{2} = \left(\frac{\alpha}{d}\right)_{n},$$

so the Davenport–Hasse relation implies that

$$G_1^{(2n)}(1,d)\,G_{n+1}^{(2n)}(1,d) = G_1^{(n)}(1,d). \tag{13}$$

For example, if $n = 2$ this is the trivial relation

$$G_1^{(4)}(1,d)\,G_3^{(4)}(1,d) = G_1^{(2)}(1,d) = 1.$$

Our main tool in establishing the analytic continuation of the $Z_i(s,w)$, $i = 1,2$, will be

Proposition 1. *For $\Re s, \Re w > 1$, both $Z_1(s,w)$ and $Z_2(s,w)$ converge absolutely. Furthermore, each has an analytic continuation for any fixed w as long as $\Re s$ is sufficiently large. In fact, the following relations hold. For n ODD*

$$\begin{aligned}
Z_1(s,w) &= Z_2(s+w-1/2,1-w) \\
&= \zeta(2ns+nw-n)\sum_{\tilde{d}_0}\frac{\overline{G(\chi_{\tilde{d}_0}^{(n)})L(1-w,(\chi_{\tilde{d}_0}^{(n)})^2)}}{(\mathbf{N}\tilde{d}_0)^{s+w/2}(\mathbf{N}d_0)^{(w-1)/2}} \\
&\quad\times\prod_{(p,d_0)=1}\left(1+\frac{\chi_{\tilde{d}_0}^{(n)}(p)}{\mathbf{N}p^{ns+(n-1)w/2-(n-1)/2}}\right) \\
&\quad\times\prod_{(p,d_0)=1}\left(1-\frac{\overline{\chi_{\tilde{d}_0}^{(n)}(p)}}{\mathbf{N}p^{ns+(n+1)w/2-(n-1)/2}}\right),
\end{aligned}$$

and for n even

$$
\begin{aligned}
Z_1(s,w) &= Z_2(s+w-1/2, 1-w) \\
&= \zeta(2ns+nw-n)\sum_{\tilde{d_0}} \frac{\overline{G\left(\left(\chi_{\tilde{d_0}}^{(2n)}\right)^{n+1}\right)} L\left(1-w, \chi_{\tilde{d_0}}^{(n)}\right)}{(\mathbf{N}\tilde{d_0})^{s+w/2}(\mathbf{N}d_0)^{(w-1)/2}} \\
&\times \prod_{(p,d_0)=1} \left(1 - \frac{1}{\mathbf{N}p^{2ns+nw-n+1}}\right).
\end{aligned}
$$

Here, $G(\psi)$ refers to the Gauss sum associated with the character ψ, normalized to have absolute value equal to 1. The sums over d_0 and $\tilde{d_0}$ are defined as follows. If n is odd, then we write $\tilde{d_0} = e_1 e_2^n$. Here, e_1 is n^{th} power free, e_2 is the square free product of all p dividing e_1 such that the exact power of p dividing e_1 is even and we sum over all such e_1. If n is even, then we sum over all $\tilde{d_0}$ that are $2n^{th}$ power free, with the proviso that if $p|\tilde{d_0}$ then an odd power of p must exactly divide $\tilde{d_0}$. We denote by d_0 the product of all the distinct primes dividing $\tilde{d_0}$.

Proposition 1 is proved by taking s, w to have large real parts, and interchanging the order of summation in $Z_1(s,w)$. A careful analysis reduces $Z_1(s,w)$ to the expressions on the right-hand side above, but with the functional equation applied to the L-series in the numerator (i.e. the argument of the L-series is w rather than $1-w$.) The sum over d then converges absolutely for any fixed w as long as the real part of s is sufficiently large. If one applies the functional equation to the L-series and uses the Davenport–Hasse relation, the sum is transformed into that given in the Proposition. Similarly, if one takes $Z_2(s+w-1/2, 1-w)$, where $\Re(1-w)$ and $\Re s$ are sufficiently large to insure absolute convergence, and interchanges the order of summation, the right-hand side of the Proposition is obtained directly.

One can alternatively take $Z_1(s,w), Z_2(s,w)$ and sum over d on the inside. If one does this, with the real parts of s, w sufficiently large, then one obtains

$$Z_1(s,w) = \sum_m \frac{D_1^{(2n)}(s,m^2)}{\mathbf{N}m^w} \tag{14}$$

and also

$$Z_2(s,w) = \sum_m \frac{D_{n-1}^{(n)}(s,m^2)}{\mathbf{N}m^w} \tag{15}$$

for n odd and

$$Z_2(s,w) = \sum_m \frac{D_{n-1}^{(2n)}(s,m^2)}{\mathbf{N}m^w} \tag{16}$$

for n even.

Applying the relations (14)–(16) and the functional equation (2), one obtains the following

Proposition 2. *For fixed s the series expressions (14)–(16) converge absolutely as long as the real part of w is sufficiently large. In the range of absolute convergence, the normalized series* $\tilde{Z}_1(s,w), \tilde{Z}_2(s,w)$ *defined in (9),(10) satisfy*

$$\tilde{Z}_1(s,w) = \tilde{Z}_1(1-s, w+2s-1)$$

and

$$\tilde{Z}_2(s,w) = \tilde{Z}_2(1-s, w+2s-1)$$

We are now in a position to obtain the analytic continuation of $\tilde{Z}_1(s,w)$ and $\tilde{Z}_2(s,w)$. First let us clear the poles by defining

$$\hat{Z}_i(s,w) = \mathscr{P}_i(s,w)\tilde{Z}_i(s,w) \tag{17}$$

for $i = 1,2$, where

$$\mathscr{P}_1(s,w) = \left(s - \frac{1}{2} - \frac{1}{2n}\right)\left(s - \frac{1}{2} + \frac{1}{2n}\right)(w)(w-1)(w+2s-2)(w+2s-1) \\ \times \left(s+w-1-\frac{1}{\delta n}\right)\left(s+w-1+\frac{1}{\delta n}\right) \tag{18}$$

and

$$\mathscr{P}_2(s,w) = \mathscr{P}_1(s+w-1/2, 1-w).$$

The factors in $\mathscr{P}$ are chosen to clear the poles in s and w in the region of absolute convergence, and also to satisfy $\mathscr{P}_i(s,w) = \mathscr{P}_i(1-s, w+2s-1)$ for $i = 1,2$. Thus, in addition to being analytic in the region of absolute convergence,

$$\hat{Z}_i(s,w) = \hat{Z}_i(1-s, w+2s-1)$$

for $i = 1,2$ and

$$\hat{Z}_1(s,w) = \hat{Z}_2(s+w-1/2, 1-w).$$

For $i = 1,2$, $\hat{Z}_i(s,w)$ converges absolutely in the region $\Re s, \Re w > 1$. The functional equation in s given above in (2) implies a polynomial bound in $|m|^s$ for the Dirichlet series in the numerators of (14)–(16) when $\Re(s) < 0$. Consequently, the Phragmen–Lindelöf principle implies a bound for these series when $0 \leq \Re(s) \leq 1$. Thus, $\hat{Z}_i(s,w)$ can be extended to a holomorphic function in the region in $\mathbb{C}^2$ given by

$$\{(s,w) \mid \Re(s) \leq 0, \Re(w) > -2\Re(s)+2\} \cup \{(s,w) \mid \Re(s) > 1, \Re(w) > 1\} \\ \cup \{(s,w) \mid 0 \leq \Re(s) \leq 1, \Re(w) > -\Re(s)+2\}$$

Arguing similarly with the L-functions appearing in the representations of $Z_i(s,w)$ given in Proposition 1, the $\hat{Z}_i(s,w)$ extend holomorphically to the region

$$\{(s,w) \mid 0 \leq \Re(w) \leq 1, \Re(s) > -\Re(w)/2 + 3/2\}$$
$$\cup \{(s,w) \mid \Re(w) \leq 0, \Re(s) > -\Re(w) + 3/2\}.$$

By Bochner's theorem, the functions $\hat{Z}_i(s,w)$ thus extend analytically to the convex closure of the union of these regions, which is the region

$$\begin{aligned} R_1 = &\{(s,w) \mid s \leq 0, \Re(w) > -2\Re(s) + 2\} \\ &\cup \{(s,w) \mid 0 \leq \Re(s) \leq 3/2, \Re(w) > -4\Re(s)/3 + 2\} \\ &\cup \{(s,w) \mid 3/2 \leq \Re(s), \Re(w) > -\Re(s) + 3/2\}. \end{aligned} \quad (19)$$

Applying the relation $\hat{Z}_1(s,w) = \hat{Z}_2(s+w-1/2, 1-w)$ we see that as the image of R_1 under the map $(s,w) \to (s+w-1/2, 1-w)$ intersects itself, we can extend both $\hat{Z}_1(s,w)$ and $\hat{Z}_2(s,w)$ to the convex hull of the union of R_1 and its image. This is the half-plane

$$R_2 = \{(s,w) \in \mathbb{C}^2 \mid \Re(w) > -2\Re(s) + 2\}.$$

Finally, applying $\hat{Z}_i(s,w) = \hat{Z}_i(1-s, w+2s-1)$ for $i = 1,2$ and taking the convex hull of the union of overlapping regions we obtain analytic continuation to $\mathbb{C}^2$.

We summarize the above discussion in

Proposition 3. *The functions $\tilde{Z}_1(s,w)$ and $\tilde{Z}_2(s,w)$ defined in (9), (10) have an analytic continuation to all of $\mathbb{C}^2$, with the exception of certain polar lines. For $\tilde{Z}_1(s,w)$, these polar lines are $s = 1/2 \pm 1/(2n)$; $w = 1,0$; $w+2s-1 = 1,0$; $s+w-1/2 = 1/2 \pm 1/(\delta n)$. For $\tilde{Z}_2(s,w)$ these polar lines are $s = 1/2 \pm 1/(\delta n)$; $w = 1,0$; $w+2s-1 = 1,0$; $s+w-1/2 = 1/2 \pm 1/(2n)$.*

4 The Residue of $\tilde{Z}_1(s,w)$ at $s = 1/2 + 1/(2n)$

Now that the analytic properties of $\tilde{Z}_1(s,w)$ have been established we can investigate the residue of this function at $s = 1/2 + 1/(2n)$. By (4), we have

$$\mathrm{Res}_{s=1/2+1/(2n)} Z_1(s,w) = \sum_m \frac{\tau_1^{(2n)}(m^2)}{\mathbf{N}m^{w+1/(2n)}}$$

and

$$\mathrm{Res}_{s=1/2+1/(2n)} \tilde{Z}_1(s,w) = \zeta(\delta n w - \delta n/2 + \delta/2 + 1)\zeta(2) \sum_m \frac{\tau_1^{(2n)}(m^2)}{\mathbf{N}m^{w+1/(2n)}}.$$

Consequently, we set $u = w + 1/(2n)$ and define

$$\tilde{L}(u) = \zeta(\delta nu - \delta n/2 + 1) \sum_m \frac{\tau_1^{(2n)}(m^2)}{\mathbf{N}m^u}. \tag{20}$$

Remark. This is one of the two Dirichlet series of interest to us. We have chosen to first derive the analytic properties of $\tilde{Z}_1(s,w)$ and then deduce the analytic properties of $\tilde{L}(u)$ by viewing this function as the residue of the two-variable Dirichlet series. It should be possible to analyze $\tilde{L}(u)$ directly by viewing it as a Rankin–Selberg convolution of the theta function on the $2n$-cover of $GL(2)$ with the quadratic theta function, but experience indicates that the two variable approach is considerably simpler to carry out.

By Proposition 3, $\tilde{L}(u)$ inherits an analytic continuation to $\mathbb{C}$ and a functional equation relating $\tilde{L}(u)$ to $\tilde{L}(1-u)$. Also, $\tilde{L}(u)$ is analytic except for possible poles at $u = 1 + 1/(2n), -1/(2n), 1 - 1/(2n), 1/(2n), 1/2 + 1/(\delta n), 1/2 - 1/(\delta n)$. Using the analytic properties of $\tilde{Z}_1(s,w)$,s corresponding properties of $\tilde{L}(u)$ are derived as follows:

$$\begin{aligned}
&\lim_{u\to 1+1/(2n)} (u - 1 - 1/(2n))\tilde{L}(u) \\
&\quad = \lim_{u\to 1+1/(2n)} (u - 1 - 1/(2n)) \lim_{s\to 1+1/(2n)} (s - 1 - 1/(2n))\tilde{Z}_1(s, u - 1/(2n)) \\
&\quad = \lim_{s\to 1+1/(2n)} (s - 1 - 1/(2n)) \lim_{w\to 1} (w-1)\tilde{Z}_1(s,w).
\end{aligned}$$

Thus, we have approached the problem by interchanging the order of two limits. Using Proposition 1 above, it is easy to compute that

$$\lim_{w\to 1} (w-1)\tilde{Z}_1(s,w) = \zeta^*(\delta ns)\zeta^*(2ns - n + 1).$$

As $w = 1$ corresponds to $u = 1 + 1/(2n)$, we see that $\tilde{L}(u)$ will have a pole at $u = 1 + 1/(2n)$ (and at $u = -1/(2n)$) if and only if $\zeta^*(\delta ns)\zeta^*(2ns - n + 1)$ has a pole at $s = 1 + 1/(2n)$. As this is not the case, the potential pole of $\tilde{L}(u)$ at $u = 1 + 1/(2n)$ does not exist.

To investigate the behavior of $\tilde{L}(u)$ at u near $1 - 1/(2n)$, we consider $\lim_{w\to 2-2s} \hat{Z}_1(s,w)$. Applying the functional equations in sequence yields

$$\hat{Z}_1(s,w) = \hat{Z}_2(s + w - 1/2, 1 - w) = \hat{Z}_2(3/2 - s - w, w + 2s - 1) = \hat{Z}_1(s, 2 - 2s - w)$$

from which we obtain

$$\begin{aligned}
\lim_{w\to 2-2s} \hat{Z}_1(s,w) = &-\left(s - \frac{1}{2} - \frac{1}{2n}\right)\left(s - \frac{1}{2} + \frac{1}{2n}\right)(2-2s)(1-2s) \\
&\times \left(s - 1 - \frac{1}{\delta n}\right)\left(s - 1 + \frac{1}{\delta n}\right)\zeta^*(\delta n - \delta ns)\zeta^*(2ns - 1).
\end{aligned} \tag{21}$$

For behavior of $\tilde{L}(u)$ at u near $1/2+1/(\delta n)$, we likewise evaluate the limit $\lim_{w\to 1+1/(\delta n)-s}\hat{Z}_1(s,w)$. Applying the functional equations in sequence we obtain

$$\hat{Z}_1(s,w)=\hat{Z}_2(s+w-1/2,2-2s-w).$$

Taking the limit as $w\to 1+1/(\delta n)-s$ yields

$$\begin{aligned}\lim_{w\to 1+1/(\delta n)-s}\hat{Z}_1(s,w)&=\left(s-\frac{1}{2}-\frac{1}{2n}\right)\left(s-\frac{1}{2}+\frac{1}{2n}\right)\left(1+\frac{1}{\delta n}-s\right)\left(\frac{1}{\delta n}-s\right)\\&\times\left(s-1+\frac{1}{\delta n}\right)\left(s+\frac{1}{\delta n}\right)\left(\frac{2}{\delta n}\right)\zeta^*(2)\zeta^*(n+1-2ns)\\&\times\lim_{s+w-\frac{1}{2}\to\frac{1}{2}+1/(\delta n)}\left(s+w-1-\frac{1}{\delta n}\right)Z_2\left(s+w-\frac{1}{2},2-2s-w\right)\\&=\left(s-\frac{1}{2}-\frac{1}{2n}\right)\left(s-\frac{1}{2}+\frac{1}{2n}\right)\left(1+\frac{1}{\delta n}-s\right)\left(\frac{1}{\delta n}-s\right)\left(s-1+\frac{1}{\delta n}\right)\\&\times\left(s+\frac{1}{\delta n}\right)\left(\frac{2}{\delta n}\right)\zeta^*(2)\zeta^*(n+1-2ns)M^{(\delta n)}_{1+(\delta-1)n}(1-s).\end{aligned}\tag{22}$$

Here,

$$M_j^{(k)}(u)=\sum\frac{\tau_j^{(k)}(m)}{\mathbf{N}m^u}\tag{23}$$

denotes the Mellin transform of the theta function on the k-fold cover of $GL(2)$, with the underlying residue symbol being the jth power of the standard one.

We have thus far computed $\hat{Z}_1(s,2-2s)$ and $\hat{Z}_1(s,1+1/(\delta n)-s)$. We will now evaluate these expressions as s approaches $1/2+1/(2n)$. Applying the relations (21) and (22) (and continuing to ignore primes dividing $2n$), we obtain for $n=2$:

$$\hat{Z}_1\left(\frac{3}{4},\frac{1}{2}\right)=\kappa^2,\tag{24}$$

and for $n=3$:

$$\hat{Z}_1\left(\frac{2}{3},\frac{2}{3}\right)=\kappa^2,\tag{25}$$

where $\kappa=\mathrm{Res}_{s=1}\zeta^*(s)$. For general $n\geq 4$, we obtain

$$\hat{Z}_1\left(\frac{1}{2}+\frac{1}{2n},1-\frac{1}{n}\right)=\zeta^*\left(\frac{\delta(n-1)}{2}\right)\zeta^*(n).\tag{26}$$

Translating back to $\tilde{L}(u)$, defined in (20) we see that as $u\to 1-1/(2n)$, for $n=2$

$$\tilde{L}(u)\sim\frac{\kappa^2}{(u-3/4)^2},\tag{27}$$

for $n = 3$

$$\tilde{L}(u) \sim \frac{\kappa^2}{(u - 5/6)^2}, \tag{28}$$

and for general $n \geq 4$

$$\tilde{L}(u) \sim \frac{\zeta^*\left(\frac{\delta(n-1)}{2}\right)\zeta^*(n)}{(u - 1 + 1/(2n))}. \tag{29}$$

In a similar manner we obtain, as $u \to 1/2 + 1/(\delta n)$, for $n \geq 4$

$$\tilde{L}(u) \sim \zeta^*(2) M_{1+(\delta-1)n}^{(\delta n)}\left(\frac{1}{2} - \frac{1}{2n}\right),$$

where $M_{1+(\delta-1)n}^{(\delta n)}(1-s)$ is defined in (23). Note that when $n = 2, n = 3$, the two poles coincide and create a double pole, while for all $n \geq 4$ these poles are separate. This may be related to the fact that the conjecture can be made consistent with the Hecke relations in only these two cases.

5 The Gamma Factors of $\tilde{L}(u)$ and a Conjecture

Recall the gamma factors associated with $\tilde{D}_1^{(n)}(s, m^2)$ defined in (1) as $\Gamma_n(s)$:

$$\Gamma_n(s) = \Gamma\left(s - \frac{1}{2} + \frac{1}{n}\right)\Gamma\left(s - \frac{1}{2} + \frac{2}{n}\right)\cdots\Gamma\left(s - \frac{1}{2} + \frac{n-1}{n}\right).$$

Applying the functional equations of Proposition 2.2 in succession, one sees that the gamma factors associated with $\tilde{Z}_1(s, w)$ are

$$\Gamma_{2n}(s)\Gamma_{\delta n}(s + w - 1/2)\Gamma(w)\Gamma(w + 2s - 1).$$

Taking the residue at $s = 1/2 + 1/(2n)$, it follows that the gamma factors associated with $\tilde{L}(u)$ are

$$\Gamma_{\delta n}(u)\Gamma\left(u - \frac{1}{2n}\right)\Gamma\left(u + \frac{1}{2n}\right). \tag{30}$$

Recall that

$$M_j^{(k)}(u) = \sum \frac{\tau_j^{(k)}(m)}{\mathbf{N}m^u}$$

denotes the Mellin transform of the theta function on the k-fold cover of $GL(2)$, where the underlying residue symbol is raised to the j power. In contrast to the situation with $\tilde{L}(u)$, it is easy to verify directly that the gamma factors associated with $M_j^{(k)}(u)$ are $\Gamma(u - 1/(2k))\Gamma(u + 1/(2k))$. We therefore define

$$\tilde{M}_j^{(k)}(u) = \Gamma\left(u - \frac{1}{2k}\right)\Gamma\left(u + \frac{1}{2k}\right)M_j^{(k)}(u).$$

It is now apparent that the gamma factors associated with $\tilde{L}(u)$, given in (30), factor into those associated to $\tilde{D}_1^{(\delta n)}(u,1)$, namely $\Gamma_{\delta_n}(u)$, times those associated to $\tilde{M}_j^{(n)}(u)$. (This is true for any j.)

Recall that for $n \geq 4$ the poles of $\tilde{L}(u)$ are simple and located at

$$u = 1 - 1/(2n), 1/(2n), 1/2 + 1/(\delta n), 1/2 - 1/(\delta n),$$

while in the cases $n = 2, 3$ they combine into double poles located at $u = 3/4, 5/6$. On the other hand, $\tilde{D}_1^{(\delta n)}(u,1)$ has simple poles at $1/2 + 1/(\delta n), 1/2 - 1/(\delta n)$, while it is easily verified that $\tilde{M}_j^{(n)}(u)$ has simple poles at $u = 1 - 1/(2n), 1/(2n)$.

Because of these observations, it is plausible to conjecture that for some value of j, $\tilde{L}(u)$ factors into a product $\tilde{M}_j^{(n)}(u)\tilde{D}_1^{(\delta n)}(u,1)$. We can investigate this more closely, by using the information provided by the Hecke operators, and conclude that a likely value for j is $j = 1$. For example, after canceling gamma factors, we might tentatively conjecture that the following Dirichlet series identities hold: for n odd

$$\zeta(nu - n/2 + 1) \sum \frac{\tau_1^{(2n)}(m^2)}{\mathbf{N}m^u} = \zeta(nu - n/2 + 1) \sum \frac{\overline{\tau_1^{(n)}(m)}}{\mathbf{N}m^u} \cdot \sum \frac{G_1^{(n)}(1,d)}{\mathbf{N}d^u}$$

and for n even

$$\zeta(2nu - n + 1) \sum \frac{\tau_1^{(2n)}(m^2)}{\mathbf{N}m^u} = \zeta(2nu - n + 1) \sum \frac{\overline{\tau_1^{(n)}(m)}}{\mathbf{N}m^u} \cdot \sum \frac{G_{n+1}^{(2n)}(1,d)}{\mathbf{N}d^u}.$$

Specializing to the case $n = 2$ and canceling $\zeta(4u - 1)$, this reduces to the relation

$$\sum \frac{\tau_1^{(4)}(m^2)}{\mathbf{N}m^u} = \sum \frac{\overline{\tau_1^{(2)}(m)}}{\mathbf{N}m^u} \cdot \sum \frac{G_3^{(4)}(1,d)}{\mathbf{N}d^u}.$$

Write $m = m_0 m_1^2$, where m_0 is square free and m_1 is unrestricted. Then by the known properties of $\tau_1^{(4)}$, it follows that

$$\tau_1^{(4)}(m_0^2 m_1^4) = \overline{G_1^{(4)}(1, m_0)} \mathbf{N}m_1^{1/2}$$

and thus the left-hand side of the expression above equals

$$\sum \frac{\tau_1^{(4)}(m_0^2 m_1^4)}{\mathbf{N}m_0^u \mathbf{N}m_1^{2u}} = \zeta(2u - 1/2) \sum \frac{G_3^{(4)}(1,d)}{\mathbf{N}d^u}.$$

As $\underline{\tau_1^{(2)}(m_0m_1^2)} = \mathbf{N}m_1^{1/2}$ if $m_0 = 1$ and vanishes otherwise, and as $G_3^{(4)}(1,d) = \overline{G_1^{(4)}(1,m_0)}$ if $d = m_0$ is square free and vanishes otherwise, the identity holds in the case $n = 2$.

The case $n = 3$ has already been discussed in Sect. 2 after the formulation of Conjecture 1. When $n \geq 4$, the highest coefficient index before periodicity which comes into play is $\mathbf{N}p^{2n-2}$. At this index, the Hecke relations confirm an equality of the left and right-hand sides. Unfortunately, they fail to confirm this equality at lower indices. The conjecture may thus need a mild modification to hold for $n \geq 4$, or it may fail completely. The question remains open.

6 A Proof of the Conjecture in the Case of a Rational Function Field and *n* Odd

In this section, we will work over the rational function field $\mathbb{F}_q(T)$. We will make crucial use of the paper [6], in the sense that we will refer to it for all notation and a number of results. We require $q \equiv 1 \bmod n$, and for convenience, we also suppose that $q \equiv 1 \bmod 4$. The conjecture is provable in this case because over a function field any Dirichlet series with a functional equation (with finite conductor) must be a ratio of polynomials. The polar behavior of the Dirichlet series determines the denominator, and a finite amount of information about the early coefficients is enough to determine the numerator.

Let $n \geq 3$ be odd. The function field analog of the series $Z_1(s,w)$ above is the Rankin–Selberg convolution of $E^{(2n)}(z,u)$ with $\theta^{(2)}(z)$. In effect, the theta function picks off the coefficients of the Eisenstein series with square index and assembles them in a Dirichlet series. The functional equation and polar behavior of the Dirichlet series are determined by the corresponding functional equations and polar behavior of the Eisenstein series in the integral:

$$\int E^{(2n)}(z,u)\,\theta^{(2)}(z)\,\overline{E^{(n)}(z,v)}\,\mathrm{d}\mu(z),$$

where the integration is taken over a truncated fundamental domain. Although the integrand is not of rapid decay, the technique of *regularizing* the integral provides the functional equation and polar behavior of a Mellin transform of the part of the product $E^{(2n)}(z,u)$ with $\theta^{(2)}(z)$ that is of rapid decay. See [17] an exposition of this. The key point for us is that all the necessary information about the Mellin transform is determined by these properties.

Denoting this Mellin transform as $R(u,v)$, we have explicitly

$$R(u,v) = \int_{\mathrm{ord}(Y)\equiv 0 \bmod n} \sum c_m^{(2n)}(u,Y)\,\tau^{(2)}(m,Y)\,|Y|^{2v-2}\,\mathrm{d}^{\times}Y, \tag{31}$$

where the sum is over $m \in A := \mathbb{F}_q[T]$ such that $-2-\deg m + 2\operatorname{ord} Y \geq 0$. Formulas for $c_m^{(2n)}$ are given in [6]. In particular, if we let

$$D_m(u,i) = \sum_{\deg c \equiv i \bmod n} |c|^{-2u} g_1^{(2n)}(m,c),$$

then

$$c_m^{(2n)}(u,Y) = q|Y|^{2-2u}\tilde{D}_m(u,Y),$$

with

$$\tilde{D}_m(u,Y) = D_m(u,0)\left(1+(1-q^{-1})\sum_{\substack{1\leq k\leq 2n\gamma-2-\deg m\\ k\equiv 0 \bmod n}} q^{k(1-2u)}\right)$$
$$+ D_m(u,1+\deg m)\, q^{2n\gamma-2-\deg m} g_{-1-\deg m}(\mu_m,T)\, q^{-2(2n\gamma-1-\deg m)u}. \tag{32}$$

Here, μ_m denotes the leading coefficient of m. Note that $\tilde{D}_m(u,Y)$ is thus a non-zero constant plus a sum of positive powers of q^{-2u} that are multiples of n.

Let $\varpi = 1/T$ be the local uniformizer. The $\tau^{(2)}(m,Y)$ are the Fourier coefficients of the quadratic theta function, described by

$$\tau^{(2)}(m,Y) = \begin{cases} |Y|^{1/2} & m = m_0^2 \text{ with } \operatorname{ord}(\varpi^{-2}m_0^2Y^2) \geq 0 \\ 0 & \text{otherwise.} \end{cases}$$

Substituting into the integral (31), we can do the Y integration, obtaining

$$R(u,v) = c\sum_{n\gamma\geq 1+\deg m_0} q^{(2u-2v-1/2)n\gamma}\tilde{D}_{m_0^2}(u,\varpi^{n\gamma}),$$

where c is a non-zero constant.

Letting $s = 2u-1/2, w = 2v-2u+1/2$, and denoting by $\tilde{R}(u,v)$, the product of $R(u,v)$ by the normalizing zeta and gamma factors of the two Eisenstein series, we have, corresponding to $\tilde{Z}_1(s,w)$,

$$\tilde{R}(u,v) = c_q q^{n-1-2ns}\zeta^*(2ns-n+1)q^{n-1-ns-nw}\zeta^*(n(s+w-1)+1)$$
$$\times \sum_{n\gamma\geq 1+\deg m_0} q^{-wn\gamma}\tilde{D}_{m_0^2}(s/2+1/4,\varpi^{n\gamma}), \tag{33}$$

where c_q is a non-zero constant.

The functional equations of the Eisenstein series imply that $\tilde{Z}_1(s,w) = \tilde{R}(u,v)$ is a rational function of $x = q^{-s}$ and $y = q^{-w}$. Also there are, at most, simple poles at

$$s = 1/2 \pm 1/2n,\ w = 0,1,\ w = 2-2s,\ w = 1-2s,\ s+w-1/2 = 1/2 \pm 1/n.$$

We therefore write

$$\tilde{Z}_1(s,w) = \frac{P(x,y)}{D(x,y)}$$

with

$$D(x,y) = (1-y^n)(1-q^ny^n)(1-q^{n-1}x^{2n})(1-q^{n+1}x^{2n})$$
$$(1-q^{n+1}x^ny^n)(1-q^{n-1}x^ny^n)(1-q^nx^{2n}y^n)(1-q^{2n}x^{2n}y^n). \quad (34)$$

Note from (33) that $\tilde{Z}_1(s,w)$ is of the form $x^{2n}(xy)^ny^n$ times a power series in x^n, y^n. Also, the functional equations of the Eisenstein series imply that

$$\tilde{Z}_1(s,w) = \tilde{Z}_1(s, 2-2s-w).$$

Combining this information with (34), we conclude that $P(x,y)$ is of the form

$$P(x,y) = x^{3n}y^{2n}\sum_{i=0}^{M}\sum_{j=0}^{N} B_{ij}x^{in}y^{jn}.$$

and satisfies the functional equation

$$P(x,y) = q^{6n}x^{6n}y^{6n}P(x, q^{-2}x^{-2}y^{-1}).$$

To go further, we consider the residue

$$R(y;q) = \lim_{x^{2n}\to q^{-n-1}}(1-q^{n+1}x^{2n})\tilde{Z}_1(s,w) = \frac{P(q^{-1/2-1/2n},y)}{\mathscr{D}(y)},$$

where

$$\mathscr{D}(y) = (1-y^n)(1-q^ny^n)(1-q^{-2})(1-q^{(n+1)/2}y^n)(1-q^{(n-3)/2}y^n)$$
$$\times(1-q^{-1}y^n)(1-q^{n-1}y^n). \quad (35)$$

Notice that $R(y;q)$ is a power series in y^n beginning with the power y^{2n}. Also, the functional equation above specializes to

$$P(q^{-1/2-1/2n},y) = q^{9n}P(q^{-1/2-1/2n}, y^{-1}q^{-1+1/n}).$$

Let us introduce for clarity the (admittedly unnecessary) variable $t = yq^{-1/2n}$. For convenience, write $\tilde{R}(t;q) = R(y;q)$, $\tilde{P}(t) = P(q^{-1/2-1/2n},y)$, and $\tilde{\mathscr{D}}(t) = \mathscr{D}(y)$, so

$$\tilde{R}(t;q) = \tilde{P}(t)\tilde{\mathscr{D}}(t).$$

The functional equation above in y becomes one sending $t \to q^{-1}t^{-1}$ and

$$\tilde{P}(t) = t^{6n}q^{3n}\tilde{P}(q^{-1}t^{-1}).$$

Thus if $\tilde{P}(t) = \sum_2^M B_i t^{ni}$, then the functional equation implies that $M = 4$ and $B_2 = q^{-n}B_4$. Also, recall that B_2 is non-zero. Thus, we arrive at the expression

$$\tilde{P}(t) = B_2 t^{2n}(1 + B_3' t^n + q^n t^{2n})$$

for certain coefficients B_2, B_3'.

Finally, we have that the residue of $\tilde{P}_1(t)$ is 0 at both $t^{-n} = q^{n+1/2}$ and at $t^{-n} = q^{-1/2}$. This forces

$$1 + B_3' t^n + q^n t^{2n} = (1 - q^{n+1/2}t^n)(1 - q^{-1/2}t^n).$$

Cancelling these two factors from the denominator $\tilde{\mathscr{D}}(t)$, we arrive at

Theorem 1. *The function $\tilde{R}(t;q)$ is of the form*

$$\tilde{R}(t;q) = \frac{c_{n,q}t^{2n}}{(1 - q^{1/2}t^n)(1 - q^{n/2+1}t^n)(1 - q^{n/2-1}t^n)(1 - q^{n-1/2}t^n)},$$

where $c_{n,q}$ is a non-zero constant.

Now we compare this to the Mellin transform computed in [6]. The function $M_n(u;q)$ introduced in (5.2) there is defined as the Mellin transform of the theta function on the n-fold cover of $GL(2)$ over the function field $\mathbb{F}_q(T)$. The Mellin transform introduces a variable w. Continuing to let $y = q^{-w}$, we have

Proposition 4. *[6] For a certain nonzero constant $c'_{n,q}$, one has*

$$M_n(y;q) = \frac{c'_{n,q}y^n}{(1 - qy^n)(1 - q^{2n-1}y^n)}.$$

Here, M_n has functional equation

$$M_n(y;q) = M_n(y^{-1}q^{-2};q).$$

We also find the Dirichlet series part $D_n(t;q)$ of the Fourier coefficient of the nth order metaplectic Eisenstein series in [6], (5.2). From this equation, we have

$$D_n(t;q) = \frac{t^n}{(1 - q^{n-1}t^n)(1 - q^{n+1}t^n)},$$

with $t = q^{-2s}$. This function has functional equation under $s \mapsto 1 - s$.

Let us compare these three expressions. We have

$$D_n(tq^{-1/2};q) = \frac{q^{-n/2}t^n}{(1-q^{n/2-1}t^n)(1-q^{n/2+1}t^n)}.$$

Also, suppose that q is an even power of the residue characteristic. Then we may compute the Mellin transform of the theta function over $\mathbb{F}_{q^{1/2}}(T)$. If we double the Mellin transform variable w to $2w$, then the resulting expression may still be expressed in terms of $y = q^{-w} = (q^{1/2})^{-2w}$. This is given by

$$M_n(y;q^{1/2}) = \frac{c'_{n,q^{1/2}}y^n}{(1-q^{1/2}y^n)(1-q^{n-1/2}y^n)}.$$

We thus find that after normalizing so that the first coefficient of every power series equals 1.

Theorem 2. *Suppose that q is an even power of the residue characteristic. Then*

$$\tilde{R}(t;q) = M_n(t;q^{1/2})D_n(tq^{-1/2};q).$$

In other words, the rational polynomial on the left-hand side, which equals the Rankin–Selberg convolution of the Eisenstein series on the $2n$-fold cover with the quadratic theta function, factors into the rational polynomial representing the Mellin transform of a theta function on the n-fold cover times the first Fourier coefficient of the Eisenstein series on the n-fold cover.

This proves the conjecture in the case of the rational function field when n is odd. Unfortunately, the conjecture is certainly not true over a number field for $n \geq 5$, as observed previously. Thus, the special nature of the rational function field seems to give rise to too many simplifications! In particular, the numerators on both sides are (after cancellations) essentially trivial in this case. In a function field of higher genus, the numerators would be polynomials, and further structure would be revealed. It remains a very interesting open question to follow through the methods of this section in the case of *any* extension of the rational function field and to see what the actual relationship is between $\tilde{R}, M_n$ and D_n.

Acknowledgements This work was supported by NSF grants DMS-0652312, DMS-0652609, DMS-0847586, DMS-1001326 and by NSA grant H98230-10-1-0183.

References

1. B. Brubaker and D. Bump. On Kubota's Dirichlet series. *J. Reine Angew. Math.*, 598:159–184, 2006.
2. B. Brubaker, D. Bump, G. Chinta, S. Friedberg, J. Hoffstein, *Weyl group multiple Dirichlet series I,* in Multiple Dirichlet Series, Automorphic Forms, and Analytic Number Theory, Proc. Sympos. Pure Math., 75, Amer. Math. Soc., Providence, RI, 2006.

3. D. Bump and J. Hoffstein. Some conjectured relationships between theta functions and Eisenstein series on the metaplectic group. In *Number theory (New York, 1985/1988)*, volume 1383 of *Lecture Notes in Math.*, pages 1–11. Springer, Berlin, 1989.
4. H. Davenport and H. Hasse, Die Nullstellen der Kongruenzzetafunktionen in gewissen zyklischen Fällen. *J. Reine Angew. Math.* 172 (1934), 151–182.
5. C. Eckhardt and S. J. Patterson, On the Fourier coefficients of biquadratic theta series, *Proc. London Math. Soc.* (3) 64 (1992), 225–264.
6. J. Hoffstein. Theta functions on the n-fold metaplectic cover of SL(2)–the function field case. *Invent. Math.*, 107(1):61–86, 1992.
7. J. Hoffstein. Eisenstein series and theta functions on the metaplectic group. In *Theta functions: from the classical to the modern*, volume 1 of *CRM Proc. Lecture Notes*, pages 65–104. Amer. Math. Soc., Providence, RI, 1993.
8. D. A. Kazhdan and S. J. Patterson. Metaplectic forms. *Inst. Hautes Études Sci. Publ. Math.*, (59):35–142, 1984.
9. T. Kubota. *On automorphic functions and the reciprocity law in a number field.* Lectures in Mathematics, Department of Mathematics, Kyoto University, No. 2. Kinokuniya Book-Store Co. Ltd., Tokyo, 1969.
10. S. J. Patterson. A cubic analogue of the theta series. *J. Reine Angew. Math.*, 296:125–161, 1977.
11. S. J. Patterson. A cubic analogue of the theta series. II *J. Reine Angew. Math.*, 296:217–220, 1977.
12. S. J. Patterson. The distribution of general Gauss sums at prime arguments. *Progress in Analytic Number Theory, vol. 2 (H. Halberstam and C. Hooley, eds.) Academic Press, New York*, 171–182, 1981.
13. S. J. Patterson. Whittaker models of generalized theta series. In *Seminar on number theory, Paris 1982–83 (Paris, 1982/1983)*, volume 51 of *Progr. Math.*, pages 199–232. Birkhäuser Boston, Boston, MA, 1984.
14. S. J. Patterson. Note on a paper of J. Hoffstein. *Glasg. Math. J.*, 49(2):243–255, 2007.
15. T. Suzuki. Metaplectic Eisenstein series and the Bump-Hoffstein conjecture. *Duke Math. J.*, 90(3):577–630, 1997.
16. G. Wellhausen, Fourierkoeffizienten von Thetafunktionen sechster Ordnung, Dissertation (Göttingen, 1996).
17. D. Zagier, The Rankin-Selberg method for automorphic functions which are not of rapid decay, *J. Fac. Sci. Univ. of Tokyo*, Sect. IA 28 (1981), 415–437.

The Patterson Measure: Classics, Variations and Applications

Manfred Denker and Bernd O. Stratmann

Abstract This survey is dedicated to S. J. Patterson's 60th birthday in recognition of his seminal contribution to measurable conformal dynamics and fractal geometry. It focuses on construction principles for conformal measures for Kleinian groups, symbolic dynamics, rational functions and more general dynamical systems, due to Patterson, Bowen-Ruelle, Sullivan and Denker-Urbański.

1 The Patterson Measure: Classics

In his pioneering work [75] Patterson laid the foundation for a comprehensive measure theoretical study of limit sets arising from (conformal) dynamical systems. Originally, his main focus was on limit sets of finitely generated Fuchsian groups, with or without parabolic elements. We begin this survey by reviewing his construction and some of its consequences in the slightly more general situation of a Kleinian group. The starting point of this construction is that to each Kleinian group G one can associate the Poincaré series $\mathscr{P}(z,s)$, given by

$$\mathscr{P}(z,s) := \sum_{g\in G} \exp(-sd(z,g(0))),$$

for $s \in \mathbb{R}$, 0 denoting the origin in the $(N+1)$-dimensional hyperbolic space $\mathbb{H}$ (throughout, we always use the Poincaré ball model for $\mathbb{H}$), z an element of $\mathbb{H}$, and

M. Denker (✉)
Mathematics Department, Pennsylvania State University, State College, PA 16802, USA
e-mail: denker@math.psu.edu

B.O. Stratmann
Fachbereich 3 – Mathematik und Informatik, Universität Bremen, Bibliothekstr. 1, 28359 Bremen, Germany
e-mail: bos@math.uni-bremen.de

V. Blomer and P. Mihăilescu (eds.), *Contributions in Analytic and Algebraic Number Theory*, Springer Proceedings in Mathematics 9, DOI 10.1007/978-1-4614-1219-9_7,

where d denotes the hyperbolic metric. The abscissa of convergence $\delta = \delta(G)$ of this series is called the Poincaré exponent of G. It is a priori not clear if $\mathscr{P}(z,s)$ converges or diverges for $s = \delta$, and accordingly, G is called of δ-divergence type if $\mathscr{P}(z,\delta)$ diverges, and of δ-convergence type otherwise. Patterson made use of this critical behaviour of $\mathscr{P}(z,s)$ at $s = \delta$ in order to build measures supported on the limit set $L(G)$ of G, that is, the set of accumulation points of the orbit $G(0)$, as follows. In order to incorporate also the δ-convergence type case, he first chooses a sequence (s_j) tending to δ from above, and then carefully crafts a slowly varying function φ such that the modified Poincaré series

$$\mathscr{P}_\varphi(z,s) := \sum_{g\in G} \varphi(d(z,g(0)))\exp(-sd(z,g(0)))$$

still has abscissa of convergence equal to δ, but diverges for $s = \delta$. With this slight alteration of the classical Poincaré series, he then defines discrete measures μ_{z,s_j} by putting weights on the orbit points in $G(0)$ according to

$$\mu_{z,s_j}(g(0)) = \frac{\varphi(d(z,g(0)))\exp(-s_j d(z,g(0)))}{\mathscr{P}_\varphi(z,s_j)}.$$

Due to the divergence of the modified Poincaré series at δ, each weak accumulation point of the resulting sequence $\left(\mu_{z,s_j}\right)$ of measures is clearly supported on $L(G)$, and each of these so obtained limit measures is what one nowadays calls a Patterson measure. One of the success stories of these measures is that if G is geometrically finite, that is, each element of $L(G)$ is either a radial limit point or else is the fixed point of some parabolic element of G, then there exists a unique measure class containing all these measures. In other words, in this situation a weak accumulation point μ_z of the sequence $\left(\mu_{z,s_j}\right)$ does not depend on the particular chosen sequence (s_j). Moreover, in this geometrically finite situation it turns out that G is of δ-divergence type. Let us now concentrate on this particular situation for a moment, that is, let us assume that G is geometrically finite. Then, a crucial property of the family $\{\mu_z : z \in \mathbb{H}\}$ is that it is δ-harmonic, meaning that for arbitrary $z,w \in \mathbb{H}$ we have, for each $x \in L(G)$,

$$\frac{\mathrm{d}\mu_z}{\mathrm{d}\mu_w}(x) = \exp(\delta b_x(z,w)),$$

where $b_x(z,w)$ denotes the signed hyperbolic distance of z to w at x, obtained by measuring the hyperbolic distance $d_x(z,w)$ between the two horocycles at x, one containing z and the other containing w, and then taking the negative of this distance if w is contained in the horoball bounded by the horocycle through z, and letting it be equal to this hyperbolic distance otherwise. Note that $d_x(z,w)$ is a Busemann function and $b_x(z,w)$ coincides with $\log(P(z,x)/P(w,x))$, for $P(\cdot,\cdot)$ denoting the Poisson kernel in $\mathbb{H}$. Let us also remark that here the wording *δ-harmonic* points towards another remarkable success story of the concept

"Patterson measure", namely, its close connection to spectral theory on the manifold associated with G. More precisely, we have that the function ϕ_0, given by

$$\phi_0(z) := \int_{\partial\mathbb{H}} P(z,x)^\delta \, d\mu_0(x),$$

is a G-invariant eigenfunction of the Laplace–Beltrami operator associated with the (smallest) eigenvalue $\delta(N-\delta)$. Moreover, ϕ_0 is always square-integrable on the convex core of $\mathbb{H}/G$, defined by forming first the convex hull of the limit set in $\mathbb{H}$, then taking a unit neighbourhood of this convex hull, and finally quotienting out G. In order to gain more geometric insight into δ-harmonicity, it is convenient to consider the measure $\mu_{\gamma(0)}$, for some arbitrary $\gamma \in G$. A straightforward computation gives that $\mu_{\gamma(0)} = \mu_0 \circ \gamma^{-1}$, and hence, the δ-harmonicity implies that

$$\frac{d\,(\mu_0 \circ \gamma^{-1})}{d\,\mu_0}(x) = P(\gamma(0),x)^\delta, \text{ for all } \gamma \in G. \tag{1}$$

This property of the Patterson measure μ_0 is nowadays called δ-conformality. Sullivan [111] was the first to recognise the geometric strength of this property, which we now briefly comment on. Let s_x denote the hyperbolic ray between $0 \in \mathbb{H}$ and $x \in \partial\mathbb{H}$, and let x_t denote the point on s_x at hyperbolic distance t from the origin. Let $B_c(x_t) \subset \mathbb{H}$ denote the $(N+1)$-dimensional hyperbolic disc centred at x_t of hyperbolic radius $c > 0$, and let $\Pi : \mathbb{H} \to \partial\mathbb{H}$ denote the shadow-projection given by $\Pi(C) := \{x \in \partial\mathbb{H} : s_x \cap C \neq \emptyset\}$. Also, if x_t lies in one of the cusps associated with the parabolic fixed points of G, let $r(x_t)$ denote the rank of the parabolic fixed point associated with that cusp, otherwise, put $r(x_t)$ equal to δ. Combining the δ-conformality of μ_0 and the geometry of the limit set of the geometrically finite Kleinian group G, one obtains the following generalized Sullivan shadow lemma [110, 111, 113]:

$$\mu_0(\Pi(B_c(x_t))) \asymp |\Pi(B_c(x_t))|_E^\delta \cdot \exp((r(x_t)-\delta)d(x_t,G(0))),$$

for all $x \in L(G)$ and $t > 0$, for some fixed sufficiently large $c > 0$, and where $|\cdot|_E$ denotes the diameter with respect to the chordal metric in $\partial\mathbb{H}$. Note that in the latter formula the "fluctuation term" $\exp((r(x_t)-\delta)d(x_t,G(0)))$ can obviously also be written in terms of the eigenfunction ϕ_0 of the Laplace–Beltrami operator. Besides, this gives a clear indication towards why the Patterson measure admits the interpretation as a "correspondence principle", which provides a stable bridge between geometry and spectral theory. However, one of the most important consequences of the generalized Sullivan shadow lemma is that it allows us to use the Patterson measure as a striking geometric tool for deriving significant geometric insights into the fractal nature of the limit set $L(G)$. For instance, it immediately follows that if G has no parabolic elements, then μ_0 coincides, up to a multiplicative constant, with the δ-dimensional Hausdorff measure on $L(G)$. Hence, in this case, the Hausdorff dimension of $L(G)$ is equal to δ. To extend this to the case in which there are parabolic elements, one first establishes the following generalization of a classical theorem of Khintchine in metrical Diophantine approximations [55]. The proof in [110] uses the generalized Sullivan shadow lemma and the techniques of Khintchine's classical result (for further results on metrical Diophantine approximations in connection with the Patterson measure, see e.g. [42,76,96–103,105], or the survey article [104]).

$$\limsup_{t\to\infty} \frac{d(x_t, G(0))}{\log t} = (2\delta(G) - r_{\max})^{-1}, \text{ for } \mu_0\text{-almost all } x \in L(G).$$

Here, $r_{\max}$ denotes the maximal rank of the parabolic fixed points of G. By combining this with the generalized Sullivan shadow lemma, an immediate application of the mass distribution principle gives that even when G has parabolic elements, we still have that δ is equal to the Hausdorff dimension of $L(G)$. Moreover, these observations immediately show that μ_0 is related to the δ-dimensional Hausdorff measure H_δ and packing measure P_δ as follows. For ease of exposition, the following table assumes that G acts on hyperbolic 3-space.

	$0 < \delta < 1$	$\delta = 1$	$1 < \delta < 2$
No cusps	$\mu_0 \asymp H_\delta \asymp P_\delta$	$\mu_0 \asymp H_1 \asymp P_1$	$\mu_0 \asymp H_\delta \asymp P_\delta$
$r_{\max} = 1$	$\mu_0 \asymp P_\delta, H_\delta = 0$	$\mu_0 \asymp H_1 \asymp P_1$	$\mu_0 \asymp H_\delta, P_\delta = \infty$
$r_{\min} = 2$	n.a.	n.a.	$\mu_0 \asymp P_\delta, H_\delta = 0$
$r_{\min} = 1, r_{\max} = 2$	n.a.	n.a.	$H_\delta = 0$, $P_\delta = \infty$

Moreover, as was shown in [105], again by applying the generalized Sullivan shadow lemma for the Patterson measure, we additionally have that δ is equal to the box-counting dimension of $L(G)$. At this point, it should also be mentioned that in [8] and [103] it was shown that in fact *every* non-elementary Kleinian group G has the property that its exponent of convergence δ is equal to the Hausdorff dimension of its uniformly radial limit set, that is, the subset of the radial limit set consisting of those limit points $x \in L(G)$ for which there exists $c > 0$ such that $d(x_t, G(0)) < c$, for all $t > 0$. The proof of this rather general result is based on an elementary geometrization of the Poincaré series and does not use any Patterson measure theory (see also [103]). These fractal geometric interpretations of the exponent of convergence are complemented by its dynamical significance. Namely, one finds that the square integrability of the eigenfunction ϕ_0 on the convex core of $\mathbb{H}/G$ implies that the invariant measure for the geodesic flow on $\mathbb{H}/G$ associated with the Patterson measure has finite total mass [113]. Using this, one then obtains that δ is equal to the measure-theoretic entropy of the geodesic flow. In particular, if there are no cusps, one can define a topological entropy for the invariant set of geodesics with both endpoints in the limit set, and this topological entropy also turns out to be equal to the critical exponent δ [111]. It is worth mentioning that in this geometrically finite situation the invariant measure for the geodesic flow is not only of finite total mass and ergodic, but it is also mixing and even Bernoulli [90]. In fact, these strong properties of the geodesic flow have been exploited intensively in the literature to derive various interesting aspects of the limit set. For instance, the marginal measure of the Patterson–Sullivan measure $|x-y|^{-2\delta}\mathrm{d}\mu_0(x)\mathrm{d}\mu_0(y)$, obtained by disintegration of the first coordinate, leads to a measure which is invariant under the Bowen–Series map. This allows us to bring standard (finite and infinite) ergodic theory into play. As an example of the effectiveness of this connection, we mention the recent result (see [53] in these Proceedings) that for a geometrically finite Kleinian group G with parabolic elements we have that, with $|\cdot|$ denoting the word metric,

$$\sum_{\substack{g\in G\\ |g|\leq n}} \exp(-\delta d(0,g(0))) = O(n^{2\delta-r_{\max}}).$$

For Kleinian groups which are not geometrically finite the Patterson measure theory is less well developed, although various promising first steps have been undertaken. Here, an interesting class is provided by finitely generated, geometrically infinite Kleinian groups acting on hyperbolic 3-space $\mathbb{H}^3$ whose limit set is not equal to the whole boundary $\partial\mathbb{H}^3$. For these groups, it had been conjectured for almost 40 years that the area of their limit sets is always equal to zero. This conjecture was named after Ahlfors and was eventually reduced to the so-called tameness-conjecture, a conjecture which was only very recently confirmed in [5] and [18]. Given the nature of this conjecture, it is perhaps not too surprising that the concept "Patterson measure" also made vital contributions to its solution.

For infinitely generated Kleinian groups, so far only the beginnings of a substantial theory have been elaborated. As Patterson showed in [77], there exist infinitely generated groups whose exponent of convergence is strictly less than the Hausdorff dimension of their limit set. Kleinian groups with this property were named in [37] as discrepancy groups. Also, an interesting class of infinitely generated Kleinian groups is provided by normal subgroups N of geometrically finite Kleinian groups G. For these groups, one always has that $L(N) = L(G)$ and $\delta(N) \geq \delta(G)/2$ (see [37]), and this inequality is in fact sharp, as was shown very recently in [11]. Moreover, by a result of Brooks in [14], one has that if G acts on hyperbolic n-space such that $\delta(G) > n/2$, then

$$N \text{ is a discrepancy group if and only if } G/N \text{ is non-amenable.}$$

This result is complemented by beautiful applications of the Patterson measure theory in [86] and [87], where it was shown for the Fuchsian case that if $G/N \cong \mathbb{Z}^k$, and hence $\delta = \delta(N) = \delta(G)$, since $\mathbb{Z}^k$ is clearly amenable, then

$$N \text{ is of } \delta\text{-divergence type} \Leftrightarrow \begin{cases} k\in\{1,2\} & \text{if } G \text{ has no parabolic elements} \\ k=1 & \text{if } G \text{ has parabolic elements.} \end{cases}$$

Finally, we mention the related work of [2], which considers the special situation of the Riemann surface $\mathbb{C}\setminus\mathbb{Z}$ uniformized by a Fuchsian group N, which is a normal subgroup of the subgroup G of index 6 of the modular group $PSL_2(\mathbb{Z})$ uniformizing the threefold punctured sphere. There it was shown that the Poincaré series $\mathscr{P}(z,s)$ associated with N has abscissa of convergence $\delta(N) = 1$ and that it has a logarithmic singularity at $s = 1$ (for further results of this type, see e.g. [4, 65, 72, 79, 80]). This result of [2] is obtained by showing that the associated geodesic flow has a factor, which is Gibbs-Markov [3] and by using a local limit theorem of Cauchy type.

2 Gibbs Measures

Rokhlin's seminal paper [88] on the foundations of measure theory, dynamical systems and ergodic theory is fundamental for our further discussion of Patterson measures and conformality of measures of the type as in (1). Let $R : \Omega_1 \to \Omega_2$ be a measurable, countable-to-one map between two Lebesgue spaces $(\Omega_i, \Sigma_i, \mu_i)$ $(i = 1,2)$ [19], where Σ_i and μ_i denote some Borel fields and measures. If R is nonsingular[1], the Jacobian J_R of R exists, meaning that, for all $E \in \Sigma_1$ such that $R|_E$ is invertible, we have that

$$\mu_2(R(E)) = \int_E J_R \mathrm{d}\mu_1. \tag{2}$$

By our assumptions, the images $R(E)$ are always measurable, in fact, throughout this section all functions and sets considered will always be assumed to be measurable. Also, note that J_R is uniquely defined, μ_1-almost everywhere. Moreover, since R is countable-to-one, the Jacobian J_R gives rise to the transfer operator $\mathscr{L}_J = \mathscr{L}_{J_R}$, given by

$$\mathscr{L}_J f(x) = \sum_{R(y)=x} f(y)/J(y), \tag{3}$$

for all measurable functions $f : \Omega_1 \to \mathbb{R}$ for which the right-hand side in (3) is well defined. (For example, the latter always holds for f bounded and R finite-to-one, and it holds, more generally, if $\|\mathscr{L}_J 1\|_\infty < \infty$.) For this type of function, we then have that (2) is equivalent to

$$\int f \mathrm{d}\mu_1 = \int \mathscr{L}_J f \mathrm{d}\mu_2. \tag{4}$$

Note that this identity can also be written in terms of the "dual operator" $\mathscr{L}_J^*$, which maps μ_2 to μ_1. If the two Lebesgue spaces agree and are equal to some Ω, then

[1] R is said to be nonsingular with respect to μ_1 and μ_2, if for each measurable set $E \subset \Omega_2$ one has that $\mu_1(R^{-1}(E)) = 0$ if and only if $\mu_2(E) = 0$.

$R : \Omega \to \Omega$ is a non-singular transformation of the Lebesgue space Ω, and in this situation we have that $\mu = \mu_1$ is a fixed point of the dual $\mathscr{L}_J^*$.

The δ-conformality of the Patterson measure in (1) can be viewed as determining the Jacobian for the transformations in the Kleinian group G. Hence, the Patterson construction in Sect. 1 solves the problem of finding a measure whose Jacobian equals a certain power of the derivative of these transformations. This naturally leads to the following question: For a given measurable function ϕ and a transformation T, when does there exist a probability measure with Jacobian equal to e^{ϕ}? It turns out that typical conditions on ϕ and T are certain kinds of conformality as well as some specific geometric and/or analytic properties. Nowadays, this type of question is well addressed, but in the mid-1970s the work in [75] paved the way for these developments (see the following sections). Here, it should also be mentioned that, parallel to this development, the theory of Gibbs measures evolved [12,122], solving the analogue question for subshifts of finite type.

Consider a compact metric space (Ω, d) and a continuous finite-to-one transformation $T : \Omega \to \Omega$. For a given continuous function $\phi : \Omega \to \mathbb{R}$, let us first identify non-singular measures m_ϕ for which (2) is satisfied with $J_T = e^{\phi}$. A good example for this situation is given by a differentiable map T of the unit interval into itself, where the Lebesgue measure satisfies the equality (2), with ϕ being equal to the logarithm of the modulus of the derivative of T.

For an expanding, open map $T : \Omega \to \Omega$ and a continuous function ϕ, Ruelle's Perron–Frobenius Theorem [12] guarantees the existence of a measure μ satisfying

$$\mu(T(A)) = \lambda \int_A \mathrm{e}^{\phi} \mathrm{d}\mu, \tag{5}$$

for some $\lambda > 0$ and for each $A \in \Sigma$ for which $T|_A$ is invertible. Each measure so obtained is called a Gibbs measure for the potential function ϕ. This type of measure represents a special case of conformal measures. An open, expanding map on a compact metric space is called R-expanding, where R refers to Ruelle. This includes subshifts of finite type (or topological Markov chains), for which the Ruelle Theorem was originally proven. In fact, such a R-expanding map has the property that the number of pre-images of all points is locally constant. Consequently, for a given $\phi \in C(X)$, the Perron–Frobenius operator (or equally, the transfer operator) $\mathscr{L}_\phi$ acts on the space $C(X)$, and is given by

$$\mathscr{L}_\phi f(x) = \sum_{y:T(y)=x} f(y) \exp(-\phi(y)).$$

In this situation, we then have that the map $m \mapsto \mathscr{L}_\phi^* m / m(\mathscr{L}_\phi 1)$ has a fixed point m_ϕ. The measure m_ϕ is a Gibbs measure whose Jacobian is equal to $\lambda \cdot \mathrm{e}^{\phi}$, for $\lambda = m_\phi(\mathscr{L}_\phi 1)$. The logarithm of the eigenvalue λ is called the pressure $P(T, -\phi)$ of $-\phi$.

The following Bowen–Ruelle–Perron–Frobenius Theorem summarised the main results in this area.

Theorem 1. *([12]) Let (Ω, T) be a topologically mixing, R-expanding dynamical system. For each Hölder continuous function $\phi : \Omega \to \mathbb{R}$, there exists a probability measure m_ϕ and a positive Hölder continuous function h such that the following hold.*

1. $\mathscr{L}_\phi^* m_\phi = \exp(P(T, \phi)) m_\phi$;
2. $\mathscr{L}_\phi h = \exp(P(T, \phi)) h$;
3. $\mathscr{L}_\phi^n f - \int f h dm_\phi$ *decreases in norm exponentially fast.*

One immediately verifies that the measure $\widetilde{m}_\phi$, given by $d\widetilde{m}_\phi = h \cdot dm_\phi$, is T-invariant, and hence, $\widetilde{m}_\phi$ is often also referred to as the invariant Gibbs measure. In fact, as the name already suggests, the existence of this type of Gibbs measures is closely related to the thermodynamic formalism for discrete time dynamical systems.

Note that the existence of m_ϕ has been derived in [52], whereas some first results in this direction were already obtained in [74]. Alternative proofs of the Bowen–Ruelle–Perron–Frobenius Theorem use, for instance, the Hilbert metric in connection with positive cones (see [38] and Sect. 5) or, for the statement in (2), the Theorem of Ionescu-Tulcea and Marinescu (see [50]). Also, note that the original version of this theorem was given in terms of subshifts of finite type. In fact, an R-expanding transformation admits a Markov partition, and therefore, the associated coding space is a subshift of finite type. Nevertheless, the theorem can also be proven directly in terms of R-expanding maps on compact metric spaces.

Finally, let us remark that the method above can be extended to systems which are neither open nor expanding. For instance, the potential function ϕ may have properties which only requires T to be expanding along certain orbits. A typical condition of this type is that the pressure function at ϕ exceeds $\sup(\phi)$, where the supremum is taken over the state space Ω. This situation arises, for instance, if T is a rational map on the Riemann sphere (see e.g. [33] and [27], or [47] for the case of a map of the interval). In this case, we still have that $\mathscr{L}_\phi$ acts on the set of continuous functions, and the proof of the existence of the invariant Gibbs measure then uses that for a Hölder continuous potential function ϕ most of the branches are contracting and that the contributions of other branches are negligible, due to the boundedness condition on ϕ. In fact, this approach turns out to be somehow characteristic for certain non-uniformly hyperbolic systems.

3 Sullivan's Conformal Measure

As already mentioned at the beginning, originally one of the main motivations for the construction of the Patterson measure was to study fractal geometric properties of limit sets of Fuchsian groups. The analogue of Patterson's construction for Julia sets of either hyperbolic or parabolic rational maps was first noticed by Sullivan in [112]. Recall that a rational function $R : S^2 \to S^2$ is called hyperbolic if its

Julia set does not contain any critical or rationally indifferent (parabolic) periodic points, whereas R is called parabolic if its Julia set contains a parabolic periodic point, but does not contain any critical point. Here, the key observation is that in these expansive cases the Julia set can be considered as being the "limit set" of the action of the rational map on its Fatou component. The elaboration of this analogue between Fuchsian groups and rational maps in [112] has led to what is nowadays called Sullivan's dictionary (for some further chapters of this dictionary, see e.g. [106–109, 112]).

The idea of a conformal measure for a rational map R appeared first in [112], Theorem 3, where the existence of a conformal measure for the function $|R'|^t$, for some $t \in \mathbb{R}$, was established. Moreover, in the same paper Sullivan showed that this measure is unique in the hyperbolic case. In fact, in this case one easily verifies that $\delta = \inf\{t > 0 : \text{a } t\text{-conformal measure exists}\}$ coincides with the Hausdorff dimension h of the Julia set. Sullivan's construction modifies the Patterson measure construction, and his method was later extended in [28] to more general classes of transformations.

Recall that the starting point of Patterson's and Sullivan's construction is to consider powers ϕ^t of the exponential of some potential function $\log\phi$, for t greater than a certain critical value, and then to proceed by letting t decrease to this critical value. However, in the case of expanding rational maps, it is much simpler to use the theory of Gibbs measures, as explained in Sect. 2.

One immediately verifies that there always exists a Gibbs measure m_t for $\phi^t = |R'|^t$, for some $t \geq 0$ (this follows from the discussion in Sect. 2). Since

$$m_t(R(A)) = \lambda_t \int_A |R'|^t \,\mathrm{d}m_t,$$

where as before $\log\lambda_t = P(R, -t\log|R'|)$, we have that the measure m_t is conformal if and only if $P(R, -t\log|R'|) = 0$. If R is expanding, it is easy to see that the pressure function is continuous and strictly decreasing, for $t \geq 0$. In particular, we have that $P(R, 0)$ $(= \log deg(R))$ is equal to the topological entropy (see [66]) and that $P(R, -t\log|R'|) \to -\infty$, for t tending to infinity. This implies that there exists a unique t for which the pressure function vanishes. In fact, this is precisely the content of the Bowen–Manning–McCluskey formula [13, 68]. Using this observation, it can then be shown that the so obtained t is equal to the Hausdorff dimension of the Julia set of R, a result due to Sullivan in [112] (see [13] for related earlier results on dynamical and geometric dimensions). Note that Sullivan's construction employs Patterson's approach, replacing the orbit under the Fuchsian group by the set of pre-images under R of some point in the Fatou set, which accumulates at the Julia set. This approach can be viewed as some kind of "external construction" (see [40]).

For more general rational functions, it is necessary to gain better control over the eigenvalues of the transfer operator. This can hardly be done by the type of functional analytic argument given above. However, for a parabolic rational map one still finds that there exists a unique non-atomic ergodic conformal measure with exponent equal to the Hausdorff dimension of the Julia set. Although there still exists such a conformal measure, in this situation one finds that every other ergodic conformal measure is concentrated on the orbit of the parabolic points (see [32, 34]). While the construction is still straightforward in this parabolic case, other cases of rational functions have to be treated with refined methods and require certain "internal constructions", of which we now recall a few (see also Sect. 4).

One of these methods is Urbański's KV-method, which considers invariant subsets of the Julia set whose closures do not contain any critical point. Given that these sets exhaust the Julia set densely, this method allows us to construct measures which converge weakly to the conformal measure in question. Here, the main work consists in showing that the obtained limit measure has no atoms at the critical orbits. This is achieved by employing a certain type of tightness argument. In a similar fashion to that outlined above, the construction leads to a conformal measure with a minimal exponent (see [31,83]). Although it is still an open problem to decide whether this measure is overall non-atomic, one nevertheless has that the minimal exponent is equal to the dynamical dimension of the system.

Another method is the constructive method of [29], which applies in the case of subexpanding rational functions and in the case of rational functions satisfying the Collet-Eckmann condition. It also applies to rational maps which satisfy the following summability condition of [40] and [84]:

$$\sum_{n=1}^{\infty} |(R^n)'(R^{n_c}(c))|^{-\alpha} < \infty,$$

for some $\alpha \geq 0$, for all critical points c in the Julia set, and for some $n_c \in \mathbb{N}$. In this case, the existence of a non–atomic conformal measure is guaranteed, given that the Julia set does not contain parabolic points and given that $\alpha < h/(h+\mu)$, where h denotes the Hausdorff dimension of the Julia set and μ the maximal multiplicity of the critical points in the Julia set.

Finally, let us also mention that for a general rational map we have that the dynamical dimension of its Julia set coincides with the minimal t for which a t-conformal measure exists [31, 83].

The following theorem summarizes the discussion above.

Theorem 2. *Let R be a rational map of the Riemann sphere, and let h denote the Hausdorff dimension of its Julia set $J(R)$. Then there exists a non-atomic h-conformal probability measure m on $J(R)$, given that one of the following conditions hold:*

(1) ([112]) R is hyperbolic. In this case, m is the unique t-conformal measure, for all $t \in \mathbb{R}$.
(2) ([32]) R is parabolic. In this case, m is the unique non–atomic t-conformal measure, for all $t \in \mathbb{R}$.
(3) ([29]) R is subexpanding (of Misiurewicz type). In this case, m is the unique non–atomic h-conformal measure.

(4) *([118]) If $J(R)$ does not contain any recurrent critical points of R, then m is the unique h-conformal measure. Moreover, m is ergodic and conservative.*
(5) *([40, 84]) R satisfies the above summability condition. In this case, m is the unique non–atomic h-conformal measure.*
(6) *([6]) R is a Feigenbaum map for which the area of $J(R)$ vanishes. In this case, m is the unique h-conformal measure and there exists a non–atomic t-conformal measure, for each $t \geq h$.*

In order to complete this list, let us also mention that Prado has shown in [82] that for certain infinitely renormalizable quadratic polynomials (originally introduced in [67]), the equality $h = \inf\{t : \exists \text{ a } t\text{-conformal measure}\}$ still holds. The ergodicity problem for the conformal measure of quadratic polynomials is treated in [81] and then extended further in [48].

An interesting new approach for obtaining the existence of conformal measures is developed by Kaimanovich and Lyubich. They study conformal streams which are defined on laminations of conformal structures. This setting is very much in the spirit of our discussion of bundle maps in Sect. 5. For further details concerning the construction of conformal streams and its application to rational functions, we refer to [51]. Moreover, note that the theory of conformal measures has also been elaborated for semigroups of rational functions (see [114–116]).

Up to now, the classification of conformal measures has not been completed. Clearly, since the space of conformal measures is compact with respect to the weak topology, we always have that there exists a conformal measure of minimal exponent. However, this measure can be either non-atomic, or purely atomic, or even a mixture of both of these types. This follows by convexity of the space of conformal measures (cf. [40]). At this point, it should be remarked that [9] contains an interesting result, which clarifies under which conditions on the critical and parabolic points one has that a conformal measure is non-atomic. Also, let us remark that an important aspect when studying conformal measures is provided by the attempts to describe the essential support of a conformal measure in greater detail (see [15, 26, 49, 71, 85]). Of course, the set of radial limit points marks the starting point for this journey.

There are various further fundamental results on the fine structure of Julia sets, which have been obtained via conformal measures. For instance, conformal measures led to the striking result that the Hausdorff dimension of the Julia set of parabolic maps of the Riemann sphere lies strictly between $p/(p+1)$ and 2 (see [1]), where p denotes the maximum of the number of petals to be found at parabolic points of the underlying rational map. Also, conformal measures have proven to be a powerful tool in studies of continuity and analyticity of the Hausdorff-dimension-function on families of rational maps [36, 125].

Recently, the existence of Sullivan's conformal measures has also been established for meromorphic functions [63]. The following theorem summarizes some of the most important cases.

Theorem 3. *Let T be a meromorphic function on $\mathbb{C}$, and let F be the projection of T onto $\{z \in \mathbb{C} : -\pi < Re(z) \leq \pi\}$. With h_T (resp. h_F) denoting the Hausdorff dimension of $J(T)$ (resp. $J(F)$), in each of the following cases we have that there exists a h_T-conformal (resp. h_F-conformal) measure.*

(1) ([57]) T is a transcendental function of the form $T(z) = R(\exp(z))$, where R is a non-constant rational function whose set of singularities consists of finitely many critical values and the two asymptotic values $R(0)$ and $R(\infty)$. Moreover, the critical values of T are contained in $J(T)$ and are eventually mapped to infinity, and the asymptotic values are assumed to have orbits bounded away from $J(T)$. In this case, there exists $t < h_F$ such that if $t > 1$ then there is only one t-conformal measure. Also, the h_F-conformal measure is ergodic, conservative and vanishes on the complement of the set of radial limit points. In particular, this h_F-conformal measure lifts to a σ-finite h_F-conformal measure for T.

(2) ([59, 61]) T is either elliptic and non-recurrent or weakly non-recurrent.[2] We then have that the h_T-conformal measure is non–atomic, ergodic and conservative, and it is unique as a non–atomic t-conformal measure.

(3) ([119, 121]) T is either exponential[3] and hyperbolic or super-growing.[4] Here, if $t > 1$ then the h_F-conformal measure is ergodic, conservative and unique as a t-conformal measure for F. Also, this conformal measure lifts to a σ-finite h_F-conformal measure for T.

(4) ([120]) T is given by $T(z) = \exp(z-1)$ (parabolic). Here, the h_F-conformal measure is non–atomic, ergodic and conservative. Also, for $t > 1$ it is the unique non–atomic t-conformal measure for F, and if $t \neq h$ then there exist discrete t-conformal measures for F, whereas no such discrete t-conformal measure for F exists for $t = h$. Again, this conformal measure lifts to a σ-finite h_F-conformal measure for T.

(5) ([94]) T is given by $T(z) = R(\exp(z))$, where R is a non-constant rational function with an asymptotic value, which eventually maps to infinity. Here, the h_F-conformal measure is non–atomic, conservative and ergodic, where h_F denotes the Hausdorff dimension of the radial Julia set of F. Also, this measure is unique as a h_F-conformal measure, and it lifts to a σ-finite h_F-conformal measure for T.

The proofs of these statements follow the general construction method, which will be described in the next section. Furthermore, the proofs use the well-known standard method of extending a finite conformal measure for an induced transformation to the full dynamics (see e.g. [35]). Note that [58] gives a finer analysis of the geometric measures appearing in part (5) of the previous theorem.

---

[2]The ω limit sets of critical points in the Fatou set are attracting or parabolic cycles and the ω limit set of critical points c in the Julia set are compact in $\mathbb{C} \setminus \{c\}$ (resp. $T^n(c) = \infty$, for some $n \geq 1$).

[3]That is of the form $T(z) = \lambda \exp(z)$.

[4]The sequence of real parts α_n (resp. the absolute value) of $T^n(0)$ is exponentially increasing, that is, $\alpha_{n+1} \geq c \exp \alpha_n$, for all $n \in \mathbb{N}$ and for some $c > 0$.

Furthermore, we would like to mention the work in [62] and [60], where one finds a discussion of the relations between different geometric measures. Also, fractal geometric properties of conformal dynamical systems are surveyed in [117] (see also the surveys in [78] and [104]).

4 Conformal Measures for Transformations

As mentioned before, the Patterson–Sullivan construction relies on approximations by discrete measures supported on points outside the limit set, and hence can be viewed as some kind of "external construction". In contrast to this, we are now going to describe an "internal construction", which uses orbits inside the limit set. The basic idea of this construction principle is inspired by the original Patterson measure construction in [75], and also by the method used for deriving equilibrium measures in the proof of the variational principle for the pressure function [73]. Note that the method does not use powers of some potential function, instead, it mimics the general construction of Gibbs measures, and one is then left to check the vanishing of the pressure function.

Throughout this section, let (X,d) be a compact metric space, equipped with the Borel σ-field $\mathscr{F}$. Also, let $T : X \to X$ be a continuous map for which the set $\mathscr{S}(T)$ of singular points $x \in X$ (that is, T is either not open at x or non-invertible in some neighbourhood of x) is finite. Furthermore, let $f : X \to \mathbb{R}$ be a given continuous function, and let $(E_n : n \in \mathbb{N})$ be a fixed sequence of finite subsets of X.

Recall that for a sequence of real numbers $(a_n : n \in \mathbb{N})$, the number $c = \limsup_{n\to\infty} a_n/n$ is called the transition parameter of that sequence. Clearly, the value of c is uniquely determined by the fact that it is the abscissa of convergence of the series $\sum_{n\in\mathbb{N}} \exp(a_n - ns)$. For $s = c$, this series may or may not converge. Similar to [75] (see also Sect. 1), an elementary argument shows that there exists a slowly varying sequence $(b_n : n \in \mathbb{N})$ of positive reals such that

$$\sum_{n=1}^{\infty} b_n \exp(a_n - ns) \begin{cases} \text{converges} & \text{for } s > c \\ \text{diverges} & \text{for } s \leq c. \end{cases} \tag{6}$$

4.1 The Construction Principle

Define $a_n = \log \sum_{x\in E_n} \exp S_n f(x)$, where $S_n f = \sum_{0\leq k<n} f \circ T^k$, and let c be the transition parameter of the sequence $(a_n : n \in \mathbb{N})$. Also, let $(b_n : n \in \mathbb{N})$ be a slowly varying sequence satisfying (6). For each $s > c$, we then define the normalized measure

$$m_s = \frac{1}{M_s} \sum_{n=1}^{\infty} \sum_{x\in E_n} b_n \exp(S_n f(x) - ns)\delta_x, \tag{7}$$

where M_s is a normalizing constant, and where δ_x denotes the Dirac measure at the point $x \in X$. A straightforward calculation then shows that, for $A \in \mathscr{F}$ such that $T|_A$ is invertible,

$$m_s(TA) = \int_A \exp(c-f)\mathrm{d}m_s + \mathrm{O}(s-c) - \frac{1}{M_s}\sum_{n=1}^{\infty}\sum_{x\in A\cap(E_{n+1}\Delta T^{-1}E_n)} b_n \exp(S_n f(T(x)) - ns). \tag{8}$$

For $s \searrow c$, any weak accumulation point of $\{m_s : s > c\}$ will be called a limit measure associated with f and $(E_n : n \in \mathbb{N})$. In order to find conformal measures among these limit measures, we now have a closer look at the terms in (8). There are two issues to discuss here. First, if A is a set which can be approximated from above by sets A_n for which $T|_{A_n}$ is invertible and for which the limit measure of their boundaries vanishes, then the outer sum on the right-hand side of (8) converges to the integral with respect to the limit measure. Obviously, this convergence depends on how the mass of m_s is distributed around the singular points. If the limit measure assigns zero measure to these points, the approximation works well. In this case, one has to check whether the second summand in (8) tends to zero as $s \searrow c$. The simplest case is that $E_{n+1} = T^{-1}(E_n)$, for all $n \in \mathbb{N}$, and then nothing has to be shown.

This discussion has the following immediate consequences.

Proposition 1. *([28]) Let T be an open map, and let m be a limit measure assigning measure zero to the set of periodic critical points. If we have*

$$\lim_{s\searrow c} \frac{1}{M_s}\sum_{n=1}^{\infty}\sum_{x\in E_{n+1}\Delta T^{-1}E_n} b_n \exp(S_n f(T(x)) - ns) = 0,$$

then there exists a $\exp(c-f)$-conformal measure μ. Moreover, if m assigns measure zero to all critical points, then $\mu = m$.

Clearly, the proposition guarantees, in particular, that for an arbitrary rational map R of the Riemann sphere we always have that there exists a $\exp(p-f)$-conformal measure supported on the associated Julia set, for some $p \in \mathbb{R}$.

Also, the above discussion motivates the following weakening of the notion of a conformal measure.

Definition 1. With the notation as above, a Borel probability measure m is called weakly $\exp(c-f)$-conformal, if

$$m(T(A)) = \int_A \exp(c-f)\,\mathrm{d}m$$

for all $A \in \mathscr{F}$ such that $T|_A$ is invertible and $A \cap \mathscr{S}(T) = \emptyset$.

The following proposition shows that these weakly conformal measures do in fact always exist.

Proposition 2. *([28]) With the notation as above, we always have that there exists a weakly* $\exp(p-f)$*–conformal Borel probability measure m, for some* $p \in \mathbb{R}$.

The following theorem addresses the question of how to find the transition parameter c, when constructing a conformal measure by means of the construction principle above. Obviously, the parameter c very much depends on the potential f as well as on the choice of the sequence $(E_n : n \in \mathbb{N})$. In most cases, the sets E_n can be chosen to be maximal separating sets, and then the parameter c is clearly equal to the pressure of ϕ. However, in general, it can be a problem to determine the value of c. The following theorem gives a positive answer for a large class of maps.

Theorem 4. *([28]) For each expansive map T, we have that there exists a weakly* $\exp(P(T,f)-f)$*-conformal measure m. If, additionally, T is an open map, then m is an* $\exp(P(T,f)-f)$*-conformal measure.*

Note that besides its fruitful applications to rational and meromorphic functions of the complex plane, the above construction principle has also been used successfully for maps of the interval (including circle maps) (see e.g. [16, 28, 43–46]). In particular, it has been employed to establish the existence of a 1-conformal measure for piecewise continuous transformations of the unit interval, which have neither periodic limit point nor wandering intervals, and which are irreducible at infinity (see [16]). Moreover, conformal measures for higher dimensional real maps appear in [17], and there they are obtained via the transfer operator method.

Currently, it is an active research area to further enlarge the class of transformations for which the existence of conformal measures can be established. This area includes the promising attempts to construct conformal measures on certain characteristic subsets of the limit set, such as on the radial limit set [26,49,85] or on certain other attractors [25]. Also, a related area of research aims to elaborate fractal geometry for systems for which weakly conformal measures exist (see e.g. [31]).

We end this section by giving two further examples of systems for which the theory of conformal measures has proven to be rather successful. The first of these is the case of expanding maps of the interval. Here, Hofbauer was one of the leading architects during the development of the general theory.

Theorem 5. *([45]) Let* $T : [0,1] \to [0,1]$ *be an expanding, piecewise monotone map of the interval which is piecewise Hölder differentiable. Let* $A \subset [0,1]$ *have the Darboux property and positive Hausdorff dimension h, and assume that the forward orbit of each element of A does not intersect the endpoints of the monotonicity intervals of T. Then we have that there exists a non-atomic h-conformal measure, which is unique as a t-conformal measure for* $t > 0$.

Also, for expansive $C^{1+\varepsilon}$-maps Gelfert and Rams obtained the following result.

Theorem 6. *([39]) Let* (X,T) *be an expansive, transitive* $C^{1+\varepsilon}$*-Markov system, whose limit set has Hausdorff dimension equal to h. Then there exists a h-conformal measure.*

In particular, we have that h is the least exponent for which a t-conformal measure exists, and h is also the smallest zero of the pressure function $P(T,-t\log|T'|)$.

Finally, let us mention that the above construction principle can obviously also be applied to iterated function systems and graph-directed Markov systems. For these dynamical systems, conformal measures are obtained by considering the inverse branches of the transformations coming with these systems. For further details we refer to [70].

5 Gibbs Measures for Bundle Maps

In this section, we give an outline of how to extend the concept of a Gibbs measure to bundles of maps over some topological (or measurable) space X (cf. Sect. 2). For this, let (X,T) be a dynamical system for which the map $T: X \to X$ factorizes over some additional dynamical system (Y,S) such that the fibres are non-trivial. Then there exists a map $\pi : X \to Y$ such that $\pi \circ T = S \circ \pi$. We will always assume that π is either continuous (if X is compact) or measurable. A system of this type is called a fibred system. Note that the set of fibred systems includes dynamical systems, which are skew products. For ease of exposition, let us mainly discuss the following two cases: (1) (Y,S) is itself a topological dynamical system and π is continuous; (2) $(Y,\mathscr{B},P,S)$ is a measurable dynamical system, with P being a probability measure on Y, S an invertible probability preserving transformation, and where π is measurable.

In the first case, one can define a family $(\mathscr{L}_\phi^{(y)} : y \in Y)$ of transfer operators, given on the space C_y of continuous functions on $\pi^{-1}(\{y\})$ (the image does not necessarily have to be a continuous function), by

$$\mathscr{L}_\phi^{(y)} f(x) = \sum_{\substack{T(z)=x \\ \pi(z)=y}} f(z)\mathrm{e}^{-\phi(z)}.$$

If the fibre maps $T_y = T|_{\pi^{-1}(y)}$ are uniformly open and expanding,[5] these operators act on the spaces of continuous functions on the fibres. In this situation, we have that an analogue of the Bowen–Ruelle–Perron–Frobenius Theorem holds. Note that it is not known whether this analogue can be obtained via some fixed point theorem. The currently known proof uses the method of invariant cones and Hilbert's projective metric (see [7, 22]). More precisely, a conic bundle $(K_y : y \in Y)$ over X is given as follows. For each $y \in Y$, let $K_y \subset C_y$ be the cone defined by

$$K_y = \{f \in C_y : \ f(x_1) \leq \rho(x_1,x_2) f(x_2);\ x_1,x_2 \in \pi^{-1}(\{y\});\ d(x_1,x_2) < a\},$$

where $\rho(x_1,x_2) = \exp(2\beta(d(x_1,x_2))^\gamma)$, and where β is chosen such that $\beta > \alpha\lambda^\gamma/2(1-\lambda^\gamma)$. Here, $0 < \gamma \leq 1$ denotes the Hölder exponent of the potential function ϕ. One then verifies that $T_y(K_y) \subset K_{S(y)}$ and that the projective diameter

[5]That is there exist $a > 0$ and $\lambda > 1$ such that for all $x,x' \in \pi^{-1}(\{y\})$ we have that $d(x,x') < a$ implies that $d(T(x),T(x')) \geq \lambda d(x,x')$.

of $K_{S(y)}$ is finite and does not depend on y. By using Birkhoff's Theorem [7], we then obtain that the fibre maps T_y are contractions. This method of employing Hilbert's projective metric in order to derive conformal measures is due to Ferrero and Schmidt, and we refer to [38] for further details. The following theorem states this so obtained analogue of the Bowen–Ruelle–Perron–Frobenius Theorem for bundle maps.

Theorem 7. *([22]) Assume that the fibre maps are uniformly expanding, open and (uniformly) exact.*[6] *For each Hölder continuous function $\phi : X \to \mathbb{R}$, we then have that there exists a unique family $\{\mu_y : y \in Y\}$ of probability measures μ_y on $\pi^{-1}(\{y\})$ and a unique measurable function $\alpha : Y \to \mathbb{R}_{>0}$ such that, for each $A \subset X$ measurable,*

$$\mu_{S(y)}(T(A)) = \alpha(y) \int_A \exp(\phi(x))\, \mathrm{d}\mu_y(x). \tag{9}$$

Moreover, the map $y \mapsto \mu_y$ is continuous with respect to the weak topology.

The family of measures obtained in this theorem represents a generalization of the concept "Gibbs measure", which also explains why such a family is called a Gibbs family. Note that the strong assumptions of the theorem are necessary in order to guarantee the continuity of the fibre measures. Moreover, note that, under some mild additional assumptions, the function α can be shown to be continuous (and in some cases, it can even be Hölder continuous) [22]. Since by changing the metric [20, 30], each expansive system can be made into an expanding system, one immediately verifies that the previous theorem can be extended such that it includes fibrewise uniformly expansive systems. The proof of this extension is given in [89].

Let us also mention that typical examples for these fibred systems are provided by Julia sets of skew products for polynomial maps in $\mathbb{C}^{\mathrm{d}}$. For these maps, it is shown in [23] that various outcomes of the usual thermodynamic formalism can be extended to the Gibbs families associated with these maps. This includes the existence of measures of maximal entropy for certain polynomial maps. Note that, alternatively, these measures can also be obtain via pluriharmonic functions.

For more general dynamical systems, the fibre measures do not have to be continuous. In fact, as observed by Bogenschütz and Gundlach, the Hilbert metric also turns out to be a useful tool for investigating the existence of Gibbs families for more general maps. One of the problems which one then usually first encounters is to locate a suitable subset of Y for which the relation in (9) is satisfied. It turns out that here a suitable framework is provided by the concept of a random dynamical system. More precisely, let us assume that the map S is invertible and that (Y, S) is equipped with a σ-algebra $\mathscr{B}$ and an S-preserving ergodic probability measure P. The following "random version" of the Bowen–Ruelle Theorem has been obtained in [10]. Note that in here we have that (9) holds P-almost everywhere. Also, note that in the special case in which S is invertible, we have that each of the operators $\mathscr{L}_\phi(y)$ is nothing else but a restriction of the transfer operator to fibres. Moreover, the theorem uses the concept of a random subshift of finite type. Such a

[6] That is for $\varepsilon > 0$ there exists some $n \geq 1$ such that $T^n(B(x, \varepsilon)) \supset \pi^{-1}(\{S^n(\pi(x))\})$.

subshift is defined by a bounded random function $l : Y \to \mathbb{N}$ and a random matrix $A(\cdot) = (a_{i,j}(\cdot))$ over Y with entries in $\{0,1\}$, such that the fibres are given by

$$\pi^{-1}(\{y\}) = \{(x_n)_{n\geq 0} : x_k \leq l(S^k(y)) \text{ and } a_{x_k,x_{k+1}}(S^k(y)) = 1, \forall k \in \mathbb{N}\}.$$

Theorem 8. *([10]) Let (X,T) be a random subshift of finite type for which $\|\log \mathscr{L}_\phi\|_\infty \in L_1(P)$, $A(\cdot)$ is uniformly aperiodic and $\phi_{|\pi^{-1}(\{y\})}$ is uniformly Hölder continuous, for each $y \in Y$. Then there exist a random variable λ with $\log\lambda \in L_1(P)$, a positive random function g with $\|\log g\|_\infty \in L_1(P)$, and a family of probability measures μ_y such that the following hold, for all $y \in Y$.*

1. $\mathscr{L}_\phi^{(y)*}\mu_{S(y)} = \lambda(y)\mu_y$;
2. $\mathscr{L}_\phi^{(y)}g = \lambda(y)g$;
3. $\int g\,d\mu_y = 1$;
4. *The system has exponential decay of correlation for Hölder continuous functions.*

Further results in this direction can be found in [41, 54, 56]. Note that none of these results makes use of the Patterson construction, but for random countable Markov shifts the construction principle of Sect. 4 has been successfully applied, and this will be discussed in the following final section of this survey.

6 Gibbs Measure on Non-Compact Spaces

Without the assumption of X being compact, the weak convergence in the Patterson construction needs some additional care in order to overcome the lack of relative compactness of the associated space of probability measures.

One of the the simplest examples, in which the quality of the whole space X does not play any role, is the following. Suppose that there exists a compact subset of X to which the forward orbit of a generic point under a given transformation $T : X \to X$ returns infinitely often. More specifically, let us assume that the map admits a countable Markov partition, and that there exists some compact atom A of this partition such that $A \subset \bigcup_{n\in\mathbb{N}} T^{-n}(A)$. We then consider the induced transformation $T_A : A \to A$, given for each $x \in A$ by

$$T_A(x) = T^{n(x)}(x),$$

where $n(x) = \inf\{k \in \mathbb{N} : T^k(x) \in A\}$. Likewise, for a given potential function ϕ on X, we define the induced potential function ϕ_A by $\phi_A(x) = \phi(x) + \cdots + \phi(T^{n(x)-1})$. In order to see in which way Gibbs measures for T_A give rise to Gibbs measures for T, let μ be a given Gibbs measure for the transformation T_A and the induced potential function ϕ_A. Then define a measure m by

$$\int f\,dm = \int \sum_{k=0}^{n(x)-1} f(T^k(x))\,d\mu(x).$$

One immediately verifies that m is a σ-finite Gibbs measure for the potential function ϕ (see [35]).

If in the absence of compactness one still wants to employ any of the general construction principles for conformal measures, discussed in Sects. 1, 2 and 4, one needs to use the concept of tightness of measures. For instance, for S-uniformal maps of the interval, tightness has been used in [25] to show that there exists a conformal measure concentrated on a dense symbolic subset of the associated limit set. Also, Urbański's KV-method, discussed in Sect. 3, appears to be very promising here, since it gives rise to conformal measures which are concentrated on non-compact subsets of X (although, strictly speaking, the construction is carried out for a compact space, where limits do of course exist). Moreover, there is ongoing research on the existence of Gibbs measures for countable topological Markov chains. In all of the results obtained thus far, tightness plays a key role. For this non-compact situation, there are various examples in the literature for which the existence of Gibbs measures is discussed. However, the first general result was derived in [69].

In the following theorem, we consider a topologically mixing Markov chain X, given by a state space Λ, a map $T : X \to X$, and a transition matrix $\Sigma = (\sigma_{ij})_{i,j\in\Lambda}$. Recall that (X,T) is said to have the big images and big pre-images property, abbreviated as (BIP), if there exist a finite set $\Lambda_0 \subset \Lambda$ of states such that for each $\ell \in \Lambda$ there exist $a,b \in \Lambda_0$ for which

$$\sigma_{a\ell}\sigma_{\ell b} = 1.$$

Note that this property is equivalent to what Mauldin and Urbański call "finitely primitive" [69]. Also, mark that the property (BIP) is more restrictive than the big image property of [1], which was there used to obtain absolutely continuous invariant measures.

The following theorem is due to Sarig. The proof of the sufficiency part of this theorem can also be found in [69].

Theorem 9. *([93]) Let (X,T) be a topologically mixing infinite topological Markov chain, and let $\phi \in C(X)$ have summable variation.*[7] *In this situation, we have that the following two statements are equivalent.*

(1) There exists an invariant Gibbs measure for ϕ.
(2) (X,T) has the property (BIP) and the Gurevic pressure $P_G(\phi)$ of ϕ is finite, that is, for some $\ell \in \Lambda$ we have

$$P_G(\phi) := \lim_{n\to\infty} \frac{1}{n} \log \sum_{T^n(x)=x} \mathbb{I}_\ell(x) \exp(\phi(x) + \cdots + \phi(T^{n-1}(x))) < \infty.$$

[7] That is $\sum_{n=1}^{\infty} V_n(\phi) < \infty$, where $V_n(\phi)$ denotes the maximal variation of ϕ over cylinders of length n.

Recently, this result has been partially extended by Stadlbauer in [95] to the case of random countable topological Markov chains. Moreover, for the situation of the theorem with the additional assumptions that a certain random (BIP) holds and that $V_1^y(\phi) < \infty$ for all $y \in Y$, it was shown in [24] that there exists an invariant measure.

Theorem 10. *([95]) Let (X,T) be a random topological Markov shift, and let ϕ be a locally fibre Hölder continuous function of index two*[8] *with finite Gurevic pressure. Also, assume that the functions $y \mapsto \sum_{k=1}^{\infty} \kappa(S^{-k}(y))r^k$, $y \mapsto \log\sup\{\mathscr{L}_\phi^{(y)} 1(x) : x \in X_{S(y)}\}$ and $y \mapsto \log\inf\{\mathscr{L}_\phi^{(y)} 1(x) : x \in X_{S(y)}\}$ are P-integrable, and let (X,T,ϕ) be of divergence type.*[9] *Then there exists a measurable function $\alpha : Y \to \mathbb{R}_+$ and a Gibbs family $\{\mu_y : y \in Y\}$ for the potential $P_G(\phi) - \phi$ such that, for all $y \in Y$ and all x in the fibre over y,*

$$\frac{\mathrm{d}\mu_{S(y)}}{\mathrm{d}\mu_y}(x) = \alpha(y)\exp(P_G(\phi) - \phi(x)).$$

In the work of Sarig in [91] and [92], which is closely related to the thermodynamic formalism, tightness is used to construct Gibbs measures via transfer operator techniques. Contrary to this approach, the results in [95] combine the Patterson measure construction with Crauel's Prohorov Theorem on tightness ([21]). To be more precise, let (X,T) be a random Markov chain over the base $(Y,\mathscr{B},R,P)$, where P is some fixed probability measure. Then Crauel's theorem states that a sequence of bundle probabilities $\{\mu_y^{(n)} : y \in Y\}$ is relatively compact with respect to the narrow topology if and only if $\{\mu_y^{(n)} : y \in Y\}$ is tight.[10] Here, convergence of the discrete fibre measures $\{\mu_y^{(n)} : y \in Y\}$ towards $\{\mu_y : y \in Y\}$ with respect to the narrow topology means that for all functions f, which are continuous and bounded as functions on fibres, we have that

$$\int\int f\mathrm{d}\mu_y^{(n)}\,\mathrm{d}P(y) = \int\int f\mathrm{d}\mu_y\,\mathrm{d}P(y).$$

The construction in Sect. 4 can then be carried out fibrewise, showing that there exist weak accumulation points with respect to the narrow topology (see also [24]). It is worth mentioning that, beyond this result, in this situation no further results on the existence of conformal measures seem to be known. Also, thorough investigations of the fractal geometry of such systems are currently still missing.

---

[8] $V_n^y(\phi) \le \kappa(y)r^n$ for $n \ge 2$ and $\int \log\kappa\,\mathrm{d}P < \infty$.

[9] For a fixed measurable family $\xi_y \in \pi^{-1}(y)$, we have that $\sum_{n:S^n(y)\in Y'} s^n(\mathscr{L}_\phi^{(y)})^n(1)(\xi_{S^n(y)})$ converges for $s < 1$ and diverges for $s = 1$, where Y' is some set of positive measure.

[10] That is for all $\varepsilon > 0$ there exists a measurable set $K \subset X$ such that $K \cap \pi^{-1}(\{y\})$ is compact, for all $y \in Y$, and $\inf_n \int \mu_y^{(n)}(K)\,\mathrm{d}P(y) > 1 - \varepsilon$.

References

1. Aaronson, J.; Denker, M.; Urbański, M.: Ergodic theory for Markov fibred systems and parabolic rational maps. Trans. Amer. Math. Soc. **337** (1993), 495–548.
2. Aaronson, J.; Denker, M.: The Poincaré series of $\mathbb{C}\setminus\mathbb{Z}$. Ergodic Theory Dynam. Systems **19**, No.1 (1999), 1–20.
3. Aaronson, J.; Denker, M.: Local limit theorems for partial sums of stationary sequences generated by Gibbs-Markov maps. Stochastics and Dynamics **1** (2001), 193–237.
4. Adachi, T.; Sunada, T.: Homology of closed geodesics in a negatively curved manifold. J. Differ. Geom. **26** (1987), 81–99.
5. Agol, I.: Tameness of hyperbolic 3-manifolds. Preprint (2004); arXiv:math.GT/0405568.
6. Avila, A.; Lyubich, M.: Hausdorff dimension and conformal measures of Feigenbaum Julia sets. J. Amer. Math. Soc. **21** (2008), 305–363.
7. Birkhoff, G.: *Lattice theory*. Third edition. AMS Colloquium Publ., Vol. XXV AMS, Providence, R.I. 1967
8. Bishop, C. J.; Jones, P. W.: Hausdorff dimension and Kleinian groups. Acta Math. **56** (1997), 1–39.
9. Blokh, A. M.; Mayer, J. C.; Oversteegen, L. G.: Recurrent critical points and typical limit sets for conformal measures. Topology Appl. **108** (2000), 233–244.
10. Bogenschütz, T.; Gundlach, M.: Ruelle's transfer operator for random subshifts of finite type. Ergodic Theory Dynam. Systems **15** (1995), 413–447.
11. Bonfert-Taylor, P.; Matsuzaki, K.; Taylor, E. C.: Large and small covers of hyperbolic manifolds. J. Geom. Anal. 21 (2011) DOI: 10.1007/s12220-010-9204-6.
12. Bowen, R.: *Equilibrium states and the ergodic theory of Anosov diffeomorphisms*. Lect. Notes in Math. **470**, Springer-Verlag, 1975.
13. Bowen, R.: Dimension of quasi circles. Publ. IHES **50** (1980), 11–25.
14. Brooks, R.: The bottom of the spectrum of a Riemannian cover. J. Reine Angew. Math. **357** (1985), 101–114.
15. Bruin, H.; Todd, M.: Markov extensions and lifting measures for complex polynomials. Ergodic Theory Dynam. Systems **27** (2007), 743–768.
16. Buzzi, J.; Hubert, P.: Piecewise monotone maps without periodic points: rigidity, measures and complexity. Ergodic Theory Dynam. Systems **24** (2004), 383–405.
17. Buzzi, J.; Paccaut, F.; Schmitt, B.: Conformal measures for multidimensional piecewise invertible maps. Ergodic Theory Dynam. Systems **21** (2001), 1035–1049.
18. Calegari, D.; Gabai, D.: Shrinkwrapping and the taming of hyperbolic manifolds. J. Amer. Math. Soc. **19** (2006), 385–446.
19. Cornfeld, I. P.; Fomin, S. V.; Sinai, Ya. G.: *Ergodic theory*. Springer Verlag New York, Heidelberg, Berlin 1982.
20. Coven, E. M.; Reddy, W. L.: Positively expansive maps of compact manifolds. Global theory of dynamical systems (Proc. Internat. Conf., Northwestern Univ., Evanston, Ill., 1979), pp. 96–110, Lecture Notes in Math., 819, Springer, Berlin, 1980.
21. Crauel, H.: *Random probability measures on Polish spaces*. Stochastics Monographs **11**, Taylor & Francis, London, 2002.
22. Denker, M.; Gordin, M.: Gibbs measures for fibred systems. Adv. Math. **148** (1999), 161–192.
23. Denker, M.; Gordin, M.; Heinemann, S.-M.: On the relative variational principle for fibre expanding maps. Ergodic Theory Dynam. Systems **22** (2002), 757–782.
24. Denker, M.; Kifer, Y.; Stadlbauer, M.: Thermodynamic formalism for random countable Markov shifts. Discrete Contin. Dyn. Syst. **22** (2008), 131–164.
25. Denker, M.; Nitecki, Z.; Urbański, M.: Conformal measures and S-unimodal maps. Dynamical systems and applications, 169–212, World Sci. Ser. Appl. Anal., 4, World Sci. Publ., River Edge, NJ, 1995.

26. Denker, M.; Mauldin, R. D.; Nitecki, Z.; Urbański, M.: Conformal measures for rational functions revisited. Dedicated to the memory of Wieslaw Szlenk. Fund. Math. **157** (1998), 161–173.
27. Denker, M.; Przytycki, F.; Urbański, M.: On the transfer operator for rational functions on the Riemann sphere. Ergodic Theory Dynam. Systems **16** (1996), 255–266.
28. Denker, M.; Urbański, M.: On the existence of conformal measures. Trans. Amer. Math. Soc. **328** (1991), 563–587.
29. Denker, M.; Urbański, M.: Hausdorff measures on Julia sets of subexpanding rational maps. Israel J. Math. **76** (1991), 193–214.
30. Denker, M.; Urbański, M.: Absolutely continuous invariant measures for expansive rational maps with rationally indifferent periodic points. Forum Math. **3** (1991), 561–579.
31. Denker, M.; Urbański, M.: On Sullivan's conformal measures for rational maps of the Riemann sphere. Nonlinearity **4** (1991), 365–384.
32. Denker, M.; Urbański, M.: Hausdorff and conformal measures on Julia sets with a rationally indifferent periodic point. J. London Math. Soc. (2) **43** (1991), 107–118.
33. Denker, M.; Urbański, M.: Ergodic theory of equilibrium states for rational maps. Nonlinearity **4** (1991), 103–134.
34. Denker, M.; Urbański, M.: Geometric measures for parabolic rational maps. Ergodic Theory Dynam. Systems **12** (1992), 53–66.
35. Denker, M.; Yuri, M.: A note on the construction of nonsingular Gibbs measures. Colloquium Mathematicum **84/85** (2000), 377–383.
36. Douady, A.; Sentenac, P.; Zinsmeister, M.: Implosion parabolique et dimension de Hausdorff. C.R. Acad. Sci. Paris Sér. I. **325** (1997), 765–772.
37. Falk, K. H.; Stratmann, B. O.: Remarks on Hausdorff dimensions for transient limit sets of Kleinian groups. Tohoku Math. Jour. **56** (4) (2004), 571–582.
38. Ferrero, P.; Schmitt, B.: Ruelle's Perron-Frobenius theorem and projective metrics. In *Random Fields Esztergom Hung., 1979*, Coll. Math. Soc. Janos Bolyai, **27**, 333–336. North-Holland, Amsterdam, 1979.
39. Gelfert, K.; Rams, M.: Geometry of limit sets for expansive Markov systems. Trans. Amer. Math. Soc. **361** (2009), 2001–2020.
40. Graczyk, J.; Smirnov, S.: Non-uniform hyperbolicity in complex dynamics. Invent. Math. **175** (2008), 335–415.
41. Gundlach, M.; Kifer, Y.: Expansiveness, specification, and equilibrium states for random bundle transformations. Discrete and Continuous Dynam. Syst. **6** (2000), 89–120.
42. Hill, R.; Velani, S.:The Jarník-Besicovitch theorem for geometrically finite Kleinian groups. Proc. London Math. Soc. (3) **77** (1998), 524 - 550.
43. Hofbauer, F.: Hausdorff and conformal measures for expanding piecewise monotonic maps of the interval. Studia Math. **103** (1992), 191–206.
44. Hofbauer, F.: Hausdorff and conformal measures for expanding piecewise monotonic maps of the interval. II. Studia Math. **106** (1993), 213–231.
45. Hofbauer, F. Hausdorff and conformal measures for weakly expanding maps of the interval. Differential Equations Dynam. Systems **2** (1994), 121–136.
46. Hofbauer, F.: The Rényi dimension of a conformal measure for a piecewise monotonic map of the interval. Acta Math. Hungar. **107** (2005), 1–16.
47. Hofbauer, F.; Keller, G.: Ergodic properties of invariant measures for piecewise monotonic transformations. Math. Z. **180** (1982), 119–140.
48. Huang, Z.; Jiang, Y.; Wang, Y.: On conformal measures for infinitely renormalizable quadratic polynomials. Sci. China Ser. A **48** (2005), 1411–1420.
49. Huang, Z.; Wang, Y. On the existence of conformal measures supported on conical points. Proc. Japan Acad. Ser. A Math. Sci. **79** (2003), 154–157.
50. Ionesu-Tulcea, C.; Marinescu, G.: Théorie ergodique pour des classes d'opérations non complètment continues. Ann. of Math. **52** (1950), 140–147.
51. Kaimanovich, V. A.; Lyubich, M.: *Conformal and harmonic measures on laminations associated with rational maps.* Mem. Amer. Math. Soc. **173** (2005) 119 pp.

52. Keane, M.: Strongly mixing g-measures. Invent. Math. **16** (1972), 309–324.
53. Kesseböhmer, M.; Stratmann, B. O.: A note on the algebraic growth rate of Poincaré series for Kleinian groups. Article dedicated to S.J. Patterson's 60th birthday (2011), in this volume.
54. K. Khanin, Y. Kifer: Thermodynamics formalism for random transformations and statistical mechanics. Amer. Math. Soc. Translations, Series 2 (1995), 107–140.
55. Khintchine, A. Y.: *Continued fractions*. Univ. of Chicago Press, Chicago and London, 1964.
56. Kifer, Y.: Equilibrium states for random expanding transformations. Random Comput. Dynam. **1** (1992/93), 1–31.
57. Kotus, J.: Conformal measures for non-entire functions with critical values eventually mapped onto infinity. Analysis (Munich) **25** (2005), 333–350.
58. Kotus, J.: Probabilistic invariant measures for non-entire functions with asymptotic values mapped onto ∞. Illinois J. Math. **49** (2005), 1203–1220.
59. Kotus, J.: Elliptic functions with critical points eventually mapped onto infinity. Monatsh. Math. **149** (2006), 103–117.
60. Kotus, J.; Urbański, M.: Conformal, geometric and invariant measures for transcendental expanding functions. Math. Ann. **324** (2002), 619–656.
61. Kotus, J.; Urbański, M.: Geometry and ergodic theory of non-recurrent elliptic functions. J. Anal. Math. **93** (2004), 35–102.
62. Kotus, J.; Urbański, M.: Geometry and dynamics of some meromorphic functions. Math. Nachr. **279** (2006), 1565–1584.
63. Kotus, J.; Urbański, M.: Fractal measures and ergodic theory for transcendental meromorphic functions. London Math. Soc. Lecture Note Series **348** (2008), 251–316.
64. Lalley, S.: Closed geodesics in homology classes on surfaces of variable negative curvature. Duke Math. J. **58** (1989), 795–821.
65. Lyons, T. J.; McKean, H. P.: Winding of the plane Brownian motion. Adv. Math. **51** (1984), 212–225.
66. Lyubich, M.: Entropy properties of rational endomorphismsof the Riemann sphere. Ergodic Theory and Dynam. Syst. **3** (1983), 351–386.
67. Lyubich, M.: Dynamics of quadratic polynomials. I, II. Acta Math. **178** (1997), 185–247, 247–297.
68. Manning, A.; McCluskey, H.: Hausdorff dimension for horseshoes. Ergodic Theory Dynam. Systems **5** (1985), 71–88.
69. Mauldin, R. D.; Urbański, M.: Gibbs states on the symbolic space over an innite alphabet. Israel J. Maths. 125 (2001), 93–130.
70. Mauldin, R. D.; Urbański, M.: *Graph directed Markov systems. Geometry and dynamics of limit sets.* Cambridge Tracts in Mathematics, **148**. Cambridge University Press, Cambridge, 2003.
71. Mayer, V.: Comparing measures and invariant line fields. Ergodic Theory Dynam. Systems **22** (2002), 555–570.
72. McKean, H. P.; Sullivan, D.: Brownian motion and harmonic functions on the class surface of the thrice punctured torus. Adv. Math. **51** (1984), 203–211.
73. Misiurewicz, M.: A short proof of the variational principle for a $\mathbb{Z}_+^n$ action on a compact space. Bull. Acad. Polon. Sci. Sér. Sci. Math. Astronom. Phys. **24** (1976), 1069–1075.
74. Parry, W.: Intrinsic Markov chains. Trans. Amer. Math. Soc. **112** (1964), 55–66.
75. Patterson, S. J.: The limit set of a Fuchsian group. Acta Math. **136** (1976), 241–273.
76. Patterson, S. J.: Diophantine approximation in Fuchsian groups. Phil. Trans. Roy. Soc. Lond. **282** (1976), 527–563.
77. Patterson, S. J.: Further remarks on the exponent of convergence of Poincaré series. Tôhoku Math. Journ. **35** (1983), 357–373.
78. Patterson, S. J.: Lectures on measures on limit sets of Kleinian groups. In *Analytical and geometric aspects of hyperbolic space*, (eds. D. B. A. Epstein); Cambridge University Press, 1987.
79. Phillips, R.; Sarnak, P.: Geodesics in homology classes. Duke Math. J. **55** (1987), 287–297.

80. Pollicott, M.; Sharp, R.: Orbit counting for some discrete groups acting on simply connected manifolds with negative curvature. Invent. Math. **117** (1994), 275–302.
81. Prado, E. A. Ergodicity of conformal measures for unimodal polynomials. Conform. Geom. Dyn. **2** (1998), 29–44.
82. Prado, E. A.: Conformal measures and Hausdorff dimension for infinitely renormalizable quadratic polynomials. Bol. Soc. Brasil. Mat. (N.S.) **30** (1999), 31–52.
83. Przytycki, F.: Lyapunov characteristic exponents are nonnegative. Proc. Amer. Math. Soc. **119** (1993), 309–317.
84. Przytycki, F.: Iterations of holomorphic Collet-Eckmann maps: conformal and invariant measures. Appendix: on non-renormalizable quadratic polynomials. Trans. Amer. Math. Soc. **350** (1998), 717–742.
85. Przytycki, F.: Conical limit set and Poincaré exponent for iterations of rational functions. Trans. Amer. Math. Soc. **351** (1999), 2081–2099.
86. Rees, M.: Checking ergodicity of some geodesic flows with infinite Gibbs measure. Ergodic Theory Dynam. Systems **1** (1981), 107–133.
87. Rees, M.: Divergence type of some subgroups of finitely generated Fuchsian groups. Ergodic Theory Dynam. Systems **1** (1981), 209–221.
88. Rokhlin, V. A.: On the fundamental ideas of measure theory. Mat. Sbornik **25** (1949), 107-150. Transl.: Amer. Math. Soc. Translations **71** (1952), 1–54.
89. Roy, M.; Urbański, M.: Conformal families of measures for fibred systems. Monatsh. Math. **140** (2003), 135–145.
90. Rudolph, D. J.: Ergodic behaviour of Sullivan's geometric mesure on a geometrically finite hyperbolic manifold. Ergodic Theory Dynam. Systems **2** (1982), 491–512.
91. Sarig, O.: Thermodynamic formalism for countable Markov shifts. Ergodic Theory Dynam. Systems **19** (1999), 1565–1593.
92. Sarig, O.: Thermodynamic formalism for null recurrent potentials. Israel J. Math. **121** (2001), 285–311.
93. Sarig, O.: Existence of Gibbs measures for countable Markov shifts. Proc. Amer. Math. Soc. **131** (2003), 1751–1758.
94. Skorulski, B.: The existence of conformal measures for some transcendental meromorphic functions. Complex dynamics, 169–201, Contemp. Math. **396**, Amer. Math. Soc., Providence, RI, 2006.
95. Stadlbauer, M.: On random topological Markov chains with big images and preimages. Stochastics & Dynamics **10** (2010), 77–95.
96. Stratmann, B. O.: Diophantine approximation in Kleinian groups. Math. Proc. Camb. Phil. Soc. **116** (1994), 57–78.
97. Stratmann, B. O.: A note on counting cuspidal excursions. Annal. Acad. Sci. Fenn. **20** (1995), 359–372.
98. Stratmann, B. O.: Fractal dimensions for Jarník limit sets; the semi-classical approach. Ark. för Mat. **33** (1995), 385–403.
99. Stratmann, B. O.: The Hausdorff dimension of bounded geodesics on geometrically finite manifolds. Ergodic Theory Dynam. Systems **17** (1997), 227–246.
100. Stratmann, B. O.: A remark on Myrberg initial data for Kleinian groups. Geometriae Dedicata **65** (1997), 257–266.
101. Stratmann, B. O.: Multiple fractal aspects of conformal measures; a survey. In *Workshop on Fractals and Dynamics*, (eds. M. Denker, S.-M. Heinemann, B. O. Stratmann); Math. Gottingensis **05** (1997), 65–71.
102. Stratmann, B. O.: Weak singularity spectra of the Patterson measure for geometrically finite Kleinian groups with parabolic elements. Michigan Math. Jour. **46** (1999), 573–587.
103. Stratmann, B. O.: The exponent of convergence of Kleinian groups; on a theorem of Bishop and Jones. Progress in Probability **57** (2004), 93–107.
104. Stratmann, B. O.: Fractal geometry on hyperbolic manifolds. In *Non-Euclidean Geometries, Janos Bolyai Memorial Volume*, (eds. A. Prekopa and E. Molnar); Springer Verlag, (2006), 227–249.

105. Stratmann, B. O.; Urbański, M.: The box-counting dimension for geometrically finite Kleinian groups. Fund. Matem. **149** (1996), 83–93.
106. Stratmann, B. O.; Urbański, M.: The geometry of conformal measures for parabolic rational maps. Math. Proc. Cambridge Philos. Soc. **128** (2000), 141–156.
107. Stratmann, B. O.; Urbański, M.: Jarnik and Julia: a Diophantine analysis for parabolic rational maps. Math. Scand. **91** (2002), 27–54.
108. Stratmann, B. O.; Urbański, M.: Metrical Diophantine analysis for tame parabolic iterated function systems. Pacific J. Math. **216** (2004), 361–392.
109. Stratmann, B. O.; Urbański, M.: Pseudo-Markov systems and infinitely generated Schottky groups. Amer. J. Math. **129** (4) (2007), 1019–1063.
110. Stratmann, B. O.; Velani, S.: The Patterson measure for geometrically finite groups with parabolic elements, new and old. Proc. Lond. Math. Soc. (3) **71** (1995), 197–220.
111. Sullivan, D.: The density at infinity of a discrete group of hyperbolic motions. IHES Publ. Math.**50** (1979), 171–202.
112. Sullivan, D.: Conformal dynamical systems. In *Geometric Dynamics*; Lect. Notes in Math. **1007** (1983), 725–752.
113. Sullivan, D.: Entropy, Hausdorff measures old and new, and limit sets of geometrically finite Kleinian groups. Acta Math. **153** (1984), 259–277.
114. Sumi, H.: On dynamics of hyperbolic rational semigroups and Hausdorff dimension of Julia sets. In *Complex dynamical systems and related areas*; (Japanese) (Kyoto, 1996). Surikaisekikenkyusho Kokyuroku **988** (1997), 98–112.
115. Sumi, H.: Dynamics of sub-hyperbolic and semi-hyperbolic rational semigroups and conformal measures of rational semigroups. In *Problems on complex dynamical systems*; (Japanese) (Kyoto, 1997). Surikaisekikenkyusho Kokyuroku **1042** (1998), 68–77.
116. Sumi, H.: On Hausdorff dimension of Julia sets of hyperbolic rational semigroups. Kodai Math. Jour. **21** (1998), 10–28.
117. Urbański, M.: Measures and dimensions in conformal dynamics. Bull. Amer. Math. Soc. (N.S.) **40** (2003), 281–321
118. Urbański, M.: Geometry and ergodic theory of conformal non-recurrent dynamics. Ergodic Theory Dynam. Systems **17** (1997), 1449–1476.
119. Urbański, M.; Zdunik, A.: The finer geometry and dynamics of the hyperbolic exponential family. Michigan Math. J. **51** (2003), 227–250.
120. Urbański, M.; Zdunik, A.: The parabolic map $f_{1/e}(z) = \frac{1}{e}e^z$. Indag. Math. (N.S.) **15** (2004), 419–433.
121. Urbański, M.; Zdunik, A.: Geometry and ergodic theory of non-hyperbolic exponential maps. Trans. Amer. Math. Soc. **359** (2007), 3973–3997.
122. Walters, P.: A variational principle for the pressure of continuous transformations. Amer. J. Math. **97** (1975), 937–971.
123. Yuri, M.: On the construction of conformal measures for piecewise C_0-invertible systems. In *Studies on complex dynamics and related topics*; (Japanese) (Kyoto, 2000). Surikaisekikenkyusho Kokyuroku **1220** (2001), 141–143.
124. Yuri, M.: Thermodynamic Formalism for countable to on Markov systems. Trans. Amer. Math. Soc. **355** (2003), 2949–2971.
125. Zinsmeister, M.: *Formalisme thermodynamique et systèmes dynamiques holomorphes.* (French) [Thermodynamical formalism and holomorphic dynamical systems] Panoramas et Synthéses [Panoramas and Syntheses], 4. Société Mathématique de France, Paris, 1996.

Moments for L-Functions for $GL_r \times GL_{r-1}$

Adrian Diaconu, Paul Garrett, and Dorian Goldfeld

Abstract We establish a spectral identity for moments of Rankin–Selberg L-functions on $GL_r \times GL_{r-1}$ over arbitrary number fields, generalizing our previous results for $r = 2$.

1 Introduction

Let k be an algebraic number field with adele ring $\mathbb{A}$. Fix an integer $r \geq 2$ and consider the general linear groups $GL_r(k)$, $GL_r(\mathbb{A})$ of $r \times r$ invertible matrices with entries in k, $\mathbb{A}$, respectively. Let Z^+ be the positive real scalar matrices in GL_r. Let π be an irreducible cuspidal automorphic representation in $L^2(Z^+ GL_r(k) GL_r(\mathbb{A}))$. Let π' run over irreducible unitary cuspidal representations in $L^2(Z^+ GL_{r-1}(k) \backslash GL_{r-1}(\mathbb{A}))$, where now Z^+ is the positive real scalar matrices in GL_{r-1}. For brevity, denote a sum over such π' by $\sum_{\pi'}$. For complex s, let $L(s, \pi \times \pi')$ denote the Rankin–Selberg convolution L–function. A *second integral moment* over the spectral family GL_{r-1} is described roughly as follows. For each irreducible cuspidal automorphic π' of GL_{r-1}, assign a constant $c(\pi') \geq 0$. Letting π_∞ be the archimedean component of π and π'_∞ the archimedean factor of each π', let $M(s, \pi_\infty, \pi'_\infty)$ be a function of complex s, whose possibilities will be described in more detail later. The corresponding second moment of π is

$$\sum_{\pi'} c(\pi') \int_{\mathrm{Re}\, s = \frac{1}{2}} |L(s, \pi \times \pi')|^2 \cdot M(s, \pi_\infty, \pi'_\infty)\, \mathrm{d}s.$$

A. Diaconu (✉) • P. Garrett
University of Minnesota, 127 Vincent Hall, Church St. SE, Minneapolis, MN 55455, USA
e-mail: cad@math.umn.edu; garre001@umn.edu

D. Goldfeld
422 Mathematics Building, Columbia University, New York, NY 10027, USA
e-mail: goldfeld@columbia.edu

V. Blomer and P. Mihăilescu (eds.), *Contributions in Analytic and Algebraic Number Theory*, Springer Proceedings in Mathematics 9, DOI 10.1007/978-1-4614-1219-9_8,

In fact, there are further correction terms corresponding to non-cuspidal parts of the spectral decomposition of $L^2(Z^+GL_{r-1}(k)\backslash GL_{r-1}(\mathbb{A}))$, but the cuspidal part presumably dominates.

The theory of second integral moments on $GL_2 \times GL_1$ has a long history, although the early papers treated mainly the case that the groundfield k is $\mathbb{Q}$. For example, see [2, 3, 10, 11, 14, 16, 21–24, 27, 35, 42–44]. Second integral moments of level-one holomorphic elliptic modular forms were first treated in [21, 22], the latter using an idea that is a precursor of part of the present approach. The study of second integral moments for $GL_2 \times GL_1$ with arbitrary level, groundfield, and infinity-type is completely worked out in [12].

The main aim of this paper is to establish an identity relating the second integral moment, described above, to the integral of a certain Poincaré series $\mathfrak{P}$ against the absolute value squared $|f|^2$ of a distinguished cuspform $f \in \pi$. Acknowledging that the spectral decomposition of $L^2(Z_{\mathbb{A}}GL_r(k)\backslash GL_r(\mathbb{A}))$ also has a non-cuspidal part generated by Eisenstein series and their residues, the identity we obtain takes the form

$$\int\limits_{Z_{\mathbb{A}}GL_r(k)\backslash GL_r(\mathbb{A})} \mathfrak{P}(g,\varphi_\infty)\cdot|f(g)|^2\,\mathrm{d}g$$

$$= \sum_{\pi'} |\rho(\pi')|^2 \int_{\mathrm{Re}\,s=\frac{1}{2}} |L(s,\pi\times\pi')|^2 \cdot M(s,\pi_\infty,\pi'_\infty,\varphi_\infty)\,\mathrm{d}s \;+\; \text{(non-cuspidal part)}.$$

Here, $M(s,\pi_\infty,\pi'_\infty,\varphi_\infty)$ is a weighting function depending on the complex parameter s, on the archimedean components π_∞ and π'_∞, and on archimedean data φ_∞ defining the Poincaré series. The global constants $\rho(\pi')$ are analogues of the leading Fourier coefficients of GL_2 cuspforms. The spectral expansion of the Poincaré series $\mathfrak{P}$ relates the second integral moment to automorphic spectral data. Remarkably, the cuspidal data appearing in the spectral expansion of $\mathfrak{P}$ comes only from GL_2.

These new identities have some similarities to the Kuznetsov trace formula [1, 19, 49, 50], in that they are derived via the spectral resolution of a Poincaré series, but they are clearly of a different nature. We have in mind application not only to cuspforms, but also to truncated Eisenstein series or wave packets of Eisenstein series, thus applying harmonic analysis on GL_r to L-functions attached to GL_1, touching upon higher integral moments of the zeta function $\zeta_k(s)$ of the ground field k.

In connection to this work, we mention the recent mean-value result of [51],

$$\int_{-T^{1-\varepsilon}}^{T^{1-\varepsilon}} \sum_{T<t_j\le 2T} \left|L(\tfrac{1}{2}+it,u_j\times\varphi)\right|^2 \mathrm{d}t \ll T^{3+\varepsilon} \quad \text{for } \varepsilon>0,$$

where φ is on GL_3, and where u_j on GL_2 has spectral data t_j, as usual. From this, the t-aspect convexity bound can be recovered. Also, [37] obtains a t-aspect subconvexity bound for standard L-functions for $GL_3(\mathbb{Q})$ for Gelbart–Jacquet lifts.

For context, we review the [15] treatment of spherical waveforms f for $GL_2(\mathbb{Q})$. In that case, the sum of moments is a single term

$$\int_{Z_{\mathbb{A}} GL_2(\mathbb{Q})\backslash GL_2(\mathbb{A})} \mathfrak{P}(g,z,w)\,|f(g)|^2\,dg$$
$$= \frac{1}{2\pi i}\int_{\mathrm{Re}\,(s)=\frac{1}{2}} L(z+s,f)\cdot \overline{L}(s,f)\cdot \Gamma(s,z,w,f_\infty)\,ds,$$

where $\Gamma(s,z,w,f_\infty)$ is a ratio of products of gammas, with arguments depending upon the archimedean data of f. Here, the Poincaré series $\mathfrak{P}(g) = \mathfrak{P}(g,z,w)$ is specified completely by complex parameters z,w, and has a *spectral expansion*

$$\mathfrak{P}(g,z,w) = \frac{\pi^{\frac{1-w}{2}}\Gamma\left(\frac{w-1}{2}\right)}{\pi^{-\frac{w}{2}}\Gamma\left(\frac{w}{2}\right)}\cdot E_{1+z}(g) + \tfrac{1}{2}\sum_{F\text{ on }GL_2} \rho_{\overline{F}}\cdot L\left(\tfrac{1}{2}+z,\overline{F}\right)\cdot \mathcal{G}\left(\tfrac{1}{2}-it_F,z,w\right)\cdot F(g)$$
$$+\frac{1}{4\pi i}\int_{\mathrm{Re}\,(s)=\frac{1}{2}} \frac{\zeta(z+s)\,\zeta(z+1-s)}{\xi(2-2s)}\,\mathcal{G}(1-s,z,w)\cdot E_s(g)\,ds,$$

for $\mathrm{Re}\,(z) \gg \frac{1}{2}$, $\mathrm{Re}\,(w) \gg 1$, where $\xi(s) = \pi^{-s/2}\Gamma(s/2)\zeta(s)$, where $\mathcal{G}$ is essentially a product of gamma function values

$$\mathcal{G}(s,z,w) = \pi^{-(z+\frac{w}{2})}\frac{\Gamma\left(\frac{z+1-s}{2}\right)\Gamma\left(\frac{z+s}{2}\right)\Gamma\left(\frac{z-s+w}{2}\right)\Gamma\left(\frac{z+s-1+w}{2}\right)}{\Gamma\left(z+\frac{w}{2}\right)},$$

and F is summed over (an orthogonal basis for) spherical (at finite primes) cuspforms on GL_2 with Laplacian eigenvalues $\frac{1}{4}+t_F^2$, and E_s is the usual spherical Eisenstein series. The continuous part, the *integral* of Eisenstein series E_s, cancels the pole at $z=1$ of the leading term, and when evaluated at $z=0$ is

$$\mathfrak{P}(g,0,w) = (\text{holomorphic at } z{=}0) + \tfrac{1}{2}\sum_{F\text{ on }GL_2} \rho_{\overline{F}}\cdot L\left(\tfrac{1}{2},\overline{F}\right)\cdot \mathcal{G}\left(\tfrac{1}{2}-it_F,0,w\right)\cdot F(g)$$
$$+\frac{1}{4\pi i}\int_{\mathrm{Re}\,(s)=\frac{1}{2}} \frac{\zeta(s)\,\zeta(1-s)}{\xi(2-2s)}\,\mathcal{G}(1-s,0,w)\cdot E_s(g)\,ds.$$

In this spectral expansion, the coefficient in front of a cuspform F includes $\mathcal{G}$ evaluated at $z=0$ and $s=\frac{1}{2}\pm \mathrm{i}t_F$, namely

$$\mathcal{G}\left(\tfrac{1}{2}-it_F,0,w\right) = \pi^{-\frac{w}{2}}\frac{\Gamma\left(\frac{\frac{1}{2}-it_F}{2}\right)\Gamma\left(\frac{\frac{1}{2}+it_F}{2}\right)\Gamma\left(\frac{w-\frac{1}{2}-it_F}{2}\right)\Gamma\left(\frac{w-\frac{1}{2}+it_F}{2}\right)}{\Gamma\left(\frac{w}{2}\right)}.$$

The gamma function has poles at $0,-1,-2,\ldots$, so this coefficient has poles at $w=\frac{1}{2}\pm \mathrm{i}t_F,\ -\frac{3}{2}\pm it_F,\ \ldots$. Over $\mathbb{Q}$, among spherical cuspforms (or for any fixed level) these values have no accumulation point. The continuous part of the spectral side at $z=0$ is

$$\frac{1}{4\pi i}\int_{\mathrm{Re}(s)=\frac{1}{2}} \frac{\xi(s)\xi(1-s)}{\xi(2-2s)} \frac{\Gamma(\frac{w-s}{2})\Gamma(\frac{w-1+s}{2})}{\Gamma(\frac{w}{2})}\cdot E_s\, ds,$$

with gamma factors grouped with corresponding zeta functions, to form the completed L-functions ξ. Thus, the evident pole of the leading term at $w=1$ can be exploited, using the continuation to $\mathrm{Re}(w)>1/2$. A contour-shifting argument shows that the continuous part of this spectral decomposition has a meromorphic continuation to $\mathbb{C}$ with poles at $\rho/2$ for zeros ρ of ζ, in addition to the poles from the gamma functions.

Already for GL_2, over general ground fields k, infinitely many Hecke characters enter both the spectral decomposition of the Poincaré series and the moment expression. This naturally complicates isolation of literal moments, and complicates analysis of poles via the spectral expansion. Suppressing constants, the moment expansion is a sum of twists by Hecke characters χ, of the form

$$\int_{Z_{\mathbb{A}}GL_2(k)\backslash GL_2(\mathbb{A})} \mathfrak{P}(s,z,w,\varphi_\infty)\cdot|f(g)|^2$$
$$=\sum_\chi\int_{\mathrm{Re}(s)=\frac{1}{2}} L(z+s,f\otimes\chi)\cdot L(1-s,\overline{f}\otimes\overline{\chi})\cdot M(s,z,\chi_\infty,\varphi_\infty)\, ds,$$

where $M(s,z,\chi_\infty,\varphi_\infty)$ depends upon complex parameters s,z and archimedean components χ_∞, f_∞, and upon auxiliary archimedean data φ_∞ defining the Poincaré series. Again suppressing constants, the spectral expansion is

$$\mathfrak{P}(g,z,\varphi_\infty) = (\infty-\text{part})\cdot E_{1+z}(g) + \sum_{F\text{ on }GL_2}(\infty-\text{part})\cdot\rho_{\overline{F}}\cdot L\left(\tfrac{1}{2}+z,\overline{F}\right)\cdot F(g)$$
$$+\sum_\chi\int_{\mathrm{Re}(s)=\frac{1}{2}}\frac{L(z+s,\overline{\chi})L(z+1-s,\chi)}{L(2-2s,\overline{\chi}^2)}\mathcal{G}(s,\chi_\infty)\cdot E_{s,\chi}(g)\, ds,$$

where the factor denoted ∞-part depends only upon the archimedean data, as does $\mathcal{G}(s,\chi_\infty)$.

In the simplest case beyond GL_2, take f a spherical cuspform for $GL_3(\mathbb{Q})$ generating an irreducible cuspidal automorphic representation $\pi=\pi_f$. We can construct a weight function $\Gamma(s,z,w,\pi_\infty,\pi'_\infty)$ with explicit asymptotic behavior, depending upon complex parameters s, z, and w, and upon the *archimedean* components π_∞ for π and for π' irreducible cuspidal automorphic on GL_2, such that the *moment expansion* has the form

$$\int_{Z_{\mathbb{A}}GL_3(\mathbb{Q})\backslash GL_3(\mathbb{A})}\mathfrak{P}(g,z,w)\cdot|f(g)|^2\, dg$$
$$=\sum_{\pi'\text{ on }GL_2}|\rho(\pi')|^2\frac{1}{2\pi i}\int_{\mathrm{Re}(s)=\frac{1}{2}}|L(s,\pi\times\pi')|^2\cdot\Gamma(s,0,w,\pi_\infty,\pi'_\infty)\, ds$$
$$+\frac{1}{4\pi i}\frac{1}{2\pi i}\sum_{k\in\mathbb{Z}}\int_{\mathrm{Re}(s_1)=\frac{1}{2}}\int_{\mathrm{Re}(s_2)=\frac{1}{2}}\frac{|L(s_1,\pi\times\pi_{E^{(k)}_{1-s_2}})|^2}{|\xi(1-2it_2)|^2}\Gamma(s_1,0,w,\pi_\infty,E^{(k)}_{1-s_2,\infty})\, ds_1\, ds_2,$$

where π' runs over (an orthogonal basis for) all level-one cuspforms on GL_2, with *no* restriction on the right K_∞-types, $E_s^{(k)}$ is the usual level-one Eisenstein series of K_∞-type k, and the notation $E_{1-s_2,\infty}^{(k)}$ means that the dependence is only upon the archimedean component. Here and throughout, for $\mathrm{Re}\,(s) = 1/2$, use $1-s$ in place of $\overline{s}$, to maintain holomorphy in complex-conjugated parameters.

More generally, for an irreducible cuspidal automorphic representation π on GL_r with $r \geq 3$, whether over $\mathbb{Q}$ or over a numberfield, the *moment expansion* includes an infinite sum of terms $|L(s, \pi \times \pi')|^2$ for π' ranging over irreducible cuspidal automorphic representations on GL_{r-1}, as well as *integrals* of products of L-functions $L(s, \pi \times \pi_1') \ldots L(s, \pi \times \pi_\ell')$ for $\pi_1', \ldots, \pi_\ell'$ ranging over ℓ-tuples of cuspforms on $GL_{r_1} \times \ldots \times GL_{r_\ell}$ for all partitions $(r_1, \ldots, r_\ell)$ of r.

Generally, the spectral expansion of the Poincaré series for GL_r is an induced-up version of that for GL_2. Suppressing constants, using groundfield $\mathbb{Q}$ to skirt Hecke characters, the spectral expansion has the form

$$\mathfrak{P} = (\infty - \text{part}) \cdot E_{z+1}^{r-1,1} + \sum_{F \text{ on } GL_2} (\infty - \text{part}) \cdot \rho_{\overline{F}} \cdot L(\tfrac{rz+r-2}{2} + \tfrac{1}{2}, \overline{F}) \cdot E_{\frac{z+1}{2}, F}^{r-2,2}$$

$$+ \int_{\mathrm{Re}\,(s)=\frac{1}{2}} (\infty - \text{part}) \cdot \frac{\zeta(\frac{rz+r-2}{2} + \frac{1}{2} - s) \cdot \zeta(\frac{rz+r-2}{2} + \frac{1}{2} + s)}{\zeta(2-2s)} \cdot E_{z+1, s-\frac{z+1}{2}, -s-\frac{z+1}{2}}^{r-2,1,1}\, ds,$$

where F is summed over an orthonormal basis for spherical cuspforms on GL_2, and where the Eisenstein series are naively normalized spherical, with $E_s^{r-1,1}$ a degenerate Eisenstein series attached to the parabolic corresponding to the partition $r-1,1$, and $E_{s_1,s_2,s_3,\chi}^{r-2,1,1}$ a degenerate Eisenstein series attached to the parabolic corresponding to the partition $r-2,1,1$.

Again over $\mathbb{Q}$, the *most-continuous* part of the moment expansion for GL_r is of the form

$$\int_{\mathrm{Re}\,(s)=\frac{1}{2}} \int_{t \in \Lambda} |L(s, \pi \times \pi_{E_{\frac{1}{2}+it}})|^2 M_t(s)\, ds\, dt$$

$$= \int \int_\Lambda \left| \frac{\Pi_{1 \leq \ell \leq r-1} L(s + \mathrm{i}t_\ell, \pi)}{\Pi_{1 \leq j < \ell < n} \zeta(1 - \mathrm{i}t_j + it_\ell)} \right|^2 M_t(s)\, ds\, dt,$$

where

$$\Lambda = \{t \in \mathbb{R}^{r-1} : t_1 + \cdots + t_{r-1} = 0\}$$

and where M_t is a weight function depending upon π. More generally, let $r - 1 = m \cdot b$. For π' irreducible cuspidal automorphic on GL_m, let

$$\pi'^{\Delta} = \pi' \otimes \ldots \otimes \pi'$$

on $GL_m \times \cdots \times GL_m$. Inside the moment expansion, we have (recall Langlands-Shahidi)

$$\int_{\mathrm{Re}(s)=\frac{1}{2}} \int_{\Lambda} |L(s, \pi \times \pi_{E_{\pi'^{\Delta}, \frac{1}{2}+it}})|^2 M_{\pi',t}(s)\, \mathrm{d}s\, \mathrm{d}t$$

$$= \int\int \left| \frac{\Pi_{1 \le \ell \le b} L(s + \mathrm{i}t_\ell, \pi \times \pi')}{\Pi_{1 \le j < \ell \le b} L(1 - \mathrm{i}t_j + it_\ell, \pi' \times \pi'^{\vee})} \right|^2 M\, \mathrm{d}s\, \mathrm{d}t.$$

Replacing the cuspidal representation π on $GL_r(\mathbb{Q})$ by a (truncated) minimal-parabolic Eisenstein series E_α with $\alpha \in \mathbb{C}^{n-1}$, the most-continuous part of the moment expansion contains a term

$$\int\int_{\Lambda} \left| \frac{\Pi_{1 \le \mu \le n,\, 1 \le \ell \le r-1}\, \zeta(\alpha_\mu + s + \mathrm{i}t_\ell)}{\Pi_{1 \le j < \ell < r} |\zeta(1 - \mathrm{i}t_j + \mathrm{i}t_\ell)} \right|^2 \mathrm{d}s\, \mathrm{d}t.$$

Taking $\alpha = 0 \in \mathbb{C}^{r-1}$ gives

$$\int\int_{\Lambda} \left| \frac{\Pi_{1 \le \ell \le r-1}\ \zeta(s + it_\ell)^r}{\Pi_{1 \le j < \ell < r}\, \zeta(1 - \mathrm{i}t_j + it_\ell)} \right|^2 M\, \mathrm{d}s\, \mathrm{d}t.$$

For example, for GL_3, where $\Lambda = \{(t, -t)\} \approx \mathbb{R}$,

$$\int\int_{\mathbb{R}} \left| \frac{\zeta(s + \mathrm{i}t)^3 \cdot \zeta(s - \mathrm{i}t)^3}{\zeta(1 - 2it)} \right|^2 M\, \mathrm{d}s\, \mathrm{d}t,$$

and for GL_4

$$\int_{(s)}\int_{\Lambda} \left| \frac{\zeta(s + \mathrm{i}t_1)^4 \cdot \zeta(s + \mathrm{i}t_2)^4 \cdot \zeta(s + \mathrm{i}t_3)^4}{\zeta(1 - it_1 + it_2)\, \zeta(1 - it_1 + it_3)\, \zeta(1 - it_2 + it_3)} \right|^2 M\, \mathrm{d}s\, \mathrm{d}t.$$

2 Background and Normalizations

We recall some facts concerning Whittaker models and Rankin–Selberg integral representations of L-functions, and spectral theory for automorphic forms, on GL_r. To compare zeta local integrals formed from vectors in cuspidal representations to local L-functions attached to the representations, we specify distinguished vectors in irreducible representations of p-adic and archimedean groups. Locally at both p-adic and archimedean places, Whittaker models with spherical vectors have a natural choice of distinguished vector, namely, the spherical vector taking value 1 at the identity element of the group.

Even in general, for the specific purposes here, at finite places the facts are clear. At archimedean places, the facts are more complicated, and, further, the situation dictates choices of data, and we are not free to make ideal choices. See [6–8] for detailed surveys, and references to the literature, mostly papers of Jacquet, Piatetski-Shapiro and Shalika. The spectral theory is due to [36,41], and proof of conjectures of [29] in [40].

Fix an integer $r \geq 2$ and consider the general linear group $G = GL_r$ over a fixed algebraic number field k. For a positive integer ℓ, in the following we use the notation "$\ell \times \ell$" to denote an ℓ-by-ℓ matrix, and let 1_ℓ denote the $\ell \times \ell$ identity matrix. Then $G = GL_r$ has the following standard subgroups:

$$P = P^{r-1,1} = \left\{ \begin{pmatrix} (r-1)\times(r-1) & * \\ 0 & 1\times 1 \end{pmatrix} \right\},$$

$$U = \left\{ \begin{pmatrix} 1_{r-1} & * \\ 0 & 1 \end{pmatrix} \right\}, \quad H = \left\{ \begin{pmatrix} (r-1)\times(r-1) & 0 \\ 0 & 1 \end{pmatrix} \right\},$$

$$\begin{aligned} N &= \{\text{upper-triangular unipotent elements in } H\} \\ &= (\text{unipotent radical of standard minimal parabolic in } H), \\ Z &= \text{center of } GL_r. \end{aligned}$$

Let $\mathbb{A} = \mathbb{A}_k$ be the adele ring of k. For a place v of k let k_v be the corresponding completion, with ring of integers $\mathfrak{o}_v$ for finite v. For an algebraic group defined over k, let G_v be the k_v-valued points of G. For $G = GL_r$ over k, let K_v be the standard maximal compact subgroup of G_v: for $v < \infty$, $K_v = GL_r(\mathfrak{o}_v)$ for $v \approx \mathbb{R}$, $K_v = O_r(\mathbb{R})$, and for $v \approx \mathbb{C}$, $K_v = U(r)$.

A standard choice of non-degenerate character on $N_k U_k \backslash N_\mathbb{A} U_\mathbb{A}$ is

$$\psi(n \cdot u) = \psi_0(n_{12} + n_{23} + \cdots + n_{r-2,r-1}) \cdot \psi_0(u_{r-1,r}),$$

where ψ_0 is a fixed non-trivial character on $\mathbb{A}/k$. A cuspform f has a Fourier-Whittaker expansion along NU

$$f(g) = \sum_{\xi \in N_k \backslash H_k} W_f(\xi g) \quad \text{where} \quad W_f(g) = \int_{N_k U_k \backslash N_\mathbb{A} U_\mathbb{A}} \overline{\psi}(nu) f(nug)\, dn\, du.$$

The Whittaker function $W_f(g)$ factors over primes, and a careful normalization of this factorization is set up below. Cuspforms F on H have corresponding Fourier-Whittaker expansions

$$F(h) = \sum_{\xi \in N'_k \backslash H'_k} W_F(\xi h) \quad \text{where} \quad W_F(g) = \int_{N'_k \backslash N'_\mathbb{A}} \overline{\psi}(n) F(nh)\, dn,$$

where $H' \approx GL_{r-2}$ sits inside H as H sits inside G, $N' = N \cap H'$, and ψ is restricted from NU to N. This Whittaker function also factors $W_F = \bigotimes_v W_{F,v}$.

At finite places v, given an irreducible admissible representation π_v of G_v admitting a Whittaker model, [31] shows that there is an essentially unique *effective vector* $W_{\pi_v}^{\text{eff}}$, generalizing the characterization of *newform* in [4], as follows. For π_v

spherical, $W^{\text{eff}}_{\pi_v}$ is the usual unique spherical Whittaker vector taking value 1 at the identity element of the group, as in [5, 46]. For non-spherical local representations, define *effective vector* as follows. Let

$$U_v^{\text{opp}}(\ell) \;=\; \left\{ \begin{pmatrix} 1_{r-1} & 0 \\ x & 1 \end{pmatrix} \;:\; x = 0 \bmod \mathfrak{p}^\ell \right\}.$$

Let $K_v^H \approx GL_{r-1}(\mathfrak{o}_v)$ be the standard maximal compact of H_v. Define a congruence subgroup of K_v by

$$K_v(\ell) \;=\; K_v^H \cdot (U_v \cap K_v) \cdot U_v^{\text{opp}}(\ell).$$

For a non-spherical Whittaker model π_v, there is a unique positive integer ℓ_v, the *conductor* of π_v, such that π_v has *no* non-zero vectors fixed by $K_v(\ell')$ for $\ell' < \ell_v$, and has a one-dimensional space of vectors fixed by $K_v(\ell_v)$. The remaining ambiguous constant is completely specified by requiring that local Rankin–Selberg integrals

$$Z_v(s, W^{\text{eff}}_{\pi_v} \times W^o_{\pi'_v}) \;=\; \int_{N_v \backslash H_v} |\det Y|^s \, W^{\text{eff}}_{\pi_v}\begin{pmatrix} Y & \\ & 1 \end{pmatrix} W^o_{\pi'_v}(Y)\, \mathrm{d}Y$$

produce the correct local factors $L_v(s, \pi_v \times \pi'_v)$ of $GL_r \times GL_{r-1}$ Rankin-Selberg L–functions for every *spherical* representation π'_v of the local GL_{r-1}, with normalized spherical Whittaker vector $W^o_{\pi'_v}$ in π'_v. That is,

$$Z_v(s, W^{\text{eff}}_{\pi_v} \times W^o_{\pi'_v}) \;=\; L_v(s, \pi_v \times \pi'_v),$$

with no additional factor on the right-hand side. See Sect. 4 of [32], and comments below. Cuspidal automorphic representations $\pi \approx \bigotimes'_v \pi_v$ of $G_{\mathbb{A}}$ admit local Whittaker models at all finite places, so locally at all finite places have a unique effective vector.

Facts concerning archimedean local Rankin–Selberg integrals for $GL_m \times GL_n$ for general m, n are more complicated than the non-archimedean cases. See [9, 47, 48], as well as the surveys [6–8]. The *spherical* case for $GL_r \times GL_{r-1}$ admits fairly explicit treatment, but this is insufficient for our purposes. Fortunately, for us there is no compulsion to attempt to specify the archimedean local data for Rankin–Selberg integrals. Indeed, the local archimedean Rankin–Selberg integrals will be *deformed* into shapes essentially unrelated to the corresponding L-factor, in any case. Thus, in the *moment expansion* in the theorem below we can use *any* systematic specification of distinguished vectors e_{π_v} in irreducible representations π_v of G_v, and $e_{\pi'_v}$ in π'_v of H_v, for v archimedean. For $v|\infty$ and π_v, a Whittaker model representation of G_v with a spherical vector, let the distinguished vector e_{π_v} be the spherical vector normalized to take value 1 at the identity element of the group. Similarly, for π'_v a Whittaker model representation of H_v with a spherical vector, let the distinguished vector $e_{\pi'_v}$ be the normalized spherical vector. Anticipating that cuspforms generating spherical representations at archimedean places make up the bulk of the space of automorphic forms, we do not give an explicit choice of distinguished vector in other archimedean representations. Rather, we formulate the normalizations below, and the moment expansion, in a fashion applicable to *any* choice of distinguished vectors in archimedean representations.

Let π be an automorphic representation of $G_{\mathbb{A}}$, factoring over primes as $\pi \approx \bigotimes_v' \pi_v$ admitting a global Whittaker model. Each local representation π_v has a Whittaker model, since π has a global Whittaker model. At each finite place v, let $W_{\pi_v}^{\text{eff}}$ be the normalized effective vector, and e_{π_v} the distinguished vector at $v|\infty$. Let $f \in \pi$ be a moderate-growth automorphic form on $G_{\mathbb{A}}$ corresponding to a monomial tensor in π, consisting of the effective vector at all finite primes, and the distinguished vector e_{π_v} at $v|\infty$. Then the global Whittaker function of f is a globally determined constant multiple of the product of the local functions:

$$W_f = \rho_f \cdot \bigotimes_{v|\infty} e_{\pi_v} \otimes \bigotimes_{v<\infty} W_{\pi_v}^{\text{eff}},$$

where ρ_f is a general analogue of the leading Fourier coefficient $\rho_f(1)$ in the $GL_2(\mathbb{Q})$ theory.

Let π' be an automorphic representation of $H_{\mathbb{A}}$ *spherical* at all finite primes, admitting a global Whittaker model. Let π' factor as $j : \bigotimes_v' \pi_v' \to \pi'$. Certainly each π_v' admits a Whittaker model. At each finite v, let $W_{\pi_v'}^o$ be the normalized spherical vector in π_v', and at archimedean v let $e_{\pi_v'}$ be the distinguished vector. For a moderate-growth automorphic form $F \in \pi'$ corresponding to a monomial vector in the factorization of π', at all finite places corresponding to the spherical Whittaker function $W_{\pi_v'}^o$, and to the distinguished vector $e_{\pi_v'}$ at archimedean places, again specify a constant ρ_F by

$$W_F = \rho_F \cdot \bigotimes_{v|\infty} e_{\pi_v'} \otimes \bigotimes_{v<\infty} W_{\pi_v'}^o.$$

When π' occurs discretely in the space of L^2 automorphic forms on H, each of the local factors of π' is unitarizable, and uniquely so up to a constant, by irreducibility. For an arbitrary vector $\varepsilon = \varepsilon_\infty$ in π_∞', let F^ε be the automorphic form corresponding to $\bigotimes_{v<\infty} W_{\pi_v}^o \otimes \varepsilon$ by the isomorphism j. Define ρ_{F^ε} by

$$W_{F^\varepsilon} = \rho_{F^\varepsilon} \cdot \bigotimes_{v<\infty} W_{\pi_v}^o \otimes \varepsilon.$$

By Schur's Lemma, the comparison of ρ_F and ρ_{F^ε} depends only upon the comparison of archimedean data, namely,

$$\frac{\rho_{F^\varepsilon}}{\rho_F} = \frac{|\varepsilon|_{\pi_\infty'}}{|\otimes_{v|\infty} e_{\pi_v'}|_{\pi_\infty'}},$$

with Hilbert space norms on the representation π_∞' at archimedean places. The ambiguity of these norms by a constant disappears in taking ratios.

Indeed, the global constants ρ_F and ρ_{F^ε} can be compared by a similar device (and induction) for F and F_ε in any irreducible π' occurring in the L^2 automorphic spectral expansion for H. We do not carry this out explicitly, since this would require setting up normalizations for the full spectral decomposition, while our main interest is in the cuspidal (hence, discrete) part.

With f cuspidal and F moderate growth, corresponding to distinguished vectors, as above, the Rankin–Selberg zeta integral is the finite-prime Rankin–Selberg L-function, with global constants ρ_f and ρ_F, and with archimedean local Rankin–Selberg zeta integrals depending upon the distinguished vectors at archimedean places:

$$\int_{H_k\backslash H_{\mathbb{A}}} |\det Y|^{s-\frac{1}{2}} F(Y) f\begin{pmatrix} Y & \\ & 1 \end{pmatrix} dY = \rho_f \cdot \rho_F \cdot L(s, \pi \times \pi') \cdot \prod_{v|\infty} Z_v(s, e_{\pi_v} \times e_{\pi'_v}).$$

The finite-prime part of the Rankin-Selberg L–function appears regardless of the archimedean local data. The global constants ρ_f and ρ_F do depend partly upon the local archimedean choices, but are global objects.

We need a spectral decomposition of part of $L^2(H_k\backslash H_{\mathbb{A}})$, as follows. Let K^H_{fin} be the standard maximal compact $GL_{r-1}(\widehat{\mathfrak{o}})$ of H_{fin}, where as usual $\widehat{\mathfrak{o}}$ is $\prod_{v<\infty} \mathfrak{o}_v$, with $\mathfrak{o}_v$ the local integers at the finite place v of k. First, there is the Hilbert direct-integral decomposition by characters ω on the *central archimedean split component* Z^+ of H: let

$$i : y \longrightarrow (y^{\frac{1}{d}}, \ldots, y^{\frac{1}{d}}, 1, 1, \ldots) \quad \text{for } y > 0, \text{ with } \quad d = [k : \mathbb{Q}]$$

be the diagonal imbedding of the positive real numbers in the archimedean factors of the ideles of k. The central archimedean split component is

$$Z^+ = \{j(y) = \begin{pmatrix} i(y)^{1/(r-1)} & & \\ & \ddots & \\ & & i(y)^{1/(r-1)} \end{pmatrix} \in H_{\mathbb{A}} : y > 0\}.$$

The point of our parametrization is that (with idele norms)

$$|\det j(y)| = |i(y)| = y \quad \text{with} \quad y > 0.$$

The corresponding spectral decomposition is

$$L^2(H_k\backslash H_{\mathbb{A}}) \approx \int_{\mathbb{R}}^{\oplus} L^2(Z^+ H_k\backslash H_{\mathbb{A}}, \omega_{\mathrm{it}})\, dt,$$

where $L^2(Z^+ H_k\backslash H_{\mathbb{A}}, \omega_{\mathrm{it}})$ is the isotypic component of functions Φ with $|\Phi|$ in $L^2(Z^+ H_k\backslash H_{\mathbb{A}})$ transforming by

$$\Phi(j(y) \cdot h) = y^{\mathrm{it}} \cdot \Phi(h) \quad \text{for} \quad y > 0 \text{ and } h \in H_{\mathbb{A}}$$

under Z^+. The projections and spectral synthesis along Z^+ can be written as

$$F(h) = \int_{\mathbb{R}} \left(\int_0^\infty F(j(y) \cdot h) y^{-it} \frac{dy}{y} \right) dt.$$

Each isotypic component $L^2(Z^+H_k \backslash H_{\mathbb{A}}, \omega_{\text{it}})$ has a direct integral decomposition as a representation of $H_{\mathbb{A}}$, of the form

$$L^2(Z^+H_k \backslash H_{\mathbb{A}}, \omega_{\text{it}}) \approx \int_{\Xi}^{\oplus} \pi' \otimes |\det|^{\text{it}} \, d\pi',$$

where Ξ is the set of irreducibles π' occurring in $L^2(Z^+H_k \backslash H_{\mathbb{A}}, \omega_0)$. That is, the irreducibles for general archimedean split-component character ω_{it} differ merely by a determinant twist from the trivial split-component character case. The measure is not described explicitly here, apart from the observation that the discrete part of the decomposition, including the cuspidal part, has counting measure.

For our applications, we are concerned with the subspaces $L^2(Z^+H_k \backslash H_{\mathbb{A}}/K_{\text{fin}}^H, \omega)$ of right K_{fin}^H-invariant functions. Since each π' factors over primes as a restricted tensor product $\pi' \approx \bigotimes_v' \pi_v'$ of irreducibles π_v' of the local points H_v, the decomposition of $L^2(Z^+H_k \backslash H_{\mathbb{A}}/K_{\text{fin}}^H, \omega)$ only involves the subset Ξ^o consisting of irreducibles $\pi' \in \Xi$ such that for every *finite* place v the local representation π_v' is *spherical*. Let π_∞' be the archimedean factor of π', and π_{fin}' the finite-place factor, so $\pi' \approx \pi_\infty' \otimes \pi_{\text{fin}}'$. Let $\pi_{\text{fin}}'^o$ be the one-dimensional space of K_{fin}^H-fixed vectors in π_{fin}'. As a representation of the archimedean part H_∞ of $H_{\mathbb{A}}$,

$$L^2(Z^+H_k \backslash H_{\mathbb{A}}/K_{\text{fin}}^H, \omega_{\text{it}}) \approx \int_{\Xi^o}^{\oplus} (\pi_\infty' \otimes \pi_{\text{fin}}'^o) \otimes |\det|^{\text{it}} \, d\pi'.$$

An automorphic spectral decomposition for F in $L^2(Z^+H_k \backslash H_{\mathbb{A}}/K_{\text{fin}}^H, \omega_{\text{it}})$ can be written in the form

$$F = \int_{\Xi^o} \sum_j \langle F, \Phi_{\pi' j} \otimes |\det|^{\text{it}} \rangle \cdot \Phi_{\pi' j} \otimes |\det|^{\text{it}} \, d\pi',$$

where Ξ^o indexes spherical automorphic representations π' with trivial archimedean split-component character entering the spectral expansion, for each of these j indexes an orthonormal basis in the archimedean component π_∞', and $\Phi_{\pi' j}$ is the corresponding moderate-growth spherical automorphic form in the global π'. The pairing is the natural one, namely,

$$\langle F, \Phi_{\pi' j} \otimes |\det|^{\text{it}} \rangle = \int_{H_k \backslash H_{\mathbb{A}}} F(h) \overline{\Phi}_{\pi' j}(h) \, |\det h|^{-it} \, dh.$$

3 Moment Expansion

We define a Poincaré series $\mathfrak{P} = \mathfrak{P}_{z,\varphi_\infty}$ depending on archimedean data φ_∞ and a complex *equivariance* parameter z. With various simplifying choices of archimedean data φ_∞ depending only on a complex parameter w, the Poincaré series $\mathfrak{P} = \mathfrak{P}_{z,w}$ is a function of the two complex parameters z, w. By design, for a cuspform f of conductor ℓ on $G = GL_r$ over a number field k, for *any* choice of data for the Poincaré series sufficient for convergence, the integral

$$\int_{Z_{\mathbb{A}} G_k \backslash G_{\mathbb{A}}} |f|^2 \cdot \mathfrak{P}$$

is an *integral moment* of L–functions attached to f, in the sense that it is a sum and integral over a spectral family, namely, a weighted average over spectral components with respect to $L^2(GL_{r-1}(k)\backslash GL_{r-1}(\mathbb{A}))$. Subsequently, we will obtain a spectral expansion of the more-simply parametrized Poincaré series $\mathfrak{P} = \mathfrak{P}_{z,w}$, giving the meromorphic continuation of this integral in the complex parameters z, w.

For $z \in \mathbb{C}$, let

$$\varphi = \bigotimes_v \varphi_v,$$

where $z \in \mathbb{C}$ specifies an equivariance property of φ, as follows. For v finite,

$$\varphi_v(g) = \begin{cases} \left|(\det A)/d^{r-1}\right|_v^z & (\text{for } g = mk \text{ with } m = \begin{pmatrix} A & 0 \\ 0 & d \end{pmatrix} \text{ in } Z_v H_v \text{ and } k \in K_v), \\ 0 & (\text{otherwise}). \end{cases}$$

For v archimedean require right K_v-invariance and left equivariance

$$\varphi_v(mg) = \left|\frac{\det A}{d^{r-1}}\right|_v^z \cdot \varphi_v(g) \quad \text{for } g \in G_v, \text{ for } m = \begin{pmatrix} A & 0 \\ 0 & d \end{pmatrix} \in Z_v H_v.$$

Thus, for $v|\infty$, the further data determining φ_v consists of its values on U_v. A simple useful choice of archimedean data parametrized by a single complex parameter w is

$$\varphi_v \begin{pmatrix} 1_{r-1} & x \\ 0 & 1 \end{pmatrix} = (1 + |x_1|^2 + \ldots + |x_{r-1}|^2)^{-[k_v:\mathbb{R}]w/2}, \text{ where } x = \begin{pmatrix} x_1 \\ \vdots \\ x_{r-1} \end{pmatrix},$$

and $w \in \mathbb{C}$. The norm $|x_1|^2 + \ldots + |x_{r-1}|^2$ is normalized to be invariant under K_v. Thus, φ is left $Z_{\mathbb{A}} H_k$-invariant. We attach to any such φ a *Poincaré series*

$$\mathfrak{P}(g) = \mathfrak{P}_\varphi(g) = \sum_{\gamma \in Z_k H_k \backslash G_k} \varphi(\gamma g).$$

Remark. There is an essentially unique choice of (parametrized) archimedean data $\varphi_\infty = \varphi_{z,w,\infty}$ such that the associated Poincaré series at $z = 0$ has a functional equation (as in [13]). For instance, when $G = GL_3$ over $\mathbb{Q}$ this choice is

$$\varphi_\infty \begin{pmatrix} I_2 & u \\ & 1 \end{pmatrix} = \varphi_{0,w,\infty} \begin{pmatrix} I_2 & u \\ & 1 \end{pmatrix} = 2^{-w} \sqrt{\pi}\, \frac{\Gamma(\frac{w}{2})\,(1+||u||^2)^{-\frac{w}{2}}\, F(\frac{w}{2},\frac{w}{2};w;\frac{1}{1+||u||^2})}{\Gamma(\frac{w+1}{2})},$$

for $z = 0$, with F the usual hypergeometric function

$$F(\alpha,\beta;\gamma;x) = \frac{\Gamma(\gamma)}{\Gamma(\alpha)\Gamma(\beta)} \cdot \sum_{m=0}^{\infty} \frac{1}{m!} \frac{\Gamma(\alpha+m)\Gamma(\beta+m)}{\Gamma(\gamma+m)}\, x^m \quad (\text{for } |x| < 1).$$

The functional equation of the Poincaré series $\mathfrak{P}_{0,w}(g)$ attached to this choice of $\varphi = \varphi_w$ when $z = 0$ is: *the function*

$$\sin\left(\frac{\pi w}{2}\right) \mathfrak{P}_{0,w}(g) + \frac{\pi\,\zeta(w)\,\zeta(2-w)}{2\,(1-w)\,\pi^{\frac{1}{2}-w}\,\Gamma(w-\frac{1}{2})\,\zeta(2w-1)} \cdot E^{1,1,1}\left(g, \frac{w}{3}, 1-\frac{2w}{3}\right)$$

is invariant as $w \to 2-w$, where $E^{1,1,1}(g,s_1,s_2) = E^{1,1,1}_{s_1,s_2}(g)$ is the minimal parabolic Eisenstein series. After our discussion of the spectral expansion of the Poincaré series, we give a general prescription for archimedean data producing Poincaré series admitting a functional equation: with suitable archimedean data, the functional equation is visible from the spectral expansion.

With subscripts ∞ denoting the archimedean parts of various objects, for $h, m \in H_\infty$, define

$$\mathscr{K}(h,m) = \mathscr{K}_{\varphi_\infty}(h,m) = \int_{U_\infty} \varphi_\infty(u)\, \psi_\infty(huh^{-1})\, \overline{\psi}_\infty(mum^{-1})\, du.$$

Let $\pi \approx \otimes' \pi_v$ be a cuspidal automorphic representation of G, with finite set S of finite primes such that π_v is spherical for finite $v \notin S$, and π_v has conductor ℓ_v for $v \in S$. We say a cuspform f in π is a *newform* if it is spherical at finite $v \notin S$ and is right $K_v(\ell_v)$-fixed for $v \in S$. As above, the global Whittaker function W_f of f factors as

$$W_f = \rho_f \cdot \bigotimes_{v<\infty} W^{\text{eff}}_{\pi_v} \otimes \bigotimes_{v|\infty} e_{\pi_v}.$$

Let $e_{\pi_\infty} = \otimes_{v|\infty} e_{\pi_v}$. Let π' be an automorphic representation of H admitting a global Whittaker model, with unitarizable archimedean factor π'_∞, with orthonormal basis $\varepsilon_{\pi' j}$ for π'_∞. Recalling that $\mathscr{K}(h,m) = \mathscr{K}_{z,\varphi_\infty}(h,m)$ depends on the parameter z and the data φ_∞, the gamma factors appearing in the moment expansion below are

$$\Gamma(e_{\pi_\infty},\pi'_\infty,s) = \Gamma_{z,\varphi_\infty}(e_{\pi_\infty},\pi'_\infty,s) = \sum_j \int_{N_\infty\backslash H_\infty} \int_{N_\infty\backslash H_\infty} \int_{K_\infty} e_{\pi_\infty}(hk)\varepsilon_{\pi' j}(h)|\det h|^{z+s-\frac{1}{2}}$$

$$\times \overline{e}_{\pi_\infty}(mk)\overline{\varepsilon}_{\pi' j}(m)|\det m|^{\frac{1}{2}-s}\mathscr{K}(h,m)\, dm\, dh\, dk.$$

The sum over the orthonormal basis for π'_∞ is simply an expression for a projection operator, so is necessarily independent of the orthonormal basis indexed by j. Thus, the sum indeed depends only on the archimedean Whittaker model π'_∞.

For each automorphic representation π' of H occurring (continuously or discretely) in the automorphic spectral expansion for H, and admitting a global Whittaker model, and *spherical* at all finite primes, let $F_{\pi'}$ be an automorphic form in π' corresponding to the spherical vector at all finite places and to the *distinguished vector* $e_{\pi'_\infty}$ in the archimedean part.

Theorem 1. *Let f be a cuspform, as just above. For* $\mathrm{Re}(z) \gg 1$ *and* $\mathrm{Re}(w) \gg 1$, *we have the moment expansion*

$$\int_{Z_\mathbb{A} G_k \backslash G_\mathbb{A}} |f|^2 \cdot \mathfrak{P}$$

$$= |\rho_f|^2 \int_{\Xi^o} |\rho_{F_{\pi'}}|^2 \int_\mathbb{R} L(\tfrac{1}{2}+it+z, \pi\otimes\pi')\, L(\tfrac{1}{2}-it, \overline{\pi}\otimes\overline{\pi}')\, \Gamma(e_{\pi_\infty}, \pi'_\infty, \tfrac{1}{2}+it)\, \mathrm{d}t\, \mathrm{d}\pi'.$$

Proof. The typical first unwinding is

$$\int_{Z_\mathbb{A} G_k \backslash G_\mathbb{A}} \mathfrak{P}(g)\, |f(g)|^2\, \mathrm{d}g = \int_{Z_\mathbb{A} H_k \backslash G_\mathbb{A}} \varphi(g)\, |f(g)|^2\, \mathrm{d}g.$$

Express f in its Fourier–Whittaker expansion, and unwind further:

$$\int_{Z_\mathbb{A} H_k \backslash G_\mathbb{A}} \varphi(g) \sum_{\eta \in N_k \backslash H_k} W_f(\eta g)\, \overline{f}(g)\, \mathrm{d}g \;=\; \int_{Z_\mathbb{A} N_k \backslash G_\mathbb{A}} \varphi(g)\, W_f(g)\, \overline{f}(g)\, \mathrm{d}g.$$

Use an Iwasawa decomposition $G = (HZ)UK$ everywhere locally to rewrite the whole integral as

$$\int_{N_k \backslash H_\mathbb{A} \times U_\mathbb{A} \times K_\mathbb{A}} \varphi(huk)\, W_f(huk)\, \overline{f}(huk)\, \mathrm{d}h\, \mathrm{d}u\, \mathrm{d}k.$$

At finite primes $v \notin S$, the right integral over K_v can be dropped, since all the functions in the integrand are right K_v-invariant. At finite primes $v \in S$, using the newform assumption on f, the one-dimensionality of the $K_v(\ell_v)$-fixed vectors in π_v implies that the K_v-type in which the effective vector lies is *irreducible*. Thus, by Schur orthogonality and inner product formulas, a diagonal integral of $f(xk_v) \cdot \overline{f}(yk_v)$ over $k_v \in K_v$ is a positive constant multiple of $f(x)\overline{f}(y)$, for all $x, y \in G_\mathbb{A}$. Thus, the integrals over K_v for v finite can be dropped entirely, and, up to a positive constant depending only upon the right K_v-type of f at $v \in S$, the whole integral is

$$\int_{N_k \backslash H_\mathbb{A} \times U_\mathbb{A} \times K_\infty} \varphi(huk)\, W_f(huk)\, \overline{f}(huk)\, \mathrm{d}h\, \mathrm{d}u\, \mathrm{d}k.$$

Since $\overline{f}$ is left H_k-invariant, it decomposes along $H_k\backslash H_{\mathbb{A}}$. The function $h \to f(huk)$ with $u \in U_{\mathbb{A}}$ and $k \in K_\infty$ is right K_{fin}^H-invariant. Thus, $\overline{f}$ decomposes as

$$\overline{f}(huk) = \int_{\mathbb{R}} \int_{\Xi^o} \sum_j \Phi_{\pi' j}(h) \,|\det h|^{\text{it}} \int_{H_k\backslash H_{\mathbb{A}}} \overline{\Phi}_{\pi' j}(m) \,|\det m|^{-it} \overline{f}(muk)\, \mathrm{d}m \, \mathrm{d}\pi' \, \mathrm{d}t.$$

Unwind the Fourier–Whittaker expansion of $\overline{f}$

$\overline{f}(huk)$

$$= \int_{\Xi^o} \sum_j \Phi_{\pi' j}(h) |\det h|^{\text{it}} \int_{H_k\backslash H_{\mathbb{A}}} \overline{\Phi}_{\pi' j}(m) \,|\det m|^{-it} \sum_{\eta \in N_k\backslash H_k} \overline{W}_f(\eta muk)\, \mathrm{d}m\, \mathrm{d}k\, \mathrm{d}\pi'$$

$$= \int_{\Xi^o} \Phi_{\pi' j}(h) \,|\det h|^{\text{it}} \int_{N_k\backslash H_{\mathbb{A}}} \overline{\Phi}_{\pi' j}(m) \,|\det m|^{-it} \,\overline{W}_f(muk)\, \mathrm{d}m\, \mathrm{d}k\, \mathrm{d}\pi'.$$

Then the whole integral is

$$\int_{Z_{\mathbb{A}} G_k\backslash G_{\mathbb{A}}} \mathfrak{P}(g) \,|f(g)|^2 \,\mathrm{d}g$$

$$= \int_{\mathbb{R}} \int_{\Xi^o} \sum_j \int_{N_k\backslash H_{\mathbb{A}}} \int_{U_{\mathbb{A}}} \int_{K_\infty} \varphi(huk) \Phi_{\pi' j}(h) |\det h|^{\text{it}} W_f(huk)$$

$$\times \int_{N_k\backslash H_{\mathbb{A}}} \overline{W}_f(muk) \overline{\Phi}_{\pi' j}(m) |\det m|^{-it} \mathrm{d}m\, \mathrm{d}h\, \mathrm{d}u\, \mathrm{d}k\, \mathrm{d}\pi'\, \mathrm{d}t.$$

The part of the integrand that depends upon $u \in U$ is

$$\int_{U_{\mathbb{A}}} \varphi(huk)\, W_f(huk)\, \overline{W}_f(muk)\, \mathrm{d}u$$

$$= \varphi(h)\, W_f(hk)\, \overline{W}_f(mk) \cdot \int_{U_{\mathbb{A}}} \varphi(u)\, \psi(huh^{-1})\, \overline{\psi}(mum^{-1})\, \mathrm{d}u.$$

The latter integrand and integral visibly factor over primes. We need the following:

Lemma 1. *Let v be a finite prime. For $h, m \in H_v$ such that $W_{\pi_v}^{\text{eff}}(h) \neq 0$ and $W_{\pi_v}^{\text{eff}}(m) \neq 0$, we have*

$$\int_{U_v} \varphi_v(h)\, \psi_v(huh^{-1})\, \overline{\psi}_v(mum^{-1})\, \mathrm{d}u = \int_{U_v \cap K_v} 1 \,\mathrm{d}u.$$

Proof. At a finite place v, $\varphi_v(u) \neq 0$ if and only if $u \in U_v \cap K_v$, and for such u

$$\psi_v(huh^{-1}) \cdot W_{\pi_v}(h) = W_{\pi_v}^{\text{eff}}(huh^{-1} \cdot h) = W_{\pi_v}^{\text{eff}}(hu) = W_{\pi_v}^{\text{eff}}(h) \cdot 1$$

by the right $U_v \cap K_v$-invariance, since f is a *newform*, in our present sense. Thus, for $W^{\rm eff}_{\pi_v}(h) \neq 0$, $\psi_v(huh^{-1}) = 1$, and similarly for $\psi_v(mum^{-1})$. Thus, the finite-prime part of the integral over U_v is just the integral of 1 over $U_v \cap K_v$, as indicated.

Returning to the proof of the theorem, the archimedean part of the integral does not behave as the previous lemma indicates the finite-prime components do, because of its non-trivial deformation by φ_∞. Thus, with subscripts ∞ denoting the infinite-adele part of various objects, for $h, m \in H_\infty$, as above, let

$$\mathscr{K}(h,m) = \int_{U_\infty} \varphi_\infty(u)\, \psi_\infty(huh^{-1})\, \overline{\psi}_\infty(mum^{-1})\, du.$$

The whole integral is

$$\int_{Z_{\mathbb{A}} G_k \backslash G_{\mathbb{A}}} \mathfrak{P}(g)\, |f(g)|^2 \, dg = \int_{\mathbb{R}} \int_{\Xi^o} \sum_j \int_{K_\infty} \int_{N_k \backslash H_{\mathbb{A}}} \int_{N_k \backslash H_{\mathbb{A}}} \mathscr{K}(h,m) \varphi(h)$$

$$\times W_f(hk) \Phi_{\pi' j}(h) |\det h|^{\mathrm{it}} \overline{W}_f(mk) \overline{\Phi}_{\pi' j}(m) |\det m|^{-it} dm\, dh\, d\pi'\, dk\, dt.$$

Normalize the volume of $N_k \backslash N_{\mathbb{A}}$ to 1. For a left N_k-invariant function Φ on $H_{\mathbb{A}}$, using the left $N_{\mathbb{A}}$-equivariance of W by ψ, and the left $N_{\mathbb{A}}$-invariance of φ,

$$\int_{N_k \backslash N_{\mathbb{A}}} \varphi(nh)\, \Phi(nh)\, W_f(nhk)\, dn = \varphi(h)\, W_f(h)$$

$$\int_{N_k \backslash N_{\mathbb{A}}} \psi(n)\, \Phi(nh)\, dn = \varphi(h)\, W_f(hk)\, W_\Phi(h),$$

where

$$W_\Phi(h) = \int_{N_k \backslash N_{\mathbb{A}}} \psi(n)\, \Phi(nh)\, dn.$$

(The integral is not against $\overline{\psi}(n)$, but $\psi(n)$.) That is, the integral over $N_k \backslash H_{\mathbb{A}}$ is equal to an integral against (up to an alteration of the character) the Whittaker function W_Φ of Φ, which factors over primes for suitable Φ. Thus, the whole integral is

$$\int_{Z_{\mathbb{A}} G_k \backslash G_{\mathbb{A}}} \mathfrak{P}(g)\, |f(g)|^2 \, dg = \int_{\mathbb{R}} \int_{\Xi^o} \sum_j \int_{N_{\mathbb{A}} \backslash H_{\mathbb{A}}} \int_{N_{\mathbb{A}} \backslash H_{\mathbb{A}}} \int_{K_\infty} \mathscr{K}(h,m)$$

$$\times W_f(hk) W_{\Phi_{\pi' j}}(h) |\det h|^{\mathrm{it}} \overline{W}_f(mk) \overline{W}_{\Phi_{\pi' j}}(m) |\det m|^{-it} dm\, dh\, d\pi'\, dk\, dt.$$

For fixed π', j, t, the integral over m, h, k is a product of two Euler products, since the Whittaker functions factor over primes, normalized by global constants ρ_f and $\rho_{\Phi_{\pi' j}}$. The functions $\{\Phi_{\pi', j} : j\}$ correspond to an orthonormal basis $\{\varepsilon_{\pi' j}\}$ in the local archimedean part π'_∞ of π', so, as noted earlier, by Schur's lemma the global constant $\rho_{\Phi_{\pi' j}}$ is independent of j. For each π', let $F_{\pi'}$ be the finite-prime spherical automorphic form corresponding to distinguished vectors at archimedean places. The $\Phi_{\pi' j}$'s are normalized spherical at all finite places. Thus, for each π' and j,

$$\int_{N_{\mathbb{A}}\backslash H_{\mathbb{A}}} \int_{N_{\mathbb{A}}\backslash H_{\mathbb{A}}} \int_{K_\infty} \varphi(h) W_f(hk) W_{\Phi_{\pi' j}}(h) |\det h|^{\mathrm{it}} \overline{W}_f(mk) \overline{W}_{\Phi_{\pi' j}}(m) |\det m|^{-it} \, \mathrm{d}m \, \mathrm{d}h \, \mathrm{d}k$$

$$= |\rho_f|^2 \cdot |\overline{\rho}_{F_{\pi'}}|^2 \cdot L(\tfrac{1}{2} + it + z, \pi \times \pi') L(\tfrac{1}{2} - it, \pi \times \pi')$$

$$\times \int_{N_\infty\backslash H_\infty} \int_{N_\infty\backslash H_\infty} \int_{K_\infty} \int_{K_\infty} e_{\pi_\infty}(huk) \varepsilon_{\pi' j}(h) |\det h|^{\mathrm{it}} \overline{\varepsilon}_{\pi' j}(m) \overline{e}_{\pi_\infty}(muk) |\det m|^{-it} \, \mathrm{d}m \, \mathrm{d}h \, \mathrm{d}k.$$

This gives the assertion of the theorem.

Remark. Automorphic forms not admitting Whittaker models do not enter this expansion.

4 Spectral Expansion of Poincaré Series

The Poincaré series admits a spectral expansion facilitating its meromorphic continuation. The only cuspidal data appearing in this expansion is from GL_2, right K_v-invariant everywhere locally.

In the Poincaré series $\mathfrak{P}$, let φ_∞ be the archimedean data, and z, w the two complex parameters. For a spherical GL_2 cuspform F, let

$$\Phi_{s,F}\Big(\begin{pmatrix} A & * \\ 0 & D \end{pmatrix} \cdot \theta\Big) = |\det A|^{2s} \cdot |\det D|^{-(r-2)s} \cdot F(D) \qquad \text{(where } \theta \in K_{\mathbb{A}})$$

and define an Eisenstein series

$$E^{r-2,2}_{s,F}(g) = \sum_{\gamma \in P_k^{r-2,2}\backslash G_k} \Phi_{s,F}(\gamma \cdot g).$$

Also, for a Hecke character $\overline{\chi}$, with

$$\Phi_{s_1,s_2,s_3,\chi}\Big(\begin{pmatrix} A & * & * \\ 0 & m_2 & * \\ 0 & 0 & m_3 \end{pmatrix} \cdot \theta\Big) = |\det A|^{s_1} \cdot |m_2|^{s_2} \chi(m_2) \cdot |m_3|^{s_3} \overline{\chi}(m_3),$$

for $\theta \in K_{\mathbb{A}}$, $A \in GL_{r-2}$, define an Eisenstein series

$$E^{r-2,1,1}_{s_1,s_2,s_3,\chi}(g) = \sum_{\gamma \in P_k^{r-2,1,1}\backslash G_k} \Phi_{s_1,s_2,s_3,\chi}(\gamma g).$$

Theorem 2. *With Eisenstein series as just above, the Poincaré series* $\mathfrak{P}$ *has a spectral expansion*

$$\mathfrak{P} = \Big(\int_{N_\infty} \varphi_\infty\Big) E^{r-1,1}_{z+1} + \sum_F \Big(\int_{PGL_2(k_\infty)} \widetilde{\varphi}_\infty W_{\overline{F},\infty}\Big) \cdot \rho_{\overline{F}} \cdot L(\tfrac{rz+r-2}{2} + \tfrac{1}{2}, \pi_{\overline{F}}) \cdot E^{r-2,2}_{\frac{z+1}{2},F}$$

$$+ \sum_\chi \frac{\chi(\mathfrak{d})}{4\pi i \kappa} \int_{\mathrm{Re}\,(s)=\frac{1}{2}} \Big(\Big(\int_{PGL_2(k_\infty)} \widetilde{\varphi}_\infty \cdot W_{E_{1-s,\overline{\chi}},\infty}\Big) \cdot \frac{L(\frac{rz+r-2}{2} + 1 - s, \overline{\chi}) \cdot L(\frac{rz+r-2}{2} + s, \chi)}{\Lambda(2-2s, \overline{\chi}^2)}$$

$$\times |\mathfrak{d}|^{-(\frac{rz+r-2}{2} + s - \frac{1}{2})} \cdot E^{r-2,1,1}_{z+1, s - \frac{(r-2)(z+1)}{2}, -s - \frac{(r-2)(z+1)}{2}, \chi}\Big)\, ds,$$

where F runs over an orthonormal basis for everywhere-spherical cuspforms for GL_2*,* $\overline{\rho}_F$ *is the* GL_2 *leading Fourier coefficient of* $\overline{F}$*,* χ *runs over unramified grossencharacters,* $\mathfrak{d}$ *is the differental ideal of k,* κ *is the residue of* $\zeta_k(s)$ *at* $s = 1$*,* $W_{F,\infty}$ *and* $W_{E_{s,\chi}}$ *are the normalized archimedean Whittaker functions for* GL_2*,* $\pi_{\overline{F}}$ *is the representation generated by* $\overline{F}$*,* $L(s,\chi)$ *is the usual grossencharacter L–function, and* $\Lambda(s,\chi)$ *is the grossencharacter L–function with its gamma factor.*

Remark. Notably, the spectral expansion of $\mathfrak{P}$ contains nothing beyond the main term, the cuspidal GL_2 part induced up to GL_r, and the continuous GL_2 part induced up to GL_r.

Proof. Rewrite the Poincaré series as summed in two stages, and apply Poisson summation, namely

$$\mathfrak{P}(g) = \sum_{Z_k H_k \backslash G_k} \varphi(\gamma g) = \sum_{Z_k H_k U_k \backslash G_k} \sum_{\beta \in U_k} \varphi(\beta\gamma g) = \sum_{Z_k H_k U_k \backslash G_k} \sum_{\psi \in (U_k \backslash U_{\mathbb{A}})^\frown} \widehat{\varphi}_{\gamma g}(\psi),$$

where

$$\widehat{\varphi}_g(\psi) = \int_{U_{\mathbb{A}}} \overline{\psi}(u)\, \varphi(ug)\, du \qquad \text{(for } g \in G_{\mathbb{A}}\text{)}.$$

The inner summand for ψ *trivial* gives the leading term in the spectral expansion of the Poincaré series. Specifically, it gives a vector from which a degenerate Eisenstein series for the $(r-1,1)$ parabolic $P^{r-1,1} = ZHU$ is formed by the outer sum. That is,

$$g \to \int_{U_{\mathbb{A}}} \varphi(ug)\, du$$

is left equivariant by a character on $P^{r-1,1}_{\mathbb{A}}$, and is left invariant by $P^{r-1,1}_k$, namely,

$$\int_{U_{\mathbb{A}}} \varphi(upg)\, du = \int_{U_{\mathbb{A}}} \varphi(p \cdot p^{-1}up \cdot g)\, du = \delta_{P^{r-1,1}}(m) \cdot \int_{U_{\mathbb{A}}} \varphi(m \cdot u \cdot g)\, du$$

$$= \left|\frac{\det A}{d^{r-1}}\right|^{z+1} \int_{U_{\mathbb{A}}} \varphi(ug)\, du \qquad \Big(\text{where } p = \begin{pmatrix} A & * \\ 0 & d \end{pmatrix}, m = \begin{pmatrix} A & 0 \\ 0 & d \end{pmatrix}, A \in GL_{r-1}\Big).$$

The normalization is explicated by setting $g = 1$:

$$\int_{U_{\mathbb{A}}} \varphi(u)\,\mathrm{d}u = \int_{U_\infty} \varphi_\infty \cdot \int_{U_{\mathrm{fin}}} \varphi_{\mathrm{fin}} = \int_{U_\infty} \varphi_\infty \cdot \mathrm{meas}\,(U_{\mathrm{fin}} \cap K_{\mathrm{fin}}) = \int_{U_\infty} \varphi_\infty .$$

A natural normalization is that this be 1, so the Eisenstein series includes the archimedean integral and finite-prime measure constant as factors:

$$\int_{U_\infty} \varphi_\infty \;\cdot\; E^{r-1,1}_{z+1}(g) \;=\; \sum_{\gamma \in P_k^{r-1,1} \backslash G_k} \left(\int_{U_{\mathbb{A}}} \varphi(u\gamma g)\,\mathrm{d}u \right).$$

The group H_k is transitive on non-trivial characters of $U_k \backslash U_{\mathbb{A}}$. For fixed non-trivial character ψ_0 on $k \backslash \mathbb{A}$, let

$$\psi^\xi(u) = \psi_0(\xi \cdot u_{r-1,r}) \qquad\qquad (\text{for } \xi \in k^\times).$$

The spectral expansion of $\mathfrak{P}$ with its leading term removed is

$$\sum_{\gamma \in P_k^{r-1,1} \backslash G_k} \; \sum_{\alpha \in P_k^{r-2,1} \backslash H_k} \left(\sum_{\xi \in k^\times} \widehat{\varphi}_{\alpha\gamma g}(\psi^\xi) \right),$$

where $P^{r-2,1}$ is the corresponding parabolic subgroup of $H \approx GL_{r-1}$. Let

$$U' = \left\{ \begin{pmatrix} 1_{r-2} & & * \\ & 1 & \\ & & 1 \end{pmatrix} \right\}, \quad U'' = \left\{ \begin{pmatrix} 1_{r-2} & & \\ & 1 & * \\ & & 1 \end{pmatrix} \right\}.$$

Let

$$\Theta = \left\{ \begin{pmatrix} 1_{r-2} & & \\ & * & * \\ & * & * \end{pmatrix} \right\} \approx GL_2.$$

Regrouping the sums, the expansion of the Poincaré series with its leading term removed is

$$\sum_{\gamma \in P_k^{r-2,1,1} \backslash G_k} \left(\sum_{\xi \in k^\times} \int_{U''_{\mathbb{A}}} \overline{\psi}^\xi(u'') \int_{U'_{\mathbb{A}}} \varphi(u'u''\gamma g)\,\mathrm{d}u'\,\mathrm{d}u'' \right)$$

$$= \sum_{\gamma \in P_k^{r-2,2} \backslash G_k} \; \sum_{\alpha \in P^{1,1} \backslash \Theta_k} \left(\sum_{\xi \in k^\times} \int_{U''_{\mathbb{A}}} \overline{\psi}^\xi(u'') \int_{U'_{\mathbb{A}}} \varphi(u'u''\alpha\gamma g)\,\mathrm{d}u'\,\mathrm{d}u'' \right).$$

Letting

$$\widetilde{\varphi}(g) = \int_{U'_{\mathbb{A}}} \varphi(u'g)\,\mathrm{d}u'$$

the expansion becomes

$$\sum_{\gamma \in P_k^{r-2,2} \backslash G_k} \; \sum_{\alpha \in P^{1,1} \backslash \Theta_k} \; \sum_{\xi \in k^\times} \int_{U''_{\mathbb{A}}} \overline{\psi}^{\xi}(u'')\, \widetilde{\varphi}(u''\alpha\gamma g)\,\mathrm{d}u''.$$

We claim the equivariance

$$\widetilde{\varphi}(pg) = |\det A|^{z+1} \cdot |a|^z \cdot |d|^{-(r-1)z-(r-2)} \cdot \widetilde{\varphi}(g),$$

for $p = \begin{pmatrix} A & * & * \\ & a & \\ & & d \end{pmatrix} \in G_{\mathbb{A}}$, with $A \in GL_{r-2}$.

This is verified by changing variables in the defining integral: let $x \in \mathbb{A}^{r-2}$ and compute

$$\begin{pmatrix} 1_{r-2} & & x \\ & 1 & \\ & & 1 \end{pmatrix} \begin{pmatrix} A & b & c \\ & a & \\ & & d \end{pmatrix} = \begin{pmatrix} A & b & c+xd \\ & a & \\ & & d \end{pmatrix} = \begin{pmatrix} A & b & c \\ & a & \\ & & d \end{pmatrix} \begin{pmatrix} 1_{r-2} & & A^{-1}xd \\ & 1 & \\ & & 1 \end{pmatrix}.$$

Thus, $|\det A|^z \cdot |a|^z \cdot |d|^{-(r-1)z}$ comes out of the definition of φ, and another $|\det A| \cdot |d|^{2-r}$ from the change-of-measure in the change of variables replacing x by Ax/d in the integral defining $\widetilde{\varphi}$ from φ. Note that

$$|a|^z \cdot |d|^{-(r-1)z-(r-2)} = |\det \begin{pmatrix} a & \\ & d \end{pmatrix}|^{-\frac{(r-2)}{2}\cdot(z+1)} \cdot |a/d|^{\frac{rz+(r-2)}{2}}.$$

Thus, letting

$$\Phi(g) = \sum_{\alpha \in P_k^{1,1} \backslash \Theta_k} \left(\sum_{\xi \in k^\times} \int_{U''_{\mathbb{A}}} \overline{\psi}^{\xi}(u'')\, \widetilde{\varphi}(u''\alpha g)\,\mathrm{d}u'' \right),$$

we can write

$$\mathfrak{P}(g) - \sum_{\gamma \in P_k^{r-1,1} \backslash G_k} \int_{U_{\mathbb{A}}} \varphi(u\gamma g)\,\mathrm{d}u = \sum_{\gamma \in P_k^{r-2,2} \backslash G_k} \Phi(\gamma g).$$

The right-hand side of the latter equality is not an Eisenstein series for $P^{r-2,2}$ in the strictest sense.

Define a GL_2 kernel $\varphi^{(2)}$ for a Poincaré series as follows. As in the general case, we require right invariance by the maximal compact subgroups locally everywhere, and left equivariance

$$\varphi^{(2)}\left(\begin{pmatrix} a & * \\ & d \end{pmatrix} \cdot D\right) = |a/d|^{\beta} \cdot \varphi^{(2)}(D).$$

The remaining ambiguity is the archimedean data $\varphi_\infty^{(2)}$, completely specified by giving its values on the archimedean part of the standard unipotent radical, namely,

$$\varphi_\infty^{(2)}\begin{pmatrix} 1 & x \\ & 1 \end{pmatrix} = \widetilde{\varphi}\begin{pmatrix} 1_{r-2} & & \\ & 1 & x \\ & & 1 \end{pmatrix} \qquad (\widetilde{\varphi} \text{ as above}).$$

Let $U^{1,1}$ be the unipotent radical of the standard parabolic $P^{1,1}$ in GL_2. Express $\varphi^{(2)}$ in its Fourier expansion along $U^{1,1}$, and remove the constant term: let

$$\varphi^*(\beta, D) = \varphi^{(2)}(\beta, D) - \int_{U^{1,1}_{\mathbb{A}}} \varphi^{(2)}(\beta, uD)\, du = \sum_{\xi \in k^\times} \int_{U^{1,1}_{\mathbb{A}}} \overline{\psi}^{\xi}(u)\, \varphi^{(2)}(\beta, uD)\, du.$$

The corresponding GL_2 Poincaré series with leading term removed is

$$\mathfrak{Q}(\beta, D) = \sum_{\alpha \in P_k^{1,1} \backslash GL_2(k)} \varphi^*(\beta, \alpha D).$$

Thus, for

$$g = \begin{pmatrix} A & * \\ & D \end{pmatrix} \qquad (\text{with } A \in GL_{r-2}(\mathbb{A}) \text{ and } D \in GL_2(\mathbb{A})),$$

the inner integral

$$g \to \int_{U''_{\mathbb{A}}} \overline{\psi}(u'')\, \widetilde{\varphi}(u''g)\, du''$$

is expressible in terms of the kernel φ^* for $\mathfrak{Q}$, namely,

$$\sum_{\xi \in k^\times} \int_{U''_{\mathbb{A}}} \overline{\psi}^{\xi}(u'')\, \widetilde{\varphi}(u''g)\, du'' = |\det A|^{z+1} \cdot |\det D|^{-\frac{(r-2)}{2}\cdot(z+1)} \cdot \varphi^*\left(\tfrac{rz+r-2}{2}, D\right).$$

Thus,

$$\sum_{\alpha \in P_k^{1,1} \backslash \Theta_k} \sum_{\xi \in k^\times} \int_{U''_{\mathbb{A}}} \overline{\psi}^{\xi}(u'')\widetilde{\varphi}(u''\alpha g)\, du'' = |\det A|^{z+1} |\det D|^{-\frac{(r-2)}{2}(z+1)} \mathfrak{Q}\left(\tfrac{rz+r-2}{2}, D\right).$$

Thus, letting

$$\Phi\begin{pmatrix} A & * \\ & D \end{pmatrix} = |\det A|^{z+1} \cdot |\det D|^{-(r-2)\cdot\frac{z+1}{2}} \cdot \mathfrak{Q}(\tfrac{rz+r-2}{2}, D),$$

with $A \in GL_{r-2}$ and $D \in GL_2$, we have

$$\mathfrak{P}(g) = \left(\int_{U_\infty} \varphi_\infty\right) \cdot E^{r_1,1}_{z+1}(g) + \sum_{\gamma \in P_k^{r-2,2}\backslash G_k} \Phi(\gamma g).$$

To obtain a spectral decomposition of the Poincaré series $\mathfrak{P}$ for GL_r, we first recall from [11] the spectral decomposition of $\mathfrak{Q}$ for $r=2$, and then form $P^{r-2,2}$ Eisenstein series from the spectral fragments.

As in [11], a direct computation shows that the spectral expansion of the GL_2 Poincaré series with constant term removed is

$$\mathfrak{Q}(\beta, D) = \sum_F \left(\int_{PGL_2(k_\infty)} \widetilde{\varphi}_\infty \cdot W_{\overline{F},\infty}\right) \cdot \overline{\rho}_F \cdot L(\beta + \tfrac{1}{2}, \pi_{\overline{F}}) \cdot F$$

$$+ \sum_\chi \frac{\chi(\mathfrak{d})}{4\pi i \kappa} \int_{\mathrm{Re}\,(s)=\frac{1}{2}} \left(\int_{PGL_2(k_\infty)} \widetilde{\varphi}_\infty \cdot W_{E_{1-s,\overline{\chi}},\infty}\right)$$

$$\times \frac{L(\beta+1-s, \overline{\chi}) \cdot L(\beta+s, \chi)}{L(2-2s, \overline{\chi}^2)} \cdot |\mathfrak{d}|^{-(\beta+s-1/2)} \cdot E_{s,\chi}(D)\, ds,$$

where F runs over an orthonormal basis of everywhere-spherical cuspforms, $\overline{\rho}_F$ is the general GL_2 analogue of the leading Fourier coefficient, $\pi_{\overline{F}}$ is the cuspidal automorphic representation generated by $\overline{F}$, $W_{\overline{F},\infty}$ and $W_{E_{s,\chi},\infty}$ are the normalized spherical vectors in the corresponding archimedean Whittaker models, $\Lambda(s,\chi)$ is the standard L-function completed by adding the archimedean factors, and $\mathfrak{d}$ is the differental idele. Thus, the individual spectral components of Φ are of the form

$$\Phi_{\frac{z+1}{2},\Psi}\left(\begin{pmatrix} A & * \\ 0 & D \end{pmatrix} \cdot \theta\right) = (\text{constant}) \cdot |\det A|^{z+1} \cdot |\det D|^{-(r-2)\frac{z+1}{2}} \cdot \Psi(D),$$

where $\theta \in K_{\mathbb{A}}$ and Ψ is either a spherical GL_2 cuspform or a spherical GL_2 Eisenstein series, in either case with trivial central character.

For Ψ a spherical GL_2 cuspform F averaging over $P_k^{r-2,2}\backslash G_k$ produces a half-degenerate Eisenstein series

$$E^{r-2,2}_{\frac{z+1}{2},F}(g) = \sum_{\gamma \in P_k^{r-2,2}\backslash G_k} \Phi_{\frac{z+1}{2},F}(\gamma \cdot g).$$

As in the appendix, the half-degenerate Eisenstein series $E^{r-2,2}_{s,F}$ has *no poles* in $\mathrm{Re}\,(s) \geq 1/2$. With $s = (z+1)/2$ this assures absence of poles in $\mathrm{Re}\,(z) \geq 0$.

The continuous spectrum part of $\mathfrak{Q}$ produces degenerate Eisenstein series on G, as follows. With $\Psi = E_{s,\chi}$ the usual spherical, trivial central character, Eisenstein series for GL_2, define an Eisenstein series

$$E^{r-2,2}_{\frac{z+1}{2},E_{s,\chi}}(g) = \sum_{\gamma \in P_k^{r-2,2} \backslash G_k} \Phi_{\frac{z+1}{2},E_{s,\chi}}(\gamma g).$$

As usual, for $\mathrm{Re}(s) \gg 0$ and $\mathrm{Re}(z) \gg 0$, this iterated formation of Eisenstein series is equal to a single-step Eisenstein series. That is, let

$$\Phi_{s_1,s_2,s_3,\chi}\left(\begin{pmatrix} A & * & * \\ 0 & m_2 & * \\ 0 & 0 & m_3 \end{pmatrix} \cdot \theta\right) = |\det A|^{s_1} \cdot |m_2|^{s_2} \chi(m_2) \cdot |m_3|^{s_3} \overline{\chi}(m_3),$$

for $\theta \in K_{\mathbb{A}}$, $A \in GL_{r-2}$, and let

$$E^{r-2,1,1}_{s_1,s_2,s_3,\chi}(g) = \sum_{\gamma \in P_k^{r-2,1,1} \backslash G_k} \Phi_{s_1,s_2,s_3,\chi}(\gamma g).$$

Taking $s_1 = 2 \cdot \frac{z+1}{2}$, $s_2 = s - \frac{(r-2)(z+1)}{2}$, and $s_3 = -s - \frac{(r-2)(z+1)}{2}$,

$$E^{r-2,2}_{\frac{z+1}{2},E_{s,\chi}} = E^{r-2,1,1}_{z+1,s-\frac{(r-2)(z+1)}{2},-s-\frac{(r-2)(z+1)}{2},\chi}.$$

Adding up these spectral components yields the spectral expansion of the Poincaré series.

Remark. Suitable archimedean data to give the Poincaré series a functional equation is best described in the context of the spectral expansion, and, due to the form of the spectral expansion, essentially reduces to GL_2. It is useful to describe the data through a *differential equation*, since this explains the outcome of the computation more transparently. Since each archimedean place affords its own opportunity for data choices, we simplify this aspect of the situation by taking groundfield $k = \mathbb{Q}$.

First, for $G = GL_2(\mathbb{Q})$, let Δ be the usual invariant Laplacian on the upper half-plane $\mathfrak{H}$, and consider the partial differential equations

$$(\Delta - s(s-1))^{\nu} u^{\beta}_{s,\nu} = \text{the distribution } f \to \int_0^{\infty} y^{\beta} \cdot f\begin{pmatrix} y & 0 \\ 0 & 1 \end{pmatrix} \frac{dy}{y},$$

for $1 \le \nu \in \mathbb{Z}$ and $s, \beta \in \mathbb{C}$, on $\mathfrak{H}$. Further, require that $u^{\beta}_{s,\nu}$ have the same equivariance as the target distribution, namely,

$$u^{\beta}_{s,\nu}(t \cdot z) = t^{\beta} \cdot u^{\beta}_{s,\nu}(z) \qquad (\text{for } t > 0 \text{ and } z \in \mathfrak{H}).$$

Then $u^{\beta}_{s,\nu}(x+iy) = y^{\beta} \cdot \varphi^{\beta}_{s,\nu}(x/y)$ for a function $\varphi^{\beta}_{s,\nu}$ on $\mathbb{R}$ satisfying the corresponding differential equation

$$\left((1+x^2)\frac{\partial^2}{\partial x^2} + 2x(1-\beta)\frac{\partial}{\partial x} + \big(\beta(\beta-1) - s(s-1)\big)\right)^{\nu} f \;=\; \delta,$$

with Dirac δ at 0.

The generalized function δ is in the L^2 Sobolev space on $\mathbb{R}$ with index $-\frac{1}{2} - \varepsilon$ for every $\varepsilon > 0$. By elliptic regularity, solutions f to this differential equation are in the local Sobolev space with index $2\nu - \frac{1}{2} - \varepsilon$, and by Sobolev's lemma are locally at least $C^{2\nu-1-2\varepsilon} \subset C^{2\nu-2}$. That is, by increasing ν solutions are made as differentiable as desired, and their Fourier transforms will have corresponding decay, giving convergence of the Poincaré series (for suitable s, β), as in [11].

The spectral expansion of the GL_2 Poincaré series $\mathfrak{P}^{\beta}_{s,\nu}$ formed with this archimedean data $\varphi^{\beta}_{s,\nu}$ is a special case of the computation in [11], recalled above, but in fact gives a much simpler outcome. For example, the *cuspidal* components are directly computed by unwinding, *integrating by parts*, and applying the characterization of $\varphi^{\beta}_{s,\nu}$ by the differential equation:

$$\langle \mathfrak{P}^{\beta}_{s,\nu}, F\rangle \;=\; \frac{\overline{\rho}_F(1) \cdot \Lambda(\beta+\frac{1}{2}, \overline{F})}{\big(s_F(s_F-1) - s(s-1)\big)^{\nu}} \qquad \text{where } \Delta F = s_F(s_F-1),$$

where $\Lambda(\cdot, F)$ is the L-function completed with its gamma factors. Thus,

$$\mathfrak{P}^{\beta}_{s,\nu} \;=\; \sum_F \frac{\overline{\rho}_F(1) \cdot \Lambda(\beta+\frac{1}{2}, \overline{F}) \cdot F}{\big(s_F(s_F-1) - s(s-1)\big)^{\nu}} \;+\; (\text{non-cuspidal}),$$

summing over an orthonormal basis of cuspforms F. Granting convergence for ν sufficiently large and $\mathrm{Re}(s), \mathrm{Re}(\beta)$ large, the cuspidal part has a meromorphic continuation in s with poles at the values s_F, as expected. Visibly, the *cuspidal* part of $\mathfrak{P}^{\beta}_{s,\nu}$ is invariant under $s \leftrightarrow 1-s$, and in these coordinates the map $\beta \to -\beta$ maps F to $\overline{F}$ (whether or not F is self-contragredient).

The *leading term* of the spectral expansion of $\mathfrak{P}^{\beta}_{s,\nu}$, via Poisson summation, is a constant multiple $C^{\beta}_{s,\nu} \cdot E_{\beta+1}$ of the spherical Eisenstein series $E_{\beta+1}$. This happens regardless of the precise choice of archimedean data, simply due to the homogeneity we have required of the archimedean data throughout.

Similarly, the *continuous* part of this Poincaré series on GL_2 is

$$\frac{1}{4\pi i}\int_{\mathrm{Re}(s_e)=\frac{1}{2}} \frac{\xi(\beta+s_e)\,\xi(\beta+1-s_e) \cdot E_{s_e}}{\xi(2s_e) \cdot \big((s_e(s_e-1) - s(s-1)\big)^{\nu}}\, ds_e,$$

where ξ is the ζ-function completed with its gamma factor. In analogy with the cuspidal discussion, the product $\xi(\beta+s_e) \cdot \xi(\beta+1-s_e)$ is invariant under $\beta \to -\beta$, since $\xi(1-z) = \xi(z)$. The *visual* symmetry in $s \leftrightarrow 1-s$ is slightly deceiving, since the meromorphic continuation (in s) through the critical line $\mathrm{Re}(s_e) = \frac{1}{2}$ (over which the Eisenstein series is integrated) introduces extra terms from residues at $s_e = s$ and $s_e = 1-s$. Indeed, parts of these extra terms cancel a pole in the leading term $C^{\beta}_{s,v} \cdot E_{\beta+1}$ at $\beta = 0$. Despite this subtlety in the continuous spectrum, the special choice of archimedean data makes meromorphic continuation in s, β visible.

In summary, for $GL_2(\mathbb{Q})$, the special choice of archimedean data makes the *cuspidal* part of the Poincaré series have a visible meromorphic continuation, and satisfy obvious functional equations. The *continuous* part of the Poincaré series satisfies functional equations modulo explicit leftover terms.

The ideal choice of archimedean data φ_∞ for the Poincaré series for $GL_r(\mathbb{Q})$ is such that the *averaged* version of it, denoted $\widetilde{\varphi}_\infty$ in the proof above, restricts to the function $\varphi^{\beta}_{s,v}$ for GL_2 just discussed: we want

$$\int_{\mathbb{R}^{r-2}} \varphi_\infty \begin{pmatrix} 1_{r-2} & 0 & u \\ 0 & 1 & x \\ 0 & 0 & 1 \end{pmatrix} du \;=\; \widetilde{\varphi}_\infty \begin{pmatrix} 1_{r-2} & 0 & 0 \\ 0 & 1 & x \\ 0 & 0 & 1 \end{pmatrix} \;=\; \varphi^{\beta}_{s,v}(x) \quad \text{for } x \in \mathbb{R}.$$

It is not obvious that, given a reasonable (even) function f on $\mathbb{R}$, there is a rotationally symmetric function u on $\mathbb{R}^{r-2}$ such that

$$\int_{\mathbb{R}^{r-2}} u(y + xe_{r-1})\, dy \;=\; f(x) \quad e_i \text{ the standard basis for } \mathbb{R}^{r-1},$$

with $\mathbb{R}^{r-2}$ sitting in the first $r-2$ coordinates in $\mathbb{R}^{r-1}$. Fourier inversion clarifies this, as follows. Supposing the integral identity just above holds, integrate further in the $(r-1)^{\text{th}}$ coordinate x, against $e^{2\pi i \xi x}$, to obtain

$$\widehat{u}(\xi e_{r-1}) \;=\; \widehat{f}(\xi) \quad \text{for } \xi \in \mathbb{R},$$

where the Fourier transform on the left-hand side is on $\mathbb{R}^{r-1}$, on the right-hand side is on $\mathbb{R}$. For u rotationally invariant, $\widehat{u}$ is also rotationally invariant, and the latter equality can be rewritten as

$$\widehat{u}(\xi) \;=\; \widehat{f}(|\xi|) \quad \text{for } \xi \in \mathbb{R}^{r-1}.$$

By Fourier inversion,

$$u(x) \;=\; \int_{\mathbb{R}^{r-1}} e^{2\pi i \langle \xi, x \rangle}\, \widehat{f}(|\xi|)\, d\xi \quad \text{for } x \in \mathbb{R}^{r-1}.$$

That is, given an even function f on $\mathbb{R}$, the latter formula yields a rotationally invariant function on $\mathbb{R}^{r-1}$, whose averages along $\mathbb{R}^{r-2}$ are the given f. This proves existence of an essentially unique φ_∞ yielding the prescribed $\varphi^{\beta}_{s,v}$.

Then the functional equation of the most-cuspidal part of the special-data Poincaré series on GL_r is inherited from the functional equation of the cuspidal part of the special-data Poincaré series on GL_2.

5 Appendix: Half-Degenerate Eisenstein Series

Take $q > 1$, and let f be a cuspform on $GL_q(\mathbb{A})$, in the strong sense that f is in $L^2(GL_q(k)\backslash GL_q(\mathbb{A})^1)$, and f meets the Gelfand–Fomin–Graev conditions

$$\int_{N_k\backslash N_{\mathbb{A}}} f(ng)\,\mathrm{d}n = 0 \qquad \text{(for almost all } g\text{)}$$

and f generates an irreducible representation of $GL_q(k_v)$ locally at all places v of k. For a Schwartz function Φ on $\mathbb{A}^{q\times r}$ and Hecke character χ, let

$$\varphi(g) = \varphi_{\chi,f,\Phi}(g) = \chi(\det g)^q \int_{GL_q(\mathbb{A})} f(h^{-1})\,\chi(\det h)^r\,\Phi(h\cdot[0_{q\times(r-q)}\ 1_q]\cdot g)\,\mathrm{d}h.$$

This function φ has the same central character as f. It is left invariant by the adele points of the unipotent radical

$$N = \left\{\begin{pmatrix} 1_{r-q} & * \\ & 1_r \end{pmatrix}\right\} \qquad \text{(unipotent radical of } P = P^{r-q,q}\text{)}.$$

The function φ is left invariant under the *k-rational* points M_k of the standard Levi component of P,

$$M = \left\{\begin{pmatrix} a & \\ & d \end{pmatrix} \ :\ a \in GL_{r-q},\ d \in GL_r\right\}.$$

To understand the normalization, observe that

$$\xi(\chi^r, f, \Phi(0,*)) = \varphi(1) = \int_{GL_q(\mathbb{A})} f(h^{-1})\,\chi(\det h)^r\,\Phi(h\cdot[0_{q\times(r-q)}\ 1_q])\,\mathrm{d}h$$

is a zeta integral as in [18] for the standard L-function attached to the cuspform f. Thus, the Eisenstein series formed from φ includes this zeta integral as a factor, so write

$$\xi(\chi^r, f, \Phi(0,*))\cdot E^P_{\chi,f,\Phi}(g) \;=\; \sum_{\gamma\in P_k\backslash GL_r(k)} \varphi(\gamma g) \quad \text{(convergent for } \mathrm{Re}\,(\chi) \gg 1\text{)}.$$

The meromorphic continuation follows by Poisson summation:

$$\begin{aligned}\xi(\chi^r, f, \Phi(0,*)) \cdot E^P_{\chi,f,\Phi}(g) \\ &= \chi(\det g)^q \sum_{\gamma \in P_k \backslash GL_r(k)} \int_{GL_q(k)\backslash GL_q(\mathbb{A})} f(h)\,\chi(\det h)^{-r} \sum_{\alpha \in GL_q(k)} \Phi(h^{-1}\cdot[0\ \alpha]\cdot g)\,\mathrm{d}h \\ &= \chi(\det g)^q \int_{GL_q(k)\backslash GL_q(\mathbb{A})} f(h)\,\chi(\det h)^{-r} \sum_{y \in k^{q\times r},\ \text{full rank}} \Phi(h^{-1}\cdot y \cdot g)\,\mathrm{d}h.\end{aligned}$$

The Gelfand–Fomin–Graev condition on f fits the full-rank constraint. Anticipating that we can drop the rank condition suggests that we define

$$\Theta_\Phi(h,g) = \sum_{y\in k^{q\times r}} \Phi(h^{-1}\cdot y\cdot g).$$

As in [18], the non-full-rank terms integrate to 0:

Proposition 1. *For f a cuspform, less-than-full-rank terms integrate to* 0, *that is,*

$$\int_{GL_q(k)\backslash GL_q(\mathbb{A})} f(h)\,\chi(\det h)^{-r} \sum_{y\in k^{q\times r},\ \text{rank} < q} \Phi(h^{-1}\cdot y\cdot g)\,\mathrm{d}h = 0.$$

Proof. Since this is asserted for arbitrary Schwartz functions Φ, we can take $g = 1$. By linear algebra, given $y_0 \in k^{q\times r}$ of rank ℓ, there is $\alpha \in GL_q(k)$ such that

$$\alpha\cdot y_0 = \begin{pmatrix} y_{\ell\times r} \\ 0_{(q-\ell)\times r}\end{pmatrix} \qquad \text{(with } \ell\text{-by-}r \text{ block } y_{\ell\times r} \text{ of rank } \ell).$$

Thus, without loss of generality fix y_0 of the latter shape. Let Y be the orbit of y_0 under left multiplication by the rational points of the parabolic

$$P^{\ell,q-\ell} = \left\{\begin{pmatrix} \ell\text{-by-}\ell & * \\ 0 & (q-\ell)\text{-by-}(q-\ell)\end{pmatrix}\right\} \subset GL_q.$$

This is some set of matrices of the same shape as y_0. Then the subsum over $GL_q(k)\cdot y_0$ is

$$\begin{aligned}&\int_{GL_q(k)\backslash GL_q(\mathbb{A})} f(h)\,\chi(\det h)^{-r} \sum_{y\in GL_q(k)\cdot y_0} \Phi(h^{-1}\cdot y)\,\mathrm{d}h \\ &= \int_{P_k^{\ell,q-\ell}\backslash GL_q(\mathbb{A})} f(h)\,\chi(\det h)^{-r} \sum_{y\in Y} \Phi(h^{-1}\cdot y)\,\mathrm{d}h.\end{aligned}$$

Let N and M be the unipotent radical and standard Levi component of $P^{\ell,q-\ell}$,

$$N = \begin{pmatrix} 1_\ell & * \\ 0 & 1_{q-\ell}\end{pmatrix}, \qquad M = \begin{pmatrix} \ell\text{-by-}\ell & 0 \\ 0 & (q-\ell)\text{-by-}(q-\ell)\end{pmatrix}.$$

Then the integral can be rewritten as an iterated integral

$$\int_{N_k M_k \backslash GL_q(\mathbb{A})} f(h)\,\chi(\det h)^{-r} \sum_{y\in Y} \Phi(h^{-1}\cdot y)\,dh$$

$$= \int_{N_{\mathbb{A}} M_k \backslash GL_q(\mathbb{A})} \sum_{y\in Y} \int_{N_k\backslash N_{\mathbb{A}}} f(nh)\,\chi(\det nh)^{-r}\,\Phi((nh)^{-1}\cdot y)\,dn\,dh$$

$$= \int_{N_{\mathbb{A}} M_k \backslash GL_q(\mathbb{A})} \sum_{y\in Y} \chi(\det h)^{-r}\,\Phi(h^{-1}\cdot y)\left(\int_{N_k\backslash N_{\mathbb{A}}} f(nh)\,dn\right)dh,$$

since all fragments but $f(nh)$ in the integrand are left invariant by $N_{\mathbb{A}}$. The inner integral of $f(nh)$ is 0, by the Gelfand–Fomin–Graev condition, so the whole is 0.

Let ι denote the transpose-inverse involution. Poisson summation gives

$$\Theta_\Phi(h,g) = \sum_{y\in k^{q\times r}} \Phi(h^{-1}\cdot y\cdot g) = |\det(h^{-1})^\iota|^r\,|\det g^\iota|^q \sum_{y\in k^{q\times r}} \widehat{\Phi}((h^\iota)^{-1}\cdot y\cdot g^\iota)$$

$$= |\det(h^{-1})^\iota|^r\,|\det g^\iota|^q\,\Theta_{\widehat{\Phi}}(h^\iota, g^\iota).$$

As with Θ_Φ, the lower-rank summands in $\Theta_{\widehat{\Phi}}$ integrate to 0 against cuspforms. Thus, letting

$$GL_q^+ = \{h\in GL_q(\mathbb{A}) \,:\, |\det h|\geq 1\} \qquad GL_q^- = \{h\in GL_q(\mathbb{A}) \,:\, |\det h|\leq 1\},$$

we have

$$\xi(\chi^r, f, \Phi(0,*))\cdot E^P_{\chi,f,\Phi}(g) = \chi(\det g)^q \int_{GL_q(k)\backslash GL_q(\mathbb{A})} f(h)\,\chi(\det h)^{-r}\,\Theta_\Phi(h,g)\,dh$$

$$= \chi(\det g)^q \int_{GL_q(k)\backslash GL_q^+} f(h)\,\chi(\det h)^{-r}\,\Theta_\Phi(h,g)\,dh$$

$$+\chi(\det g)^q \int_{GL_q(k)\backslash GL_q^-} f(h)\,\chi(\det h)^{-r}\,\Theta_\Phi(h,g)\,dh$$

$$= \chi(\det g)^q \int_{GL_q(k)\backslash GL_q^+} f(h)\,\chi(\det h)^{-r}\,\Theta_\Phi(h,g)\,dh$$

$$+\,\chi(\det g)^q \int_{GL_q(k)\backslash GL_q^-} |\det(h^{-1})^\iota|^r\,|\det g^\iota|^q\,f(h)\,\chi(\det h)^{-r}\,\Theta_{\widehat{\Phi}}(h^\iota, g^\iota)\,dh.$$

By replacing h by h^ι in the second integral, convert it to an integral over $GL_q(k)\backslash GL_q^+$, and the whole is

$$\xi(\chi^r, f, \Phi(0,*)) \cdot E^P_{\chi,f,\Phi}(g) = \chi(\det g)^q \int_{GL_q(k)\backslash GL_q^+} f(h)\, \chi(\det h)^{-r}\, \Theta_\Phi(h,g)\, dh$$

$$+ \chi^{-1}(\det g^\iota)^q \int_{GL_q(k)\backslash GL_q^+} f(h^\iota)\, \chi^{-1}(\det h^\iota)^{-r}\, \Theta_{\widehat{\Phi}}(h, g^\iota)\, dh.$$

Since $f \circ \iota$ is a cuspform, the second integral is entire in χ. Thus, we have proven

$$\xi(\chi^r, f, \Phi(0,*)) \cdot E^P_{\chi,f,\Phi} \quad \text{is entire.}$$

Acknowledgements The authors were partially supported by NSF grant DMS-0652488.

References

1. R.W. Bruggeman, *Fourier coefficients of cusp forms*, Inv. Math. **45** no. 1 (1978), 1–18.
2. R.W. Bruggeman, Y. Motohashi, *Fourth power moment of Dedekind zeta-functions of real quadratic number fields with class number one*, Funct. Approx. Comment. Math. **29** (2001), 41–79.
3. R.W. Bruggeman, Y. Motohashi, *Sum formula for Kloosterman sums and fourth moment of the Dedekind zeta-function over the Gaussian number field*, Funct. Approx. Comment. Math. **31** (2003), 23–92.
4. W. Casselman, *On some results of Atkin and Lehner*, Math. Ann. **206** (1973), 311–318.
5. W. Casselman, J. Shalika, *The unramified principal series of p-adic groups, II, the Whittaker function*, Comp. Math. **41** (1980), 207–231.
6. J. Cogdell, *L-functions and Converse Theorems*, in *Automorphic Forms and Applications*, P. Sarnak, F. Shahidi eds, IAS/Park City Mathematics Series, vol. 12, AMS, Providence, 2007, 95–178.
7. J. Cogdell, *Analytic theory of L-functions for GL_n*, in *An Introduction to the Langlands Program*, J. Bernstein and S. Gelbart eds, Birhauser, 2003, 197–228.
8. J. Cogdell, *Lectures on L-functions, Converse Theorems, and Functoriality for GL_n*, Fields Institute Notes, in *Lectures on Automorphic L-functions*, Fields Institute Monographs no. 20, AMS, Providence, 2004.
9. J. Cogdell, I. Piatetski-Shapiro, *Remarks on Rankin-Selberg convolutions*, in *Contributions to Automorphic Forms, Geometry, and Number Theory (Shalikafest 2002)*, H. Hida, D. Ramakrishnan, and F. Shahidi eds., Johns Hopkins University Press, Baltimore, 2005.
10. J.B. Conrey, D.W. Farmer, J.P. Keating, M.O. Rubinstein, N.C. Snaith, *Integral Moments of L-functions*, Proc. London Math. Soc. **91** (2005), 33–104.
11. A. Diaconu, P. Garrett, *Integral moments of automorphic L-functions*, J. Math. Inst. Jussieu **8** no. 2 (2009), 335–382.
12. A. Diaconu, P. Garrett, *Subconvexity bounds for automorphic L-functions*, J. Math. Inst. Jussieu **9** no. 1 (2010), 95–124.
13. A. Diaconu, P. Garrett, D. Goldfeld, *Natural boundaries and integral moments of L-functions*, to appear, Proceedings of 2008 Edinburgh Conference on Multiple Dirichlet Series.
14. A. Diaconu, D. Goldfeld, *Second moments of quadratic Hecke L-series and multiple Dirichlet series, I*, Proc. Symp. Pure Math. **75**, AMS, Providence, 2006, 59–89.

15. A. Diaconu, D. Goldfeld, *Second moments of GL_2 automorphic L–functions*, in proceedings of Gauss-Dirichlet Conference (Göttingen, 2005), Clay Math. Proc., AMS (2007), 77–105.
16. A. Diaconu, D. Goldfeld, J. Hoffstein, *Multiple Dirichlet series and moments of zeta and L-functions*, Comp. Math. **139** no. 3 (2003), 297–360.
17. H. Donnelly, *On the cuspidal spectrum for finite volume symmetric spaces*, J. Diff. Geom. **17** (1982), 239–253.
18. R. Godement, H. Jacquet, *Zeta functions of simple algebras*, SLN 260, Springer-Verlag, Berlin, 1972.
19. D. Goldfeld, *Automorphic forms and L–functions for the group $GL(n,\mathbb{R})$*, Cambridge Studies in Adv. Math. vol. 99, Cambridge Univ. Press, 2006.
20. A. Good, *Cusp forms and eigenfunctions of the Laplacian*, Math. Ann. **255** (1981), 523–548.
21. A. Good, *The square mean of Dirichlet series associated with cusp forms*, Mathematika **29** (1983), 278–95.
22. A. Good, *The convolution method for Dirichlet series*, in *Selberg Trace Formula and Related Topics (Brunswick, Maine, 1984)*, Contemp. Math. **53**, AMS, Providence, 1986, 207–214.
23. G.H. Hardy, J.E. Littlewood, *Contributions to the theory of the Riemann zeta-function and the theory of the distribution of primes*, Acta Math **41** (1918), 119–196.
24. D.R. Heath-Brown, *The fourth power moment of the Riemann zeta function*, Proc. London Math. Soc. **38** no. 3 (1979), 385–422.
25. J. Hoffstein, P. Lockhart, *Coefficients of Maass forms and the Siegel zero*, with appendix *An effective zero-free region*, by D. Goldfeld, J. Hoffstein, D. Lieman, Ann. of Math. **140** (1994), 161–181.
26. J. Hoffstein and D. Ramakrishnan, *Siegel zeros and cuspforms*, Int. Math. Research Notices **6** (1995), 279–308.
27. A.E. Ingham, *Mean-value theorems in the theory of the Riemann zeta function*, Proc. London Math. Soc. **27** no. 2 (1926), 273–300.
28. H. Iwaniec and P. Sarnak, *Perspectives on the analytic theory of L–functions*, GAFA special volume (2000), 705–741.
29. H. Jacquet, *On the residual spectrum of GL_n*, in *Lie Group Representations, II*, SLN 1041.
30. H. Jacquet, I. Piatetski-Shapiro, J. Shalika, *Automorphic forms on GL_3, I, II*, Ann. of Math. **109** (1979), 169–212, 213–258.
31. H. Jacquet, I. Piatetski-Shapiro, J. Shalika, *Conducteur des représentations du groupe linéaire*, Math. Ann. **256** (1981), 199–214.
32. H. Jacquet, I. Piatetski-Shapiro, J. Shalika, *Rankin-Selberg convolutions*, Amer. J. Math. **105** (1983), 367–464.
33. H. Jacquet, J. Shalika, *On Euler products and the classification of automorphic representations I, II*, Amer. J. Math. **103** (1981), 499–588, 777–815.
34. H. Jacquet, J. Shalika, *Rankin-Selberg convolutions: archimedean theory*, in *Festschrift in Honor of I.I. Piatetski-Shapiro, I*, Weizmann Science Press, Jerusalem, 1990, 125–207.
35. M. Jutila, *Mean values of Dirichlet series via Laplace transforms*, London Math. Soc. Lecture Notes, vol. 247, Cambridge Univ. Press, 1997, 169–207.
36. R. Langlands, *On the functional equations satisfied by Eisenstein series*, SLN 544, 1975.
37. X. Li, *Bounds for GL(3)XGL(2) L-functions and GL(3) L-functions*, Ann. of Math. (2) **173** (2011), no. 1, 301–336.
38. E. Lindenstrauss and A. Venkatesh *Existence and Weyl's law for spherical cusp forms*, GAFA **17** vol 1 (2007), 220–251.
39. P. Michel, *Analytic number theory and families of automorphic L–functions*, in IAS/Park City Summer Institute Series Vol 12 *Automorphic forms and applications*, (2007), 179–296.
40. C. Moeglin, J.-L. Waldspurger, *Le spectre résiduel de GL_n*, with appendix *Poles des fonctions L de pairs pour GL_n*, Ann. Sci. École Norm. Sup. **22** (1989), 605–674.
41. C. Moeglin, J.-L. Waldspurger, *Spectral Decompositions and Eisenstein series*, Cambridge Univ. Press, Cambridge, 1995.
42. Y. Motohashi, *Spectral theory of the Riemann zeta function*, Cambridge Tracts in Mathematics 127, Cambridge Univ. Press, 1997.

43. Y. Petridis and P. Sarnak, *Quantum unique ergodicity for* $SL_2(\mathscr{O})\backslash H^3$ *and estimates for L-functions*, J. Evol. Equ. **1** (2001), 277–290.
44. P. Sarnak, *Fourth moments of Grössencharakteren zeta functions*, Comom. Pure Appl. Math. **38** no. 2 (1985), 167–178.
45. J. Shalika, *The multiplicity-one theorem for* GL_n, Ann. of Math. **100** (1974), 171–193.
46. T. Shintani, *On an explicit formula for class-one Whittaker functions on* GL_n *over p-adic fields*, Proc. Japan Acad. **52** (1976), 180–182.
47. E. Stade, *Mellin transforms of* $GL_n(\mathbb{R})$ *Whittaker functions*, Amer. J. Math. **123** (2001), 121–161.
48. E. Stade, *Archimedean L-factors on* $GL_n \times GL_n$ *and generalized Barnes integrals*, Israel J. Math. **127** (2002), 201–209.
49. N.R. Wallach, *Real reductive groups, II*, Pure and Applied Math. vol. 132, Academic Press, Boston, 1992.
50. Y. Ye, *A Kuznetsov formula for Kloosterman sums on* GL_n, Ramanujan J. **4** no. 4 (2000), 385–395.
51. M. Young, *The second moment of GL(3) X GL(2) Lfunctions, integrated*, Advances in Math., to appear.

Further Remarks on the Exponent of Convergence and the Hausdorff Dimension of the Limit Set of Kleinian Groups

Martial R. Hille

Abstract In [Patterson, Further remarks on the exponent of convergence of Poincaré series, Tôhoku Math. Journ. 35 (1983), 357–373], it was shown how to construct for a given $\varepsilon > 0$ a Kleinian group of the first kind with exponent of convergence smaller than ε.
We show the more general result that for any $m \in \mathbb{N}$ there are Kleinian groups acting on $(m+1)$-dimensional hyperbolic space for which the Hausdorff dimension of their uniformly radial limit set is less than a given arbitrary number $d \in (0,m)$ and the Hausdorff dimension of their Jørgensen limit set is equal to a given arbitrary number $j \in [0,m)$.
Additionally, our result clarifies which part of the limit set gives rise to the result of Patterson's original construction.
The key idea in our construction is to combine the previous techniques of Patterson with a description of various limit sets in terms of the coding map.

1 Introduction and Statement of Results

Let Γ be a Kleinian group acting on $(m+1)$-dimensional hyperbolic space and let $L(\Gamma)$ denote its limit set. One of the important questions in the theory of Kleinian groups is to understand the relation between the exponent of convergence $\delta(\Gamma)$ of the Poincare series associated with Γ and the Hausdorff dimension $\dim_{\mathrm{H}} L(\Gamma)$ of the limit set $L(\Gamma)$. It was eventually proved by Patterson in [11] that for geometrically finite non-elementary Kleinian groups these quantities do in fact coincide. However,

M.R. Hille (✉)
Institut für Mathematik, Humboldt-Universität zu Berlin, Rudower Chaussee 25, 12489 Berlin, Germany
e-mail: hille@math.hu-berlin.de

V. Blomer and P. Mihăilescu (eds.), *Contributions in Analytic and Algebraic Number Theory*, Springer Proceedings in Mathematics 9, DOI 10.1007/978-1-4614-1219-9_9,

for infinitely generated Kleinian groups, this does not hold in general as was shown in [12].

In fact, in [12] Patterson constructed Kleinian groups of the first kind for which $\delta(\Gamma)$ is arbitrarily small. Recall that a Kleinian group is said to be of the first kind if its limit set equals the boundary of hyperbolic space.

In [1], Bishop and Jones clarified the relationship of $\dim_{\mathrm{H}} L(\Gamma)$ and $\delta(\Gamma)$ by proving that $\delta(\Gamma) = \dim_{\mathrm{H}} L_r(\Gamma)$ for *all* non-elementary Kleinian groups Γ. Here, $L_r(\Gamma)$ denotes the radial limit set, a certain subset of $L(\Gamma)$ (see Definition 3 below). Note that if Γ is geometrically finite, then every limit point is either a radial point or a parabolic fixed point. In [5] further subsets of $L(\Gamma)$ have been introduced, in particular the Jørgensen limit set $L_{\mathrm{J}}(\Gamma)$, whose name was inspired by Sullivan's notion "Jørgensen end", which was introduced in [17, Fig. 1] (see also [10, p. 172]).

In this paper, we introduce the dynamical limit set $L_{\mathrm{dyn}}(\Gamma) := L(\Gamma) \setminus L_{\mathrm{J}}(\Gamma)$ and extend the coding map from the limit set $L(\Gamma)$ to the code space associated with a set of generators of Γ. We then prove the following result. This will be done using the relationship between the code space and various limit sets via the coding map.

Main Theorem. *For every $m \in \mathbb{N}$ and for every $d, j \in (0,m)$, there exists a Kleinian group Γ acting on $(m+1)$-dimensional hyperbolic space such that*

$$\delta(\Gamma) \leq d \quad \textit{and} \quad \dim_{\mathrm{H}} L_{\mathrm{J}}(\Gamma) = j.$$

In particular, Γ can be chosen to be of Schottky type.

Note that the statement clearly holds for $j = 0$ as well.

2 Definitions and Basic Facts

Let us first recall some basic definitions and facts.

Definition 1. The ball $\mathbb{D}^{m+1} := \{x = (x_0, \ldots, x_m) \in \mathbb{R}^{m+1} \mid \|x\| < 1\}$, when equipped with the metric given by $ds^2 = dx^2/(1 - \|x\|^2)$, is called the *Poincaré model* of the $(m+1)$-dimensional hyperbolic space. Its boundary will be denoted by $\mathbb{S}^m$.

Let B be a $(m+1)$-ball in $\mathbb{R}^{m+1}$ whose boundary ∂B is orthogonal to $\mathbb{S}^m$. Then $B \cap \mathbb{D}^{m+1} \neq \emptyset$ is a hyperbolic half-space. From now on, we only consider hyperbolic half-spaces of this type. For each such $(m+1)$-dimensional hyperbolic half-space C there is a unique open Euclidean $(m+1)$-ball B_C whose boundary ∂B_C is orthogonal to $\mathbb{S}^m$ and for which we have that $B_C \cap \mathbb{D}^{m+1} = C$. Let $\mathrm{Ext}(B_C)$ and $\mathrm{Int}(B_C)$ refer to the exterior and interior of B_C, respectively. For a hyperbolic half-space C, we denote its hyperbolic boundary by ∂C, that is, $\partial C = \partial B_C \cap \mathbb{D}^{m+1}$, and its boundary in $\mathbb{S}^m$ by $\overline{\partial} C$, that is $\overline{\partial} C = \mathbb{S}^m \cap (B_C \cup \partial B_C)$. Let $\mathrm{diam}_E(C)$ denote the Euclidean diameter of B_C. We now give the definition of a Kleinian group of Schottky type.

Definition 2. A group Γ acting on $\mathbb{D}^{m+1}$ will be called a Kleinian group of Schottky type if there exists a non-empty countable set $\{C_i\}_{i\in I(\Gamma)\subset\mathbb{Z}\setminus\{0\}}$ of pairwise disjoint $(m+1)$-dimensional hyperbolic half-spaces and a set $\{\gamma_i\}_{i\in I(\Gamma)}$ of orientation preserving isometries of $\mathbb{D}^{m+1}$ such that the following hold.

- For each C_i, there is a unique open Euclidean $(m+1)$-ball B_{C_i} for which we have that $B_{C_i}\cap\mathbb{D}^{m+1}=C_i$.
- For every $i\in I(\Gamma)$, we have that the map γ_i extends to a Lipschitz continuous map g_i (with the same Lipschitz constant as γ_i) which maps $\mathrm{Ext}(B_{C_i})$ onto $\mathrm{Int}(B_{C_{-i}})$. Here, Lipschitz continous is meant with respect to the Euclidean metric.
- The group Γ is generated by $\{\gamma_i\}_{i\in I(\Gamma)}$.
- There exist $\varepsilon>0$ such that the following holds. For each C_i, there exist finitely many $C_j\in\{C_k\}_{k\in I(\Gamma)}$ such that $\mathrm{diam}_E(C_j)>\mathrm{diam}_E(C_i)$. For these C_j, we then have $B_{C_j}\cap(1+\varepsilon)B_{C_i}=\emptyset$. Here, $(1+\varepsilon)B_{C_i}$ refers to the Euclidean ball with centre equal to the centre of B_{C_i} and with diameter $(1+\varepsilon)\,\mathrm{diam}_E(C_i)$.

With this notation, let $D:=\bigcap_{i\in I(\Gamma)}C_i^c$. Here, C_i^c denotes the complement of C_i in $\mathbb{D}^{m+1}$. Note that it was shown in [9] that D is a Dirichlet fundamental domain constructed with respect to the origin.

In other words, a group Γ will be called a Kleinian group of Schottky type if and only if Γ is a free discrete subgroup of the group of orientation preserving isometries of the $(m+1)$-dimensional hyperbolic space generated by a countable set of hyperbolic (or loxodromic) elements. For further details on Kleinian groups of Schottky type, we refer to [9].

Let us quickly recall the following types of limit sets for a Kleinian group of Schottky type.

Definition 3. Let Γ be a Kleinian group of Schottky type acting on $\mathbb{D}^{m+1}$. We then define the following types of limit sets of Γ.

- For an arbitrary $x\in\mathbb{D}^{m+1}$ the set $\bigcup_{\gamma\in\Gamma}\gamma(x)$ has accumulation points exclusively at the boundary $\partial\mathbb{D}^{m+1}=\mathbb{S}^m$ of hyperbolic space. The set $L(\Gamma)$ of these accumulation points is called the limit set of Γ. (Note that $L(\Gamma)$ is independent of the choice of x [9, p. 22, D. 3]).
- An element $x\in L(\Gamma)$ is called a uniformly radial limit point if for some $c>0$ the ray from $0\in\mathbb{D}^{m+1}$ to x is fully contained in $\bigcup_{\gamma\in\Gamma}b(\gamma(0),c)$. Here, $b(\gamma(0),c)$ refers to the hyperbolic ball centred at $\gamma(0)$ of radius c. The set $L_{\mathrm{ur}}(\Gamma)$ of uniformly radial limit points is called the uniformly radial limit set of Γ (see e.g. [17]).
- An element $x\in L(\Gamma)$ is called a Jørgensen limit point if and only if, for some Dirichlet domain D_z of Γ based at some point $z\in\mathbb{D}^{m+1}$, there exists $\gamma\in\Gamma$ such that $\gamma(D_z)$ contains the hyperbolic geodesic ray from $\gamma(z)$ to x. The set $L_{\mathrm{J}}(\Gamma)$ of Jørgensen limit points is called the Jørgensen limit set of Γ (see e.g. [10]).
- Following [7], we define the dynamical limit set $L_{\mathrm{dyn}}(\Gamma)$ by

$$L_{\mathrm{dyn}}(\Gamma):=L(\Gamma)\setminus L_{\mathrm{J}}(\Gamma).$$

Let us assume that D_0 is a fundamental domain of Γ based at the origin, and let $\{\gamma_1, \gamma_2, \ldots\}$ be an irreducible set of generators of Γ. It is well known (cf. [6, 13]) that for a finitely generated Kleinian group Γ there is a coding of the limit set of Γ by infinite words. In particular, to any point in the limit set is associated an infinite (reduced) word in the generators of Γ. Furthermore, for a finitely generated Kleinian group of Schottky type, this map is one-to-one (cf. [6]). In a nutshell, this coding can be interpreted as coming from the tesselation of $\mathbb{D}^{m+1}$ by the Dirichlet fundamental domain D of Γ (say, at the origin), or, more precisely, the sides of the fundamental domain. Given an infinite word $\gamma_1\gamma_2\ldots$ of a Kleinian group of Schottky type and its associated limit point x, one easily verifies that the geodesic ray from the origin to x passes through the images $\gamma_1(D)$, $\gamma_1\gamma_2(D)$ and so on.

Note that for a fixed set of generators $\{\gamma_1, \ldots, \gamma_k\}$ of Γ the coding map can be extended to a map $\pi : L(\Gamma) \to \{1, \ldots, k\}^\infty \subset \mathbb{N}^\infty$. For an infinitely generated Kleinian group of Schottky type, we extend this coding map to $\pi : L(\Gamma) \to \{0\} \cup \bigcup_{n\in\mathbb{N}}\{1, 2, \ldots\}^n \cup \mathbb{N}^\infty$ by allowing finite codings. Note that this means that π is not in general one-to-one, but the restriction of π to the preimage $\pi^{-1}(\mathbb{N}^\infty)$ is one-to-one. This coding map is not unknown: see for instance [3], where it has been employed in the situation of an infinitely generated Schottky group whose generators accumulate at exactly one point; hence, the coding map in [3, p. 570] is one-to-one. Also note that in [3, p. 570] it is considered to be a map from the code space to the limit set, while in this artcile we use the reverse direction (since in general the generators have more than one accumulation point).

With this coding map π, we have the following lemma.

Lemma 1. *For a Kleinian group Γ of Schottky type and for the coding map π as above we have*

$$L_{\text{ur}}(\Gamma) = \pi^{-1}\left(\{\underline{i} \in \mathbb{N}^\infty \mid \limsup\{i_k \mid k \in \mathbb{N}\} < \infty\}\right);$$

$$L_{\text{dyn}}(\Gamma) = \pi^{-1}\left(\mathbb{N}^\infty\right);$$

$$L_{\text{J}}(\Gamma) = L(\Gamma) \setminus L_{\text{dyn}}(\Gamma).$$

In particular, this implies that $L_{\text{J}}(\Gamma)$ corresponds exactly to those points in $L(\Gamma)$, which do not have an infinite (but a finite) coding. Therefore, a point x lies in $L_{\text{J}}(\Gamma)$ if and only if there is some Dirichlet domain D_z of Γ based at some point $z \in \mathbb{D}^{m+1}$ such that x is an accumulation point of sides of D_z.

Remark. The first assertion is well known, see for instance [15, p. 240]. The second and third assertions appear to be common knowledge, see for instance the comments [3, p. 570]. For the sake of completeness, we nevertheless include the proof.

Proof. In order to prove the assertion regarding L_{ur}, one proceeds as follows. First note that if $\Gamma_1 \subset \Gamma_2 \subset \ldots \subset \Gamma_k \subset \ldots$ is an increasing sequence of subgroups of the Kleinian group $\Gamma = \bigcup_k \Gamma_k$, then $L_{\text{ur}}(\Gamma) = \bigcup_k L_{\text{ur}}(\Gamma_k)$. If Γ is a Kleinian group of Schottky type, then it is freely generated, say by generators $\gamma_1, \gamma_2, \ldots$. Hence, $\Gamma_k := \langle \gamma_i \mid i \leq k \rangle$ gives such an increasing sequence. For each of the finitely generated groups Γ_k, one has that each limit point is coded by a unique infinite word (from the alphabet $\{1,\ldots,k\}$). Combining this observation with the fact that $L_{\mathrm{ur}}(\Gamma_k) = L(\Gamma_k)$, it follows that $L_{\mathrm{ur}}(\Gamma)$ can be symbolically described as stated in Lemma 1.

In order to see that the Jørgensen limit set of a Kleinian group of Schottky type is contained in the set of limit points which do not have an infinite coding, let $x \in L_{\mathrm{J}}(\Gamma)$ be fixed. By definition, we then have, for some Dirichlet domain D_z of Γ based at some point $z \in \mathbb{D}^{m+1}$, that there exists $\gamma \in \Gamma$ such that $\gamma(D_z)$ contains the hyperbolic geodesic ray from $\gamma(z)$ to x. Hence, the Euclidean distance from x to the set of sides of the Dirichlet domain D_z must be equal to zero. That is, x must be an accumulation point of sides of D_z, since a Kleinian group of Schottky type is by definition a free group generated by loxodromic elements (in particular, a Kleinian group of Schottky type has no parabolic elements). Note that if x is an accumulation point of sides of some Dirichlet domain D_z, then there exists a geodesic ray as above. Hence, $L_{\mathrm{J}}(\Gamma)$ is equal to the Γ-orbit of the accumulation points of sides of D_z. Recall that a word $i_1 i_2 \ldots$ can be interpreted as a coding obtained by listing the fundamental domains in the Γ-orbit of D_z, which are passed when one travels along the ray from 0 to x. In particular, this shows that a Jørgensen limit point x can only be coded by a finite word, since the ray from 0 to x intersects at most finitely many fundamental domains.

In order to show that the set of limit points $x \in L(\Gamma)$ which do not have an infinite coding is contained in $L_{\mathrm{J}}(\Gamma)$, we use the contrapositive method and proceed as follows. Assume that $x \notin L_{\mathrm{J}}(\Gamma)$. For each $\gamma \in \Gamma$, we then have that the hyperbolic geodesic ray from $\gamma(z)$ to x is not completely contained in $\gamma(D_z)$. Now, if x would be coded by a finite word, then this would mean that the geodesic ray from 0 to x eventually stays in one of the images of the fundamental domain, say $g(D_z)$. By convexity of $g(D_z)$, it then follows that the geodesic ray from $g(z)$ to x is fully contained in $g(D_z)$. This is a contradiction, and hence shows that the geodesic ray from 0 to x must pass through infinitely many images of the fundamental domain. This implies that there exists an infinite coding associated with x, and therefore, x is not contained in the set of limit points without infinite coding. This completes the proof of the lemma. □

Definition 4. Let Γ be a Kleinian group acting on $\mathbb{D}^{m+1}$, and let $s \in \mathbb{R}$. The series

$$\sum_{\gamma \in \Gamma} \mathrm{e}^{-sd(0,\gamma(0))}$$

will be called the *Poincaré series* associated with Γ. The exponent of convergence of this series will be denoted by $\delta(\Gamma)$ and referred to as the *Poincaré exponent associated with the Kleinian group* Γ.

Theorem 1 (Bishop, Jones). *For each non-elementary Kleinian group Γ, the Poincaré exponent $\delta(\Gamma)$ coincides with the Hausdorff dimension of the uniformly radial limit set.*

For a proof see [1]. A more detailed proof can be found in [14].

3 Proof of Main Theorem

We are now ready to prove the Main Theorem. Recall that the main statements of the theorem are as follows.

For every $m \in \mathbb{N}$ and every $d, j \in (0,m)$, there exists a Kleinian group $\Gamma \subset \mathrm{Iso}(\mathbb{D}^{m+1})$ such that

$$\dim_{\mathrm{H}} L_{\mathrm{ur}}(\Gamma) \leq d \quad \text{and} \quad \dim_{\mathrm{H}} L_{\mathrm{J}}(S) = j.$$

In particular, Γ can be chosen to be of Schottky type.

Proof. Note that the case $j = 0$ is trivial, since any finitely generated Kleinian group of Schottky type with small enough Poincare exponent might serve as an example. For the case $j = m$, we refer to [12].

Hence, let $m \in \mathbb{N}$ and $j, d \in (0,m)$ be fixed. The idea is to construct an infinitely generated Kleinian group of Schottky type Γ_∞. Fix a strictly increasing sequence $\{\underline{d}_n\}_{n \in \mathbb{N} \cup \{0\}}$ of positive real numbers such that $\lim_{n \to \infty} \underline{d}_n = d$. Now choose a finitely generated Kleinian group of Schottky type $\Gamma_0 := \langle \gamma_1, \ldots, \gamma_l \rangle$ acting on $\mathbb{D}^{m+1}$ such that $\delta(\Gamma_0) < \underline{d}_0$. For $I(\Gamma_0) := \{1, \ldots, l\} \cup \{-1, \ldots, -l\}$ let $\{C_i\}_{i \in I(\Gamma_0)}$ denote the hyperbolic half-spaces associated with the generators of Γ_0 (as in Definition 2). We then have that $D = \bigcap_{i \in I(\Gamma_0)} C_i^c$ is a Dirichlet domain for Γ_0. Recall that $\overline{\partial} D$ denotes the intersection of $\mathbb{S}^m$ with the closure $\overline{D}$ of D. Choose a closed m-dimensional ball $X \subset \mathbb{S}^m$, which is contained in an open subset of $\overline{\partial} D$. Moreover, choose some $\ell \in (0,1)$ and a set of two injective contractions $S := \{\varphi_1, \ \varphi_2 : X \to X\}$ such that $\varphi_1(X) \cap \varphi_2(X) = \emptyset$, and such that

$$L(S) := \left(\bigcap_{n \in \mathbb{N}} \ \bigcup_{i_1, \ldots, i_n \in \{1,2\}} \varphi_{i_n} \circ \ldots \circ \varphi_{i_1}(X) \right) \text{ satisfies } \dim_{\mathrm{H}} L(S) = j.$$

Such an S is called an iterated function system (IFS) and that such an S exists is well known; for further details on iterated function systems, we refer for example to the book [4].

As mentioned above, we will now inductively construct an infinitely generated Kleinian group of Schottky type Γ_∞. In order to do that we will for each $n \in \mathbb{N}$ and each $k \in \{1, \ldots, 2^n\}$ find a suitable hyperbolic isometry $\gamma_{n,k} \in \mathrm{Iso}(D^{m+1})$ and define the free groups

$$\Gamma_n := \langle \gamma_{i,j} \mid 1 \leq i \leq n, \ 1 \leq j \leq 2^n \rangle \star \Gamma_0.$$

Therefore, the start of the induction is given by Γ_0. For ease of exposition, we also use the free groups

$$\Gamma_{n,k} := \Gamma_{n-1} \star \langle \gamma_{n,j} \mid 1 \leq j \leq k \rangle = \Gamma_{n,k-1} \star \langle \gamma_{n,k} \rangle.$$

The crucial point is that for each $n \in \mathbb{N}$ and each $k \in \{1,\ldots,2^n\}$ we can inductively find $\gamma_{n,k} \in \mathrm{Iso}(D^{m+1})$ such that the following properties are satisfied:

1. $\Gamma_{n,k}$ is a Kleinian group of Schottky type.
2. The choice of $\gamma_{n,k}$ does not interfere with the induction step, that is

$$\overline{\partial} C_{\pm\gamma_{n,k}} \cap \bigcup_{\underline{e}\in\{1,2\}^n} \varphi_{\underline{e}}(X) = \emptyset.$$

3. The sides of the fundamental domain for $\Gamma_{n,k-1} \star \langle\gamma_{n,k}\rangle$ are located suitably, that is, for the unique $(k_0,k_1,\ldots,k_{n-1}) \in \{1,2\}^n$ with $k = \sum_{i=0}^{n-1}(k_i-1)2^i$, we have

$$\mathrm{dist}\left(B_{C_{\pm\gamma_{n,k}}}, \varphi_{k_{n-1}} \circ \ldots \circ \varphi_{k_0}(X)\right) \leq \ell^n.$$

   This property will be crucial when proving the fact regarding $L_{\mathrm{J}}(\Gamma)$.
4. The exponent of convergence of $\Gamma_{n,k-1} \star \langle\gamma_{n,k}\rangle$ is small enough, that is,

$$\delta\left(\Gamma_{n,k}\right) \leq \underline{d}_{n,k}.$$

Clearly, by construction (in particular condition (3) above), we have that the set of accumulation points of the set $\{\partial C_i : i \in I(\Gamma_\infty)\}$ of sides of $\partial D(\Gamma_\infty)$ in $\mathbb{S}^m$ is equal to $L(S)$. This implies that $L_{\mathrm{J}}(\Gamma_\infty) = \Gamma_\infty(L(S))$, and hence $\dim_{\mathrm{H}} L_{\mathrm{J}}(\Gamma_\infty) = \dim_{\mathrm{H}} L(S)$. Furthermore, by choice of S, we have that $\dim_{\mathrm{H}} L(S) = j$. Combining these observations, we conclude that $\dim_{\mathrm{H}} L_{\mathrm{J}}(\Gamma) = j$. This gives the equality stated in the theorem.

For the inequality $\delta(\Gamma_\infty) \leq d$, note that by Theorem 1 and the definition of $L_{\mathrm{ur}}(\Gamma_\infty)$, we have that $\delta(\Gamma_\infty) = \lim_{n\to\infty} \delta(\Gamma_n)$. (Note that this result was originally proven by means of conformal measures by Sullivan in [16], while this simple proof is e.g. contained in [5, Remark 1].) Since $\delta(\Gamma_n) \leq \underline{d}_n$ for each $n \in \mathbb{N}$, we conclude that $\lim_{n\to\infty} \delta(\Gamma_n) \leq \lim_{n\to\infty} \underline{d}_n = d$. Combining these observations, the inequality in the theorem follows.

This completes the proof of the Main Theorem. □

Remark. Note that by choosing the finitely generated Kleinian group of Schottky type Γ_0 in such a way that $d - \delta(\Gamma_0)$ is small, one can construct an infinitely generated Kleinian group Γ_∞ with $\delta(\Gamma_\infty) \in (\delta(\Gamma_0), d]$.

Remark. Finally, we would like to remark that even though in this construction we control $\dim_{\mathrm{H}} L_r(\Gamma)$ and $\dim_{\mathrm{H}} L_{\mathrm{J}}(\Gamma)$, this does not imply that we control $\dim_{\mathrm{H}} L_{\mathrm{dyn}}(\Gamma)$ and hence also not $\dim_{\mathrm{H}} L(\Gamma)$.

Namely, note that (by Theorem 1 of Bishop and Jones) one has $\delta(\Gamma) = \dim_{\mathrm{H}} L_{\mathrm{ur}}(\Gamma) = \dim_{\mathrm{H}} L_r(\Gamma)$ and that by definition $L_r(\Gamma) \subset L_{\mathrm{dyn}}(\Gamma)$. However, $\dim_{\mathrm{H}} L_r(\Gamma)$ and $\dim_{\mathrm{H}} L_{\mathrm{dyn}}(\Gamma)$ do not coincide in general. This follows from a result of Brooks in [2]. More precisely, on the one hand, it was shown in [2] that for a normal subgroup N of a convex cocompact Kleinian group Γ (with $\delta(\Gamma) > n/2$)

one has $\delta(N) < \dim_{\mathrm{H}} L(N)$ if and only if Γ/N is non-amenable. On the other hand, at least for certain such N, it is easy to show that $\dim_{\mathrm{H}} L_{\mathrm{J}}(N) < \dim_{\mathrm{H}} L(N)$. This can be seen as follows. Consider a purely hyperbolic free group $\Gamma := G \star H := \langle g \rangle \star \langle h_1, h_2 \rangle$ for fixed g, h_1, h_2 such that $\delta(\Gamma) > n/2$ and the free normal subgroup $N := \langle \{hgh^{-1} \mid h \in H\} \rangle$. Then one has $\Gamma/N \simeq H$ and $L_{\mathrm{J}}(N) = N(L(H))$. Combining these observations with the fact that $\dim_{\mathrm{H}} L(N) = \delta(\Gamma) > \delta(H)$ it follows that $\max\{\delta(N), \dim_{\mathrm{H}} L_{\mathrm{J}}(N)\} < \dim_{\mathrm{H}} L_{\mathrm{dyn}}(N) = \delta(\Gamma)$.

Furthermore, for normal subgroups N of a convex cocompact Kleinian group Γ one has (by a result of Falk and Stratmann in [5]) that $\dim_{\mathrm{H}} L_r(N) \leq \dim_{\mathrm{H}} L_{\mathrm{dyn}}(N) \leq 2 \cdot \dim_{\mathrm{H}} L_r(N)$. However, in general no such upper bound is known. Hence, in the construction in this paper we do control $\dim_{\mathrm{H}} L_r(\Gamma)$ and $\dim_{\mathrm{H}} L_{\mathrm{J}}(\Gamma)$ but not $\dim_{\mathrm{H}} L_{\mathrm{dyn}}(\Gamma)$ nor $\dim_{\mathrm{H}} L(\Gamma)$.

References

1. C.J. Bishop, P.W. Jones, 'Hausdorff dimension and Kleinian groups', *Acta Math.* **179** (1) (1997) 1–39.
2. R. Brooks, 'The bottom of the spectrum of a Riemannian covering', *J. Reine Angew. Math.* **357** (1985) 101–114.
3. F. Dal'bo, A.N. Starkov, 'On a classification of limit points of infinitely generated Schottky groups', *J.Dynam. Control Systems* **6** (2000) 561–578.
4. K.J. Falconer, *Techniques in Fractal Geometry*, J. Wiley, 1997.
5. K.H. Falk, B.O. Stratmann, 'Remarks on Hausdorff dimensions for transient limit sets of Kleinian groups', *Tohoku Math. J.* **56** (2004) 571–582.
6. W.J. Floyd, 'Group completions and limit sets of Kleinian groups', *Invent. Math.* **57** (1980) 205–218.
7. M.R. Hille, 'Resonances for graph directed Markov systems and geometry of infinitely generated dynamical systems',*Ph.D. thesis*, University of St Andrews, 2009.
8. T. Jørgensen, 'Compact 3-manifolds of constant negative curvature fibering over the circle', *Ann. of Math.* **106** (1977) 61–72.
9. B. Maskit, *Kleinian Groups*, Grundlehren d. Math. Wiss. **287**, Springer-Verlag, Berlin, 1988.
10. J.R. Parker, B.O. Stratmann, 'Kleinian groups with singly cusped parabolic fixed points', *Kodai Journal of Mathematics* **24** (2001) 169–206.
11. S.J. Patterson, 'The limit set of a Fuchsian group', *Acta Math.* **136** (1976) 241–273.
12. S.J. Patterson, 'Further remarks on the exponent of convergence of Poincaré series', *Tôhoku Math. Journ.* **35** (1983) 357–373.
13. C. Series, 'The infinite word problem and limit sets in Fuchsian groups', *Ergodic Theory and Dynamical Systems* **1** (1981), 337–360
14. B.O. Stratmann, 'The exponent of convergence of Kleinian groups; on a theorem of Bishop and Jones', *Progress in Probability* **57** (2004) 93–107.
15. B.O. Stratmann, 'Fractal geometry on hyperbolic manifolds', *Non-Euclidean Geometries, Janos Bolyai Memorial Volume*, (eds. A. Prekopa and E. Molnar), Springer Verlag, 2006, page 227–249.
16. D. Sullivan, 'The density at infinity of a discrete group of hyperbolic motions', *IHES Publ. Math.* **50** (1979) 171–202.
17. D. Sullivan, 'A finiteness theorem for cusps', *Acta Math.* **142** (1981) 289–299.

A Note on the Algebraic Growth Rate of Poincaré Series for Kleinian Groups

Marc Kesseböhmer and Bernd O. Stratmann

Abstract In this note, we employ infinite ergodic theory to derive estimates for the algebraic growth rate of the Poincaré series for a Kleinian group at its critical exponent of convergence.

Keywords Poincaré series • Infinte ergodic theory • Kleinian groups

Dedicated to S.J. Patterson on the occasion of his 60th birthday.

1 Introduction and Statements of Result

In this note, we study the Poincaré series

$$\mathscr{P}(z,w,s) := \sum_{g\in G} \mathrm{e}^{-sd(z,g(w))}$$

of a geometrically finite, essentially free Kleinian group G acting on the $(N+1)$-dimensional hyperbolic space $\mathbb{D}$, for arbitrary $z,w \in \mathbb{D}$. Here, $d(z,w)$ denotes the hyperbolic distance between z and w, and $s \in \mathbb{R}$. It is well known that a group of this type is of δ-divergence type, which means that $\mathscr{P}(z,w,s)$ diverges for s equal to the exponent of convergence $\delta = \delta(G)$ of $\mathscr{P}(z,w,s)$. We are in particular interested in the situation in which G is a zonal group, that is, we always assume that G has

M. Kesseböhmer (✉) • B.O. Stratmann
Fachbereich 3 – Mathematik und Informatik, Universität Bremen, Bibliothekstr. 1, 28359 Bremen, Germany
e-mail: mhk@math.uni-bremen.de; bos@math.uni-bremen.de

V. Blomer and P. Mihăilescu (eds.), *Contributions in Analytic and Algebraic Number Theory*, Springer Proceedings in Mathematics 9, DOI 10.1007/978-1-4614-1219-9_10,

parabolic elements. For Kleinian groups of this type, we then consider the partial Poincaré sum

$$\mathscr{P}_n(z,w,s) := \sum_{\substack{g\in G\\ |g|\leq n}} \mathrm{e}^{-sd(z,g(w))},$$

for $n \in \mathbb{N}$, and where $|\cdot|$ denotes the word metric in G. The main result of this note is the following asymptotic estimate for these partial Poincaré sums, for s equal to the exponent of convergence δ. Here, $r_{\max}$ denotes the maximal rank of the parabolic fixed points of G, and $\asymp$ denotes comparability, that is, $b_n \asymp c_n$ if and only if (b_n/c_n) is uniformly bounded away from zero and infinity, for two sequences (b_n) and (c_n) of positive real numbers.

Theorem 1. *For a geometrically finite, essentially free, zonal Kleinian group G and for each $z,w \in \mathbb{D}$, we have*

$$\mathscr{P}_n(z,w,\delta) \asymp \begin{cases} n^{2\delta - r_{\max}} & \text{for} \quad \delta < (r_{\max}+1)/2, \\ n/\log n & \text{for} \quad \delta = (r_{\max}+1)/2, \\ n & \text{for} \quad \delta > (r_{\max}+1)/2. \end{cases}$$

Note that the results in this note grew out of the authors closely related recent studies in [9] of the so-called sum-level sets for regular continued fractions. These sets are given by

$$\mathscr{C}_n := \left\{ [a_1,a_2,\ldots] \in [0,1] : \sum_{i=1}^{k} a_i = n, \text{ for some } k \in \mathbb{N} \right\},$$

where $[a_1,a_2,\ldots]$ denotes the regular continued fraction expansion. Inspired by a conjecture in [7], it was shown in [9] that for the Lebesgue measure $\lambda(\mathscr{C}_n)$ of these sets one has that, with $b_n \sim c_n$ denoting $\lim_{n\to\infty} b_n/c_n = 1$,

$$\lambda(\mathscr{C}_n) \sim \frac{1}{\log_2 n} \text{ and } \sum_{k=1}^{n} \lambda(\mathscr{C}_k) \sim \frac{n}{\log_2 n}.$$

For refinements of these results, we also refer to [10]. It is not hard to see that in here the second asymptotic estimate implies Theorem 1 for G equal to the (subgroup of index 3 of the) modular group $PSL_2(\mathbb{Z})$.

2 Preliminaries

2.1 The Canonical Markov Map

As already mentioned in the introduction, throughout, we exclusively consider a geometrically finite, essentially free, zonal Kleinian group G. By definition (see [8]), a group of this type can be written as a free product $G = H * \Gamma$, where $H = \langle h_1, h_1^{-1}\rangle * \cdots * \langle h_u, h_u^{-1}\rangle$ denotes the free product of finitely many elementary, loxodromic groups, and $\Gamma = \Gamma_1 * \cdots * \Gamma_v$ denotes the free product of finitely many parabolic subgroups of G such that $\Gamma_i = \langle \gamma_{i_1}, \gamma_{i_1}^{-1}, \ldots, \gamma_{i_{r_i}}, \gamma_{i_{r_i}}^{-1}\rangle$ is the parabolic subgroup of G associated with the parabolic fixed point p_i of rank r_i. Clearly, $\Gamma_i \cong \mathbb{Z}^{r_i}$ and $\gamma_{i_j}^{\pm}(p_i) = p_i$, for all $j = 1, \ldots, r_i$ and $i = 1, \ldots, v$. Also, note that G has no relations other than those which originate from cusps of rank at least 2, that is, those Γ_i with $r_i > 1$. Without loss of generality, we can assume that G admits the choice of a Poincaré polyhedron F with a finite set $\mathscr{F}$ of faces such that if two elements s and t of $\mathscr{F}$ intersect inside $\mathbb{D}$, then the two associated generators g_s and g_t must have the same fixed point, which then, in particular, has to be a parabolic fixed point of G of rank at least 2.

Let us now first recall from [17] the construction of the relevant coding map T associated with G, which maps the radial limit set $L_r(G)$ into itself. This construction parallels the construction of the well-known Bowen-Series map (cf. [4, 13–16]).

For $\xi, \eta \in L_r(G)$, let $\gamma_{\xi,\eta} : \mathbb{R} \to \mathbb{D}$ denote to the directed geodesic from η to ξ such that $\gamma_{\xi,\eta}$ intersects the closure $\overline{F}$ of F in $\mathbb{D}$, normalized such that $\gamma_{\xi,\eta}(0)$ is the summit of $\gamma_{\xi,\eta}$. The exit time $e_{\xi,\eta}$ is defined by

$$e_{\xi,\eta} := \sup\{s : \gamma_{\xi,\eta}(s) \in \overline{F}\}.$$

Since $\xi, \eta \in L_r(G)$, we clearly have that $|e_{\xi,\eta}| < \infty$. By Poincaré's polyhedron theorem (cf. [6]), we have that the set $\mathscr{F}$ carries an involution $\mathscr{F} \to \mathscr{F}$, given by $s \mapsto s'$ and $s'' = s$. In particular, for each $s \in \mathscr{F}$ there is a unique face-pairing transformation $g_s \in G$ such that $g_s(\overline{F}) \cap \overline{F} = s'$. We then let

$$\mathscr{L}_r(G) := \{(\xi, \eta) : \xi, \eta \in L_r(G), \xi \neq \eta \text{ and } \exists t \in \mathbb{R} \ : \ \gamma_{\xi,\eta}(t) \in \overline{F}\},$$

and define the map $S : \mathscr{L}_r(G) \to \mathscr{L}_r(G)$, for all $(\xi, \eta) \in \mathscr{L}_r(G)$ such that $\gamma_{\xi,\eta}(e_{\xi,\eta}) \in s$, for some $s \in \mathscr{F}$, by

$$S(\xi, \eta) := (g_s(\xi), g_s(\eta)).$$

In order to show that the map S admits a Markov partition, we introduce the following collection of subsets of the boundary $\partial\mathbb{D}$ of $\mathbb{D}$. For $s \in \mathscr{F}$, let A_s refer to the open hyperbolic halfspace for which $F \subset \mathbb{D} \setminus A_s$ and $s \subset \partial A_s$. Also, let $\Pi : \mathbb{D} \to \partial\mathbb{D}$ denote the shadow-projection given by $\Pi(A) := \{\xi \in \partial\mathbb{D} : \sigma_\xi \cap A \neq \emptyset\}$, where σ_ξ denotes the ray from 0 to ξ. Then the projections a_s of the side s to $\partial\mathbb{D}$ is given by

$$a_s := \mathrm{Int}(\Pi(A_s)).$$

If G has exclusively parabolic fixed points of rank 1, then $a_s \cap a_t = \emptyset$, for all $s, t \in \mathscr{F}, s \neq t$. Hence, by convexity of F, we have $\gamma_{\xi,\eta}(e_{\xi,\eta}) \in s$ if and only if $\xi \in a_s$. In other words, $S(\xi, \eta) = (g_s\xi, g_s\eta)$ for all $\xi \in a_s$. This immediately gives that the projection map $\pi : (\xi, \eta) \mapsto \xi$ onto the first coordinate of $\mathscr{L}_r(G)$ leads to a canonical factor T of S, that is, we obtain the map

$$T : L_r(G) \to L_r(G), \text{ given by } T|_{a_s \cap L_r(G)} := g_s.$$

Clearly, T satisfies $\pi \circ S = T \circ \pi$. Since $T(a_s) = g_s(a_s) = \mathrm{Int}(\partial\mathbb{D} \setminus a_{s'})$, it follows that T is a non-invertible Markov map with respect to the partition $\{a_s \cap L_r(G) : s \in \mathscr{F}\}$.

If there are parabolic fixed points of rank greater than 1, then, a priori, S does not have a canonical factor. In this situation, the idea is to construct an invertible Markov map $\widetilde{S}$ which is isomorphic to S and which has a canonical factor. This can be achieved by introducing a certain rule on the set of faces associated with the parabolic fixed points of rank greater than 1, which keeps track of the geodesic movement within these cusps. This then permits to define a coding map also in this higher rank case, and, for ease of notation, this map will also be denoted by T. (For further details we refer to [17], where this construction is given for G having parabolic fixed points of rank 2 and acting on 3-dimensional hyperbolic space; the general case follows from a straightforward adaptation of this construction.) For this so obtained coding map T we then have the following result.

Proposition 1 ([17, Proposition 2, Proposition 3]). *The map T is a topologically mixing Markov map with respect to the partition generated by $\{a_s \cap L_r(G) : s \in \mathscr{F}\}$. Moreover, the map $\widetilde{S}$ is the natural extension of T.*

2.2 Patterson Measure Theory

In order to introduce the T-invariant measure on $L(G)$ relevant for us here, let us first briefly recall some of the highlights in connection with the Patterson measure and the Patterson–Sullivan measure (for detailed discussions of these measures, we refer to [11,12,18,19,21], see also [5] in these Proceedings). By now it is folklore that, given some sequence (s_n) of positive reals which tends to δ from above, the Patterson measure m_δ is a probability measure supported on $L(G)$, given by a weak accumulation point of the sequence of measures

$$\left((\mathscr{P}_\infty(0,0,\delta_n))^{-1} \sum_{g \in G} \mathrm{e}^{-\delta_n d(0,g(0))} \mathbb{1}_{g(0)} \right).$$

For geometrically finite Kleinian groups, and therefore, in particular, for the type of groups considered in this note, it is well known that the so obtained limit measure is non-atomic and does not depend on the particular choice of the sequence (s_n). Hence, in particular, m_δ is unique. Moreover, we have that m_δ is δ-conformal, that is, for all $g \in G$ and $\xi \in L(G)$, we have

$$\frac{d\,(m_\delta \circ g)}{d\,m_\delta}(\xi) = \left(\frac{1 - |g(0)|^2}{|\xi - g^{-1}(0)|^2} \right)^\delta.$$

This δ-conformality is one of the key properties of m_δ, and for geometrically finite Kleinian groups it has the following, very useful geometrization. For this, let ξ_t denote the unique point on the ray σ_ξ such that the hyperbolic distance between 0 and ξ_t is equal to t, for arbitrary $\xi \in L(G)$ and $t > 0$. Also, let $B_c(\xi_t) \subset \mathbb{D}$ denote the $(N+1)$-dimensional hyperbolic disc centred at ξ_t of hyperbolic radius $c > 0$. Moreover, if ξ_t lies in one of the cusps associated with the parabolic fixed points of G, we let $r(\xi_t)$ denote the rank of the parabolic fixed point associated with that cusp, otherwise, we put $r(\xi_t)$ equal to δ. We then have the following generalization of Sullivan's shadow lemma, where "diam" denotes the Euclidean diameter in $\partial\mathbb{D}$.

Proposition 2 ([18, 20]). *For fixed, sufficiently large $c > 0$, and for all $\xi \in L(G)$ and $t > 0$, we have*

$$m_\delta(\Pi(B_c(\xi_t))) \asymp (\mathrm{diam}(\Pi(B_c(\xi_t))))^\delta \cdot \mathrm{e}^{(r(\xi_t)-\delta)d(\xi_t,G(0))}.$$

A further strength of the Patterson measure in the geometrically finite situation lies in the fact that it gives rise to a measure $\widetilde{m}_\delta$ on $(L(G)\times L(G))\setminus\{\text{diag.}\}$, which is ergodic with respect to the action of G on $(L(G)\times L(G))\setminus\{\text{diag.}\}$, given by $g((\xi,\eta)) = (g(\xi),g(\eta))$. This measure is usually called the Patterson–Sullivan measure, and it is given by

$$d\widetilde{m}_\delta(\xi,\eta) := \frac{dm_\delta(\xi)dm_\delta(\eta)}{|\xi-\eta|^{2\delta}}.$$

The (first) marginal measure of the Patterson–Sullivan measure then defines the measure μ_δ on $L(G)$, given by

$$\mu_\delta := \widetilde{m}_\delta \circ \pi^{-1}.$$

The advantage of the measure μ_δ is that it is suitable for non-trivial applications of certain results from infinite ergodic theory. In fact, for the system $(L(G),T,\mu_\delta)$ the following results have been obtained in [17].

Proposition 3. *The map T is conservative and ergodic with respect to the T-invariant, σ-finite measure μ_δ, and μ_δ is infinite if and only if $\delta \le (r_{\max}+1)/2$. Moreover, if G has parabolic fixed points of rank less than $r_{\max}$, then μ_δ gives finite mass to small neighbourhoods around these fixed points.*

2.3 Infinite Ergodic Theory

In this section, we summarize some of the infinite ergodic theoretical properties of the system $(L(G),T,\mu_\delta)$. For further details, we refer to [17].

Recall that we always assume that G is a geometrically finite, essentially free, zonal Kleinian group, and note that for our purposes here we only have to consider the parabolic subgroups of maximal rank, since, by Proposition 3, μ_δ gives infinite measure to arbitrary small neighbourhoods of a fixed point of a parabolic generator of G *only* if the parabolic fixed point is of maximal rank $r_{\max}$. Then define

$$\mathscr{D}_0 := \bigcap_{\substack{\gamma \text{ a generator of } \Gamma_i \\ i=1,\ldots,v; r_i = r_{\max}}} (\mathbb{D} \setminus \mathrm{Cl}_{\mathbb{D}}(A_{\gamma\circ\gamma})),$$

and let

$$\mathscr{D} := L_r(G) \cap \Pi(\mathscr{D}_0).$$

Recall that the induced transformation $T_{\mathscr{D}}$ on $\mathscr{D}$ is defined by $T_{\mathscr{D}}(\xi) := T^{\rho(\xi)}(\xi)$, where ρ denotes the return time function, given by $\rho(\xi) := \min\{n \in \mathbb{N} : T^n(\xi) \in \mathscr{D}\}$. One then considers the induced system $(\mathscr{D}, T_{\mathscr{D}}, \mu_{\delta,\mathscr{D}})$, where $\mu_{\delta,\mathscr{D}}$ denotes the restriction of μ_δ to $\mathscr{D}$. Using standard techniques from ergodic theory, for this induced system the following result was obtained in [17]. Here, $b_n \ll c_n$ means that (b_n/c_n) is uniformly bounded away from infinity.

Fact ([17]). *The map $T_{\mathscr{D}}$ has the Gibbs–Markov property with respect to the measure $\mu_{\delta,\mathscr{D}}$. That is, there exists $c \in (0,1)$ such that for arbitrary cylinders $[\omega_1]$ of length n and $[\omega_2]$ of length m such that $[\omega_2] \subset T^n_{\mathscr{D}}([\omega_1])$, we have for $\mu_{\delta,\mathscr{D}}$-almost every pair $\eta, \xi \in [\omega_2]$,*

$$\left| \log \frac{d\mu_{\delta,\mathscr{D}} \circ T^{-n}_{\mathscr{D},\omega_1}}{d\mu_{\delta,\mathscr{D}}}(\xi) - \log \frac{d\mu_{\delta,\mathscr{D}} \circ T^{-n}_{\mathscr{D},\omega_1}}{d\mu_{\delta,\mathscr{D}}}(\eta) \right| \ll c^m,$$

where $T^{-n}_{\mathscr{D},\omega_1}$ denotes the inverse branch of $T^n_{\mathscr{D}}$ mapping $T^n_{\mathscr{D}}([\omega_1])$ bijectively to $[\omega_1]$.

Let $\widehat{T}_{\mathscr{D}}$ denotes the dual operator of $T_{\mathscr{D}}$, given by

$$\mu_{\delta,\mathscr{D}}(f \cdot g \circ T) = \mu_{\delta,\mathscr{D}}\left(\widehat{T}_{\mathscr{D}}(f) \cdot g\right), \text{ for all } f \in L^1(\mu_{\delta,\mathscr{D}}), g \in L^\infty(\mu_{\delta,\mathscr{D}}).$$

The Gibbs–Markov property of $T_{\mathscr{D}}$ then allows to employ the following chain of implications (cf. [1, 2]):

$T_{\mathscr{D}}$ has the Gibbs–Markov property with respect to $\mu_{\delta,\mathscr{D}}$.

$\Longrightarrow$ *There exists $c_0 \in (0,1)$ such that, for all $f \in L^1(\mu_{\delta,\mathscr{D}})$ and $n \in \mathbb{N}$, we have*

$$\left\| \widehat{T}^n_{\mathscr{D}} f - \mu_{\delta,\mathscr{D}}(f) \right\|_L \ll c_0^n \|f\|_L.$$

(Here, $\|\cdot\|_L$ refers to the Lipschitz norm (cf. [2], p. 541).)

$\Longrightarrow$ *$T_{\mathscr{D}}$ is continued fraction mixing (cf. [2], p. 500).*

$\Longrightarrow$ *The set $\mathscr{D}$ is a Darling–Kac set for T. That is, there exists a sequence (v_n) (called the* return sequence *of T) such that*

$$\frac{1}{v_n} \sum_{i=0}^{n-1} \widehat{T}^i \mathbb{1}_{\mathscr{D}}(\xi) \to \mu_\delta(\mathscr{D}), \text{ uniformly for } \mu_\delta\text{-almost every } \xi \in \mathscr{D}.$$

Finally, let us also remark that the growth rate of the sequence (v_n) can be determined explicitly as follows. Recall from [1, Sect. 3.8] that the wandering rate of the Darling–Kac set $\mathscr{D}$ is defined by the sequence (w_n), which is given, for each $n \in \mathbb{N}$, by

$$w_n := \mu_\delta\left(\bigcup_{k=1}^{n} T^{-(k-1)}(\mathscr{D})\right).$$

An application of [1, Proposition 3.8.7] gives that the return sequence and the wandering rate are related through

$$v_n \cdot w_n \sim \frac{n}{\Gamma(1+\beta)\Gamma(2-\beta)},$$

where $\beta := \max\{0, 1 + r_{\max} - 2\delta\}$ coincides with the index of variation of the regularly varying sequence (w_n). Hence, we are left with to determine the wandering rate. But this has been done in [17, Theorem 1], where it was shown that

$$w_n \asymp \begin{cases} n^{r_{\max}-2\delta+1} & \text{for} \quad \delta < (r_{\max}+1)/2 \\ \log n & \text{for} \quad \delta = (r_{\max}+1)/2 \\ 1 & \text{for} \quad \delta > (r_{\max}+1)/2. \end{cases}$$

Hence, by combining these observations, it follows that

$$v_n \asymp \begin{cases} n^{2\delta-r_{\max}} & \text{for} \quad \delta < (r_{\max}+1)/2 \\ n/\log n & \text{for} \quad \delta = (r_{\max}+1)/2 \\ n & \text{for} \quad \delta > (r_{\max}+1)/2. \end{cases}$$

3 Proof of the Theorem 1

As we have seen in the previous section, we have that the set $\mathscr{D} := L_r(G) \cap \Pi(\mathscr{D}_0)$ is a Darling–Kac set. Combining this with Proposition 2, Proposition 3, and the fact that the Patterson measure m_δ and its T–invariant version μ_δ are comparable on $\mathscr{D}$, one obtains

$$\begin{aligned}\frac{1}{v_n}\sum_{\substack{g(w)\in\mathscr{D}_0\\|g|\le n}} \mathrm{e}^{-\delta d(z,g(w))} &\asymp \frac{1}{v_n}\sum_{k=0}^{n} m_\delta\left(\mathscr{D}\cap T^{-k}(\mathscr{D})\right) \asymp \frac{1}{v_n}\sum_{k=0}^{n}\mu_\delta\left(\mathscr{D}\cap T^{-k}(\mathscr{D})\right)\\ &= \frac{1}{v_n}\sum_{k=0}^{n}\mu_\delta\left(\mathbb{1}_{\mathscr{D}}\cdot\widehat{T}^k\mathbb{1}_{\mathscr{D}}\right) = \mu_\delta\left(\mathbb{1}_{\mathscr{D}}\cdot\frac{1}{v_n}\sum_{k=0}^{n}\widehat{T}^k\mathbb{1}_{\mathscr{D}}\right)\\ &\sim \left(\mu_\delta(\mathbb{1}_{\mathscr{D}})\right)^2.\end{aligned}$$

Since $\mu_\delta(\mathbb{1}_{\mathscr{D}}) \asymp 1$, it follows that

$$\sum_{\substack{g(w)\in\mathscr{D}_0\\|g|\le n}} \mathrm{e}^{-\delta d(z,g(w))} \asymp v_n.$$

To extend this estimate to the full G-orbit of w, let

$$Q_i := \bigcap_{\gamma \text{ a generator of } \Gamma_i} (\mathbb{D}\setminus \mathrm{Cl}_{\mathbb{D}}(A_\gamma))$$

denote the fundamental domain for the action of Γ_i on $\mathbb{D}$, for each $i \in \{1,\ldots,v\}$. Clearly, we can assume, without loss of generality, that z and w are contained in each of the fundamental domains Q_i. For every $\gamma \in \Gamma_i$ such that $|\gamma| = k$, for some $1 < k \le n$, we then have, with the convention $v_0 := 1$,

$$\sum_{\substack{g(w)\in\gamma(Q_i)\\|g|\le n}} \mathrm{e}^{-\delta d(z,g(w))} \asymp k^{-2\delta} v_{n-k}.$$

Also, note that

$$\mathrm{card}\{\gamma \in \Gamma_i : |\gamma| = k\} \asymp k^{r_i - 1}.$$

Combining these observations, it follows that

$$\begin{aligned}
\mathscr{P}_n(\delta,z,w) &\asymp \sum_{\substack{g(w)\in\mathscr{D}_0\\|g|\le n}} \mathrm{e}^{-\delta d(z,g(w))} + \sum_{\substack{i=1,..,v\\r_i=r_{\max}}} \sum_{\substack{\gamma\in\Gamma_i\\|\gamma|\ge 2}} \sum_{\substack{g(w)\in\gamma(Q_i)\\|g|\le n}} \mathrm{e}^{-\delta d(z,g(w))}\\
&\asymp v_n + \sum_{\substack{i=1,..,v\\r_i=r_{\max}}} \sum_{k=2}^{n} \sum_{\substack{\gamma\in\Gamma_i\\|\gamma|=k}} \sum_{\substack{g(w)\in\gamma(Q_i)\\|g|\le n}} \mathrm{e}^{-\delta d(z,g(w))}\\
&\asymp v_n + \sum_{k=2}^{n} k^{r_{\max}-1} k^{-2\delta} v_{n-k}.
\end{aligned}$$

In order to finish the proof, recall that, by a result of Beardon [3], one has that $\delta > r_{\max}/2$. Therefore, there exists $\kappa = \kappa(G) > 0$ such that $\delta > r_{\max}/2 + \kappa$. Moreover, note that we can assume, without loss of generality, that (v_n) is increasing. Using these observations, it now follows that, on the one hand,

$$\sum_{k=2}^{n} v_{n-k} k^{-2\delta + r_{\max} - 1} \ll v_n \sum_{k=2}^{n} k^{-2\delta + r_{\max} - 1} \ll v_n \sum_{k=1}^{n} k^{-1-2\kappa} \ll v_n.$$

On the other hand, we clearly have that $\sum_{k=2}^{n} v_{n-k} k^{-2\delta + r_{\max} - 1} \gg v_{n-2}$. Combining these observations with the estimate for the asymptotic growth rate of the return sequence (v_n), given in the previous section, the proof of Theorem 1 follows.

Acknowledgements We would like to thank the *Mathematische Institut der Universität Göttingen* for the warm hospitality during our research visit in Summer 2009. In particular, we would like to thank P. Mihailescu and S.J. Patterson for the excellent organization of the *International Conference: Patterson 60 ++*, which took place during the period of our visit.

References

1. J. Aaronson. *An introduction to infinite ergodic theory.* Mathematical Surveys and Monographs **50**, American Mathematical Society, 1997.
2. J. Aaronson, M. Denker, M. Urbański. Ergodic theory for Markov fibred systems and parabolic rational maps. *Trans. Amer. Math. Soc.* **337** (2): 495–548, 1993.
3. A. F. Beardon. The exponent of convergence of Poincaré series. *Proc. London Math. Soc.* (3) **18**: 461–483, 1968.
4. R. Bowen, C. Series. Markov maps associated with Fuchsian groups. *Publ. Math., Inst. Hautes Etud. Sci.* **50**:153–170, 1979.
5. M. Denker, B.O. Stratmann. Patterson measure: classics, variations and applications.
6. D. Epstein, C. Petronio. An exposition of Poincare's polyhedron theorem. *Enseign. Math., II. Ser.* **40**: 113–170, 1994.
7. J. Fiala, P. Kleban. Intervals between Farey fractions in the limit of infinite level. *Ann. Sci. Math. Qubec* **34** (1), 63–71, 2010.
8. M. Kesseböhmer, B.O. Stratmann. A multifractal formalism for growth rates and applications to geometrically finite Kleinian groups. *Ergod. Th. & Dynam. Sys.*, **24** (01):141–170, 2004.
9. M. Kesseböhmer, B.O. Stratmann. On the Lebesgue measure of sum-level sets for continued fractions. To appear in *Discrete Contin. Dyn. Syst.*
10. M. Kesseböhmer, B.O. Stratmann. A dichotomy between uniform distributions of the Stern-Brocot and the Farey sequence. To appear in *Unif. Distrib. Theory.* **7** (2), 2011.
11. P. Nicholls. *The ergodic theory of discrete groups,* London Math. Soc. Lecture Note Series **143**, Cambr. Univ. Press, Cambridge, 1989.
12. S.J. Patterson. The limit set of a Fuchsian group. *Acta Math.* **136** (3–4): 241–273, 1976.
13. C. Series. The modular surface and continued fractions. *J. London Math. Soc. (2)* **31** (1): 69–80, 1985.
14. C. Series. Geometrical Markov coding of geodesics on surfaces of constant negative curvature. *Ergod. Th. & Dynam. Sys.* **6**: 601–625, 1986.
15. M. Stadlbauer. The Bowen–Series map for some free groups. Dissertation 2002, University of Göttingen. *Mathematica Gottingensis* **5**: 1–53, 2002.
16. M. Stadlbauer. The return sequence of the Bowen–Series map associated to punctured surfaces. *Fundamenta Math.* **182**: 221–240, 2004.
17. M. Stadlbauer, B.O. Stratmann. Infinite ergodic theory for Kleinian groups. *Ergod. Th. & Dynam. Sys.* **25**: 1305–1323, 2005.
18. B.O. Stratmann, S.L. Velani. The Patterson measure for geometrically finite groups with parabolic elements, new and old. *Proc. London Math. Soc. (3)* **71** (1): 197–220, 1995.
19. D. Sullivan. The density at infinity of a discrete group of hyperbolic motions. *Inst. Haut. Études Sci. Publ. Math.* **50**: 171–202, 1979.
20. D. Sullivan. Entropy, Hausdorff measures old and new, and limit sets of geometrically finite Kleinian groups. *Acta Math.* **153**: 259–277, 1984.
21. D. Sullivan. Discrete conformal groups and measurable dynamics. *Bull. Amer. Math. Soc.* **6**: 57–73, 1982.

Appendix* SNOQIT I†: Growth of Λ-Modules and Kummer Theory

Preda Mihăilescu

One by one the guests arrive
The guests are coming through
And "Welcome, welcome" cries a voice
"Let all my guests come in!".[1]

To S. J. Patterson, at his 60th birthday.

Abstract Let $A = \varprojlim_n$ be the projective limit of the p-parts of the class groups in some $\mathbb{Z}_p$-cyclotomic extension. The main purpose of this paper is to investigate the transition $\Lambda a_n \to \Lambda a_{n+1}$ for some special $a = (a_n)_{n \in \mathbb{N}} \in A$, of infinite order. Using an analysis of the $\mathbb{F}_p[T]$-modules $\mathscr{A}_n / p\mathscr{A}_n$ and $\mathscr{A}_n[p]$, we deduce some restrictive conditions on the structure and rank of these modules. Our model can be applied also to a broader variety of cyclic p-extensions and associated modules. In particular, it applies to certain cases of subfields of Hilbert or Takagi class fields, i.e. finite cyclic extensions.

As a consequence of this taxonomy (the term *taxonometric* research was coined by Samuel Patterson; it very well applies to this work and is part of the dedication at the occasion of his 60th birthday), we can give a proof in CM fields of the conjecture of Gross concerning the non-vanishing of the p-adic regulator of p-units.

---

*Due to time constraints, the main results of this paper did not undergo a complete review process. We decided to publish them nevertheless, because of the strong connection with the theme of this book. Theorem 3 of this contribution will appear in amplified form elsewhere.

†SNOQIT = "Seminar Notes on Open Questions in Iwasawa Theory" refers to a seminar held together with S. J. Patterson in 2007/08.

[1] Leonard Cohen: *The Guests*.

P. Mihăilescu
Mathematisches Institut der Universität Göttingen, 37073 Göttingen, Germany
e-mail: preda@uni-math.gwdg.de

V. Blomer and P. Mihăilescu (eds.), *Contributions in Analytic and Algebraic Number Theory*, Springer Proceedings in Mathematics 9, DOI 10.1007/978-1-4614-1219-9,

1 Introduction

Let p be an odd prime and $\mathbb{K} \supset \mathbb{Q}[\zeta]$ be a galois extension containing the pth roots of unity, while $(\mathbb{K}_n)_{n\in\mathbb{N}}$ are the intermediate fields of its cyclotomic $\mathbb{Z}_p$-extension $\mathbb{K}_\infty$. Let $A_n = (\mathscr{C}(\mathbb{K}_n))_p$ be the p-parts of the ideal class groups of $\mathbb{K}_n$ and $\mathbf{A} = \varprojlim_n A_n$ be their projective limit. The subgroups $\mathbf{B}_n \subset A_n$ are generated by the classes containing ramified primes above p and we let

$$\begin{aligned} A'_n &= A_n/\mathbf{B}_n, \\ \mathbf{B} &= \varprojlim_n \mathbf{B}_n, \quad \mathbf{A}' = \mathbf{A}/\mathbf{B}. \end{aligned} \tag{1}$$

We denote as usual the galois group $\Gamma = \mathrm{Gal}(\mathbb{K}_\infty/\mathbb{K})$ and $\Lambda = \mathbb{Z}_p[\Gamma] \cong \mathbb{Z}_p[[\tau]] \cong \mathbb{Z}_p[[T]]$, where $\tau \in \Gamma$ is a topological generator and $T = \tau - 1$; we let

$$\omega_n = (T+1)^{p^{n-1}} - 1 \in \Lambda, \quad \nu_{n+1,n} = \omega_{n+1}/\omega_n \in \Lambda.$$

If $\mathbb{H}_n \supset \mathbb{K}_n$ are the maximal p-abelian unramified extensions of $\mathbb{K}_n$ – the p-Hilbert class fields of $\mathbb{K}_n$ – then $X_n := \mathrm{Gal}(\mathbb{H}_n/K_n) \cong A_n$ via the Artin Symbol, which we shall denote by φ. Let $\mathbb{H} = \cup_n \mathbb{H}_n$ be the maximal unramified p-extension of $\mathbb{K}_\infty$ and $X = \mathrm{Gal}(\mathbb{H}/\mathbb{K}_\infty)$. The isomorphisms $\varphi : A_n \to X_n$ are norm compatible and yield and isomorphism in the projective limit, which we shall also denote by φ:

$$\varphi(\mathbf{A}) = \varphi\left(\varprojlim_n A_n\right) = \varprojlim_n(\varphi(A_n)) = \varprojlim_n(X_n) = X. \tag{2}$$

The module $\mathbf{A}$ is a finitely generated Λ-torsion module, and we assume for simplicity in this paper that $\mu(\mathbf{A}) = 0$.

The groups $A_n, \mathbf{B}_n, A'_n$, etc. are *multiplicative* Λ-modules and we shall denote the action of Λ-multiplicatively, so $a^T = \tau(a)/a$ for $a \in \mathbf{a}$. Whenever intensive use of group ring actions makes additive preferable, we either state the related results in terms of abstract modules, written additively, or explicitly state a switch to additive notation, by slight abuse of language.

If X is a finite abelian group, we denote by X_p its p-Sylow group. The exponent of X_p is the smallest power of p that annihilates X_p; the *subexponent*

$$\mathrm{sexp}(X_p) = \min\left\{\mathrm{ord}(x) \ : \ x \in X_p \setminus X_p^p\right\}.$$

Fukuda proves in [8] (see also Lemma 4 below) that if $\mu(\mathbb{K}) = 0$, then there for the least $n_0 \geq 0$ such that $p\text{-rk}(A_{n_0+1}) = p\text{-rk}(A_{n_0})$ we also have $p\text{-rk}(\mathbf{A}) = p\text{-rk}(A_{n_0})$: the p-rank of A_n becomes stationary after the first occurrence of a stationary rank. It is a general property of finitely generated Λ-modules of finite p-rank, that their p-rank must become stationary after some fixed level – the additional fact that this already happens after the first rank stabilization is a consequence of an early theorem of Iwasawa (see Theorem 2 below), which relates the Λ-module $\mathbf{A}$ to class field theory. The theorem has a class field theoretical proof and one can show that the properties it reveals are not shared by arbitrary finitely generated Λ-modules.

The purpose of this paper is to pursue Fukuda's observation at the level of individual cyclic Λ-modules and also investigate the *prestable* segment of these modules. We do this under some simplifying conditions and focus on specific cyclic Λ-modules defined as follows:

Definition 1. Let $\mathbb{K}$ be a CM field and $a = (a_n)_{n\in\mathbb{N}} \in \mathbf{A}^-$ have infinite order. We say that a is *conic*[2] if the following conditions are fulfilled:

1. There is a Λ-submodule $\mathbf{C} \subset \mathbf{A}^-$ such that

$$\mathbf{A}^- = \mathbf{C} \oplus \Lambda a.$$

   We say in this case that Λa is Λ-complementable.
2. Let $c = (c_n)_{n\in\mathbb{N}} \in \Lambda a$. If $c_n = 1$ for some $n > 0$, then $c \in \omega_n(\Lambda a)$.
3. If $b \in \mathbf{A}^-$ and there is a power $q = p^k$ with $b^q \in \Lambda a$, then $b \in \Lambda a$.
4. If $f_a(T) \in \mathbb{Z}_p[T]$ is the exact annihilator of Λa, then $(f_a(T), \omega_n(T)) = 1$ for all $n > 0$. The common divisor here is understood in $\mathbb{Q}_p$, so f_a, ω_n are coprime as polynomials in $\mathbb{Q}_p$.

The above definition is slightly redundant, containing all the properties that we shall require. See also Sect. 2.1 for a more detailed discussion of the definition. We shall denote a conic module by $\mathscr{A} = \Lambda a$ and $\mathscr{A}_n = \Lambda a_n$.

The first purpose of this paper is to prove the following theorem.

Theorem 1. *Let p be an odd prime, $\mathbb{K}$ be a galois CM extension containing a pth root of units and let $\mathbb{K}_n, A_n$ and $\mathbf{A}$ be defined like above, such that $\mu(\mathbf{A}) = 0$. Let $a = (a_n)_{n\in\mathbb{N}} \in \mathbf{A}^- \setminus (\mathbf{A}^-)^p$ be conic, $q = \mathrm{ord}(a_1)$ and let $f_a(T)$ be the exact annihilator polynomial of a. Then there is an integer $n_0 \geq 1$ such that*

1. *For all $n \geq n_0$, we have $p\text{-}\mathrm{rk}(\mathscr{A}_n) = \lambda(\mathscr{A}) = \deg(f_a)$. Moreover, $\mathrm{ord}(a_n) = p^{n-n_0}\mathrm{ord}(a_{n_0}) = p^{n+z(a)}$, for some $z(a) \in \mathbb{Z}$.*
2. *If $p\text{-}\mathrm{rk}(\mathscr{A}_m) < \deg(\omega_m)$ for some $m \leq n_0$, then $m+1 \geq n_0$.*

For all $n > 0$, we have $\mathrm{ord}(a_{n+1}) \leq p\,\mathrm{ord}(a_n)$.

The theorem is obtained by a tedious algebraic analysis of the rank growth in the *transitions $\mathscr{A}_n \hookrightarrow \mathscr{A}_{n+1}$*.

A class of examples of conic modules is encountered for quadratic ground fields $\mathbb{K}$, such that the p-part $A_1(\mathbb{K})$ of the class group is $\mathbb{Z}_p$-cyclic. We shall give

[2]One can easily provide examples of non conic elements, by considering $\mathbb{K}_\infty$ as a $\mathbb{Z}_p$-extension of $\mathbb{K}_n$ for some $n > 1$. It is an interesting question to find some conditions related only to the field $\mathbb{K}$, which assure the existence of conic elements.

in Sect. 3.2 a series of such examples, drawn from the computations of Ernvall and Metsänkylä in [6]. A further series of applications concern the structure of the components $e_{p-2k}\mathbf{A}$ of the class group of p-cyclotomic extensions, when the Bernoulli number $B_{2k} \equiv 0 \bmod p$. If the conjecture of Kummer–Vandiver or the cyclicity conjecture holds for this component, then the respective modules are conic.

The question about the detailed structure of annihilator polynomials in Iwasawa extensions is a difficult one, and it has been investigated in a series of papers in the literature. For small, e.g. quadratic fields, a probabilistic approach yields already satisfactory results. In this respect, the Cohen-Lenstra [4] and Cohen-Martinet [5] heuristics have imposed themselves, being confirmed by a large amount of empirical results; see also Bhargava's use of these heuristics in [2] for recent developments.

At the other end, for instance in p-cyclotomic fields, computations only revealed linear annihilator polynomials. In spite of the improved resources of modern computers, it is probably still infeasible to pursue intensive numeric investigations for larger base fields. In this respect, we understand this paper as a proposal for a new, intermediate approach between empirical computations and general proofs: *empirical case distinctions* leading to some structural evidence. In this sense, the conditions on conic elements are chosen such that some structural results can be achieved with feasible effort. The results indicate that for large base fields, the repartitions of exact annihilators of elements of $\mathbf{A}^-$ can be expected to be quite structured and far from uniform repartition within all possible distinguished polynomials.

1.1 Notations

We shall fix some notations. The field $\mathbb{K}$ is assumed to be a CM galois extension of $\mathbb{Q}$ with group Δ, containing a pth root of unity ζ but no p^2th roots of unity. We let $(\zeta_{p^n})_{n\in\mathbb{N}}$ be a norm coherent sequence of p^nth roots of unity, so $\mathbb{K}_n = \mathbb{K}[\zeta_{p^n}]$. Thus, we shall number the intermediate extensions of $\mathbb{K}_\infty$ by $\mathbb{K}_1 = \mathbb{K}, \mathbb{K}_n = \mathbb{K}[\zeta_{p^n}]$. We have uniformly that $\mathbb{K}_n$ contains the p^nth but not the p^{n+1}th roots of unity. In our numbering, ω_n annihilates $\mathbb{K}_n^\times$ and all the groups related to $\mathbb{K}_n$ ($A_n, \mathscr{O}(\mathbb{K}_n)$, etc.)

Let $A = \mathscr{C}(\mathbb{K})_p$, the p-Sylow subgroup of the class group $\mathscr{C}(\mathbb{K})$. The p-parts of the class groups of $\mathbb{K}_n$ are denoted by A_n and they form a projective sequence with respect to the norms $N_{m,n} := \mathbf{N}_{\mathbb{K}_m/\mathbb{K}_n}, m > n > 0$, which are assumed to be surjective. The projective limit is denoted by $\mathbf{A} = \varprojlim_n A_n$. The submodule $\mathbf{B} \subset \mathbf{A}$ is defined by (1) and $\mathbf{A}' = \mathbf{A}/\mathbf{B}$. At finite levels $A'_n = A_n/\mathbf{B}_n$ is isomorphic to the ideal class group of the ring of the p-units in $\mathbb{K}_n$. The maximal p-abelian unramified extension of $\mathbb{K}_n$ is $\mathbb{H}_n$ and $\mathbb{H}'_n \subset \mathbb{H}_n$ is the maximal subfield that splits all the primes above p. Then $\mathrm{Gal}(\mathbb{H}'_n/\mathbb{K}_n) \cong A'_n$ (e.g. [11], Sects. 3 and 4).

If the coherent sequence $a = (a_n)_{n\in\mathbb{N}} \in \mathbf{A}^-$ is a conic element, then $p\text{-rk}(\Lambda a) < \infty$. We write $\mathscr{A} = \Lambda a$ and $\mathscr{A}_n = \Lambda a_n$: the finite groups $\mathscr{A}_n$ form a projective sequence of Λ-modules with respect to the norms. The exact annihilator polynomial of $\mathscr{A}$ is denoted by $f_a(T) \in \mathbb{Z}_p[T]$.

If $f \in \mathbb{Z}_p[T]$ is some distinguished polynomial that divides the characteristic polynomial of $\mathbf{A}$, we let $\mathbf{A}(f) = \cup_n \mathbf{A}[f^n]$ be the union of all power f-torsions in $\mathbf{A}$. Since $\mathbf{A}$ is finitely generated, this is the maximal submodule annihilated by some power of f. If $B \subset \mathbf{A}(f)$ is some Λ-module, then we let $k = \mathrm{ord}_f(B)$ be the least integer such that $f^k B = 0$.

1.2 List of Symbols

We give here a list of the notations introduced below in connection with Iwasawa theory.

p	A rational prime,
ζ_{p^n}	Primitive p^nth roots of unity with $\zeta_{p^n}^p = \zeta_{p^{n-1}}$ for all $n > 0$,
μ_{p^n}	$= \{\zeta_{p^n}^k, k \in \mathbb{N}\}$,
$\mathbb{K}$	A galois CM extension of $\mathbb{Q}$ containing the pth roots of unity,
$\mathbb{K}_\infty, \mathbb{K}_n$	The cyclotomic $\mathbb{Z}_p$ – extension of $\mathbb{K}$, and intermediate fields,
Δ	$= \mathrm{Gal}(\mathbb{K}/\mathbb{Q})$,
$A(\mathbf{K})$	The p-part of the ideal class group of the field $\mathbf{K}$,
s	The number of primes above p in $\mathbb{K}$,
Γ	$= \mathrm{Gal}(\mathbb{K}_\infty/\mathbb{K}) = \mathbb{Z}_p\tau, \quad \tau$ a topological generator of Γ,
T	$= \tau - 1$,
$*$	Iwasawa's involution on Λ induced by $T^* = (p-T)/(T+1)$,
Λ	$= \mathbb{Z}_p[[T]], \quad \Lambda_n = \Lambda/(\omega_n\Lambda)$,
ω_n	$= (T+1)^{p^{n-1}} - 1, \quad (\mathbb{K}_n^\times)^{\omega_n} = \{1\}$,
A'_n	$= A'(\mathbb{K}_n)$, the p-part of the ideal class group of the p-integers of $\mathbb{K}_n$,
A'	$= \varprojlim A'_n$,
$\mathbf{B}$	$= \langle \{b = (b_n)_{n\in\mathbb{N}} \in A : b_n = [\wp_n], \wp_n \supset (p)\}\rangle_{\mathbb{Z}_p}$,
$\mathbb{H}_\infty$	The maximal p-abelian unramified extension of $\mathbb{K}_\infty$,
$\mathbb{H}'_\infty \subset \mathbb{H}_\infty$	The maximal subextension of $\mathbb{H}_\infty$ that splits the primes above p,
φ	The Artin symbol, see also (2).

The following notations are specific for transitions:

(A,B)	$=$ A conic transition, A,B are finite $\mathbb{Z}_p[T]$-modules,
G	$= <\tau>$, a cyclic p-group acting on the modules of the transition,
T	$= \tau - 1$,
$S(X)$	$= X[p]$, the p-torsion of the p group X, or its socle,
$R(X)$	$= X/(pX)$, the "roof" of the p group X,
N, ι	$=$ The norm and the lift associated with the transition (A,B),
K	$=$ Ker $(N : B \to A)$,
ω	$=$ Annihilator of A, such that $N = p + \omega N'$,

$$\begin{aligned} d &= \deg(\omega(T)); \quad \nu = \tfrac{(\omega+1)^p-1}{\omega}, \\ \nu\omega &= \text{Annihilator of } B, \\ \mathscr{T} &= B/\iota(A), \text{ The transition module associated to } (A,B), \\ s,s' &= \text{Generators of } S(A), S(B) \text{ as } \mathbb{F}_p[T]\text{-modules}, \\ a,b &= \text{Generators of } A,B \text{ as } \mathbb{Z}_p[T]\text{-modules}, \\ r,r' &= p\text{-rk}(A), p\text{-rk}(B). \end{aligned}$$

1.3 Ramification and Its Applications

Iwasawa's Theorem 6 [11] plays a central role in our investigations. Let us recall the statement of this theorem in our context; we use here an extended version of this theorem, which is found with identical statements and proofs in [16], Lemma 13.14 and 13.15 and [13], Chap. 5, Theorem 4.2. In view of (2), the statement below is obtained by applyng φ^{-1} to the galois groups in Washington's formulation of Lemma 13.15. Note also that our numbering starts at one, while it starts at zero in Washington numeration of the intermediate levels of $\mathbb{K}_\infty$.

Theorem 2 (Iwasawa, Theorem 6 [11]). *Let $\mathbb{K}$ be a number field and $P = \{\wp_i : i = 1,2,\ldots,s\}$ be the primes of $\mathbb{K}$ above p and assume that they ramify completely in $\mathbb{K}_\infty/\mathbb{K}$. Let $\mathbb{H}/\mathbb{K}_\infty$ be the maximal p-abelian unramified extension of $\mathbb{K}_\infty$ and $H = \mathrm{Gal}(\mathbb{H}_\infty/\mathbb{K})$, while $I_i \subset H, i = 1,2,\ldots,s$ are the inertia groups of some primes of $\mathbb{H}_\infty$ above $\wp_i$. Let $a_i \in \mathbf{A}$ be such that $\varphi(\sigma_i)I_1 = I_i$, for $i = 2,3,\ldots,s$.*

Let $Y_1 = \mathbf{A}^T \cdot [a_2,\ldots,a_s]_{\mathbb{Z}_p} \subset \mathbf{A}$ and $Y_n = \nu_{n,1}Y_1$. Then $\mathbf{A}/Y_n \cong \mathbf{A}_n$.

Note that the context of the theorem is not restricted to CM extensions. In fact, Iwasawa's theorem applies also to non-cyclotomic $\mathbb{Z}_p$-extensions, but we shall not consider such extensions in this paper.

The following Theorem settles the question about Y_1^- in CM extensions:

Theorem 3. *Let $\mathbb{K}$ be a galois CM extension of $\mathbb{Q}$ and $\mathbf{A}$ be defined like above. Then $\mathbf{A}^-(T) = \mathbf{A}^-[T] = \mathbf{B}^-$.*

We prove the theorem in chapter 4. Then we derive from Theorem 3 the following result:

Corollary 1. *Let $\mathbb{K}$ be a CM s-field and $\mathbf{B}_n, A'_n$ be defined by (1). Then $(\mathbf{A}')^-[T] = \{1\}$.*

This confirms a conjecture of Gross and Kuz'min stated by Federer and Gross in [7] in the context of p-adic regulators of p-units of number fields, and earlier by Kuz'min [12] in a class field oriented statement, which was shown by Federer and Gross to be equivalent to the non-vanishing of p-adic regulators of p-units. We prove here the class field theoretic statement for the case of CM fields. The conjecture was known to be true for abelian extensions, due to previous work of Greenberg [9].

1.4 Sketch of the Proof

We start with an overview of the proof. Our approach is based on the investigation of the growth of the ranks $r_n := p\text{-rk}(\mathscr{A}_n) \to p\text{-rk}(\mathscr{A}_{n+1})$; for this, we use *transitions* $C_n := \mathscr{A}_{n-1}/\iota_{n,n+1}(\mathscr{A}_n)$, taking advantage of the fact that our assumptions assure that the ideal lift maps are injective for all n. Since we also assumed $p\text{-rk}(\mathscr{A}) < \infty$, it is an elementary fact that the ranks r_n must stabilize for sufficiently large n. Fukuda proved recently that this happens after the first n for which $r_n = r_{n+1}$. We call this value n_0: the *stabilization index*, and focus upon the *critical section* $\mathscr{A}_n : n < n_0$. In this respect, the present work is inspired by Fukuda's result and extends it with the investigation of the critical section; this reveals useful criteria for stabilization, which make that the growth of conic Λ-modules is quite controlled: at the exception of some modules with *flat* critical section, which can grow in rank indefinitely, but have constant exponent p^k, the rank is bounded by $p(p-1)$.

The idea of our approach consists in modeling the *transitions* $(A,B) = (\mathscr{A}_n, \mathscr{A}_{n+1})$ by a set of dedicated properties that are derived from the properties of conic elements. The conic transitions are introduced in Definition 3 below. Conic transitions do not only well describe the critical section of conic Λ-modules, but they also apply to sequences $A_1, A_2, \ldots, A_n$ of more general finite modules on which a p-cyclic group $< \tau >$ acts via the group ring $\mathbf{R} = \mathbb{Z}/(p^N \cdot \mathbb{Z})[\tau] = \mathbb{Z}_p[T], T = \tau - 1$, with $p^N A_n = 0$; the modules A, B are in particular assumed to be cyclic as $\mathbf{R}$-modules and they fulfill some additional properties with respect to norms and lifts. As a consequence, the same theory can be applied, for instance, to sequences of class groups in cyclic p-extensions, ramified or unramified.

The ring $\mathbf{R}$ is a local ring with maximal ideal (p,T); since this ideal is not principal, it is customary to use the Fitting ideals for the investigation of modules on which $\mathbf{R}$ acts. Under the additional conditions of conicity however, the transitions (A,B) come equipped with a wealth of useful $\mathbb{F}_p[T]$-modules. Since $\mathbb{F}_p[T]$ has a maximal ideal (T), which is principal, this highly simplifies the investigation. The most important $\mathbb{F}_p[T]$-modules related to a transition are the *socle*, $S(B) = B[p]$ and the *roof*, $R(B) = B/pB$. It is a fundamental, but not evident fact, that $S(B)$ is a cyclic $\mathbb{F}_p[T]$, and we prove this by induction in Lemma 8. With this, the transitions are caught between two pairs of cyclic $\mathbb{F}_p[T]$-modules, and the relation between these modules induces obstructions on the growth types. These obstructions are revealed in a long sequence of tedious case distinctions, which develop in a natural way.

The relation between rank growth and norm coherence reveals in Corollary 2 the principal condition for termination of the rank growth: assuming that $p\text{-rk}(A_1) = 1$, this must happen as soon as $p\text{-rk}(A_n) < p^{n-1}$. This is a simple extension of Fukuda's results, giving a condition for growth termination, for rank stabilization. A further important module associated with the transition is the kernel of the norm, $K := \text{Ker}\ (N : B \to A)$. The structure of K is an axiom of conic transitions, which is proved to hold in the case of conic Λ-modules. The analysis of growth in conic transitions is completed in the chapter 2.

In chapter 3, the analysis of transitions can be easily adapted to conic Λ-modules, yielding an inductive proof of their structure, as described in Theorem 1. In the fourth chapter, we prove the Theorem 3 and Corollary 1.

Except for the second chapter, the material of this paper is quite simple and straightforward. In particular, the main proof included in chapter 3 follows easily from the technical preparation in chapter 2. Therefore, the reader wishing to obtain first an overview of the main ideas may skip the second chapter in a first round and may even start with chapter 4, in case her interest goes mainly in the direction of the proof of the conjectures included in that chapter.

The Lemmata 4, 5 are crucial for our approach to Kummer theory. They imply the existence of some index $n_0 \geq 0$ such that for all coherent sequences $a = (a_n)_{n\in\mathbb{N}} \in \mathbf{A}^- \setminus (\mathbf{A}^-)^p$, there is a constant $z = z(a) \in \mathbb{Z}$ such that:

$$\begin{aligned} p\text{-rk}(\mathscr{A}_n) &= p\text{-rk}(\mathscr{A}_{n_0}), \\ a_{n+1}^p &= \iota_{n,n+1}(a_n), \\ \text{ord}(a_n) &= p^{n+z}. \end{aligned} \tag{3}$$

2 Growth of Λ-Modules

We start with a discussion of the definition of conicity:

2.1 *The Notion of Conic Modules and Elements*

We have chosen in this paper a defensive set of properties for conic modules, in order to simplify our analysis of the growth of Λ-modules. We give here a brief discussion of these choices. The restriction to CM fields and submodules $\Lambda a \subset \mathbf{A}^-$ is a sufficient condition for ensuring that all lift maps $\iota_{n,n+1}$ are injective. One can prove in general that for a of infinite order, these maps are injective beyond a fixed stabilization index n_0 that will be introduced below. For $n < n_0$, the question remains still open, if it suffices to assume that $\text{ord}(a) = \infty$ in order to achieve injectivity at all levels. It is conceivable that the combination of the methods developed in this paper may achieve this goal, but the question allows no simple answer, so we defer it to later investigations.

By assuming additionally that $(f_a(T), \omega_n(T)) = 1$, we obtain as a consequence of these assumptions, that for $x = (x_m)_{m\in\mathbb{N}}$ with $x_m = 1$, we have $x \in \omega_n\mathbf{A}$. In the same vein, if $x_m^{\omega_n} = 1$ for $m > n$, then $x_m \in \iota_{n,m}(\mathscr{A}_n)$. These two consequences are very practical and will be repeatedly applied below.

The fundamental requirement to conic elements $a \in \mathbf{A}$ is that the module Λa has a direct complement, which is also a Λ-module. Conic modules exist – see for instance Corrolary 1 or the case of imaginary quadratic extensions $\mathbb{K}$ with $\mathbb{Z}_p$ cyclic $(\mathscr{C}(\mathbb{K}))_p$ and only one prime above p. The simplifying assumption allows to derive interesting properties of the growth of Λ-modules that may be generalized to arbitrary modules.

This condition in fact implies the property 3. of the definition 1, a condition which we also call $\mathbb{Z}_p$-*coalescence closure* of Λa, meaning that Λa is equal to the smallest $\mathbb{Z}_p$-submodule of $\mathbf{A}$, which contains Λa and has a direct complement as a $\mathbb{Z}_p$-module. Certainly, given property 1 and using additive notation, $b = g(T)a + x, x \in \mathbf{B}, g \in \mathbb{Z}_p[T]$, and then $qb \in \Lambda a$ implies by property 1 that $qx = 0$, so b is twisted by a p-torsion element, which is inconsistent with the fact that $\mathbf{A}^-$ was assumed $\mathbb{Z}_p$-torsion free. It is also an interesting question, whether the assumption of property 1 and $a_i \in T\mathbf{A}$ are sufficient to imply property 1.

2.2 *Auxiliary Identities and Lemmata*

We shall frequently use some identities in group rings, which are grouped below. For $n > 0$ we let $\mathbf{R}_n = \mathbb{Z}/(p^N \cdot \mathbb{Z})[T]/(\omega_n)$ for some large $N > 0$, satisfying $N > \exp(A_n)$. The ring $\mathbf{R}_n$ is local with maximal ideal (ω_n) and we write $\overline{T}$ for the image of T in this ring. Since $\overline{T}^{p^n} \in p\mathbf{R}_n$, it follows that $\overline{T}^{p^{n+N}} = 0$; thus, $\overline{T} \in \mathbf{R}_n$ is nilpotent and $\mathbf{R}_n$ is a principal ideal domain.

We also consider the group ring $\mathbf{R}'_n := \mathbb{Z}/(p^N \cdot \mathbb{Z})[\omega_n]/(\nu_{n+1,n})$, which is likewise a local principal ideal domain with maximal ideal generated by the nilpotent element ω_n. From the binomial development of $\nu_{n+1,n}$, we deduce the following fundamental identities in Λ:

$$\begin{aligned}
\nu_{n+1,n} &= \frac{(\omega_n+1)^p - 1}{\omega_n} = \omega_n^{p-1} + \sum_{i=1}^{p-1} \binom{p}{i} / p \cdot \omega_n^{i-1} \\
&= p(1 + O(\omega_n)) + \omega_n^{p-1} = \omega_n^{p-1} + pu(\omega_n), \\
u(\omega_n) &= 1 + \frac{p-1}{2}\omega_n + \cdots + \omega_n^{p-2} \in \Lambda^\times, \\
\omega_n^p &= \omega_n \cdot (\nu_{n+1,n} - pu(\omega_n)) = \omega_{n+1} - p\omega_n u(\omega_n). \qquad (4)
\end{aligned}$$

The above identities are equivariant under the Iwasawa involution $* : \tau \mapsto (p+1)\tau^{-1}$. Note that we fixed the cyclotomic character $\chi(\tau) = p+1$. If $f(T) \in \Lambda$, we write $f^*(T) = f(T^*)$, the *reflected* image of $f(T)$. The reflected norms are $\nu^*_{n+1,n} = \omega^*_{n+1}/\omega^*_n$. From the definition of ω^*_n, we have the following useful identity:

$$\omega_n + t\omega_n^* = p^{n-1}c, \quad t \in \Lambda_n^\times, c \in \mathbb{Z}_p^\times. \qquad (5)$$

We shall investigate the growth of the modules A_n for $n \to n+1$. Suppose now that A is a finite abelian p-group which is cyclic as an $\mathbb{Z}_p[T]$-module, generated by $a \in A$. We say that a monic polynomial $f \in \mathbb{Z}_p[T]$ is a *minimal polynomial* for a, if f has minimal degree among all monic polynomials $g \in a^{\top} = \{x \in \mathbb{Z}_p[T] : xa = 0\} \subset \mathbb{Z}_p[T]$.

We note the following consequence of Weierstrass preparation:

Lemma 1. *Let $I = (g(T)) \subset \mathbb{Z}_p[T]$ be an ideal generated by a monic polynomial $g(T) \in \mathbb{Z}_p[T]$. If $n = \deg(g)$ is minimal amongst all the degrees of monic polynomials generating I, then $g(T) = T^n + ph(T)$, with $h(T) \in \mathbb{Z}_p[T]$ and $\deg(h) < n$.*

Proof. Let $g(T) = T^n + \sum_{i=0}^{n-1} c_i T^i$. Suppose that there is some $i < n$ such that $p \nmid c_i$. Then the Weierstrass Preparation Theorem ([16], Theorem 7.3) implies that $g(T)\mathbb{Z}_p[T] = g_2(T)\mathbb{Z}_p[T]$, for some polynomial with $\deg(g_2(T)) \leq n$, which contradicts the choice of g. Therefore, $p \mid c_i$ for all $0 \leq i < n$, which completes the proof of the lemma.

Remark 1. As a consequence, if A is a finite abelian group, which is a Λ-cyclic module of p-rank n, then there is some polynomial $g(T) = T^r - ph(T)$, which annihilates A.

We shall use the following simple application of Nakayama's Lemma:

Lemma 2. *Let X be a finite abelian p-group of p-rank r and $\mathscr{X} = \{x_1, x_2, \ldots, x_r\} \subset X$ be a system with the property that the images $\overline{x}_i \in X/pX$ form a base of this $\mathbb{F}_p$-vector space. Then $\mathscr{X}$ is a system of generators of X.*

Proof. This is a direct consequence of Nakayama's Lemma, [14], Chap. VI, Sect. 6, Lemma 6.3.

The following auxiliary lemma refers to elementary abelian p groups with group actions.

Lemma 3. *Let E be an additively written finite abelian*[3] *p-group of exponent p. Suppose there is a cyclic group $G = <\tau>$ of order p acting on E, and let $T = \tau - 1$. Then E is an $\mathbb{F}_p[T]$-module and E/TE is an $\mathbb{F}_p$-vector space. If $r = \dim_{\mathbb{F}_p}(E/TE)$, then every system $\mathscr{E} = \{e_1, e_2, \ldots, e_r\} \subset E$ such that the images $\overline{e}_i \in E/(TE)$ form a base of the latter vector space, is a minimal system of generators of E as an $\mathbb{F}_p[T]$-module. Moreover $E[T] \cong E/(TE)$ as $\mathbb{F}_p$-vector spaces and $E = \oplus_{i=1}^{r} \mathbb{F}_p[T]e_i$ is a direct sum of r cyclic $\mathbb{F}_p[T]$-modules.*

Proof. The modules $E[T]$ and E/TE are by definition annihilated by T; since $\mathbb{F}_p[T]/(T\mathbb{F}_p[T]) \equiv \mathbb{F}_p$, they are finite dimensional $\mathbb{F}_p$-vector spaces. Let $\mathscr{E}$ be defined like in the hypothesis. The ring $\mathbb{F}_p[T]$ is local with principal maximal ideal $T\mathbb{F}_p[T]$, and T is a nilpotent of the ring since $\tau^p = 1$ so we have the following

[3] These groups are sometimes denoted by *elementary abelian p-groups*, e.g. [16], Sect. 10.2.

identities in $\mathbb{F}_p[\tau] = \mathbb{F}_p[T]$: $0 = \tau^p - 1 = (T+1)^p - 1 = T^p$. It follows from Nakayama's Lemma that $\mathscr{E}$ is a minimal system of generators. The map $T : E \to E$ is a nilpotent linear endomorphism of the $\mathbb{F}_p$-vector space E, so the structure theorem for Jordan normal forms of nilpotent maps implies that $E = \oplus_{i=1}^r \mathbb{F}_p[T]e_i$. One may also read this result by considering the exact sequence

$$0 \longrightarrow E[T] \longrightarrow E \longrightarrow E \longrightarrow E/(TE) \longrightarrow 0$$

in which the arrow $E \to E$ is the map $e \mapsto Te$. The diagram indicates that $E[T] \cong E/(TE)$, hence the claim.

In the situation of Lemma 3, we denote the common $\mathbb{F}_p$-dimension of $E[T]$ and E/TE by *T-rank of E*.

2.3 Stabilization

We shall prove in this section the relations (3). First we introduce the following notations:

Definition 2. given a finite abelian p-group X, we write $S(X) = X[p]$ for its p-torsion: we denote this torsion also by *the socle* of X. Moreover, the factor $X/X^p = R(X)$ – the *roof* of X. Then $S(X)$ and $R(X)$ are $\mathbb{F}_p$-vector spaces and we have the classical definition of the p-rank given by $p\text{-rk}(X) = \text{rank}\ (S(X)) = \text{rank}\ (R(X))$, the last two ranks being dimensions of $\mathbb{F}_p$-vector spaces. We say that $x \in X$ is *p-maximal*, or simply maximal, if $x \notin X^p$.

Suppose there is a cyclic p-group $G = \tau$ acting on X, such that X is a cyclic $\mathbb{Z}_p[T]$-module with generator $x \in X$, where $T = \tau - 1$. Suppose additionally that $S(X)$ is also a cyclic $\mathbb{F}_p[T]$-module. Let $s := (\text{ord}(x)/p)x \in S(X)$. Then we say that S is *straight* if s generates $S(X)$ as an $\mathbb{F}_p[T]$-module; otherwise, $S(X)$ is *folded*.

The next lemma is a special case of Fukuda's Theorem 1 in [8]:

Lemma 4 (Fukuda). *Let $\mathbb{K}$ be a CM field and $A_n, \mathbf{A}$ be defined like above. Suppose that $\mu(\mathbf{A}^-) = 0$ and let $n_0 > 0$ be such that $p\text{-rk}(A^-_{n_0}) = p\text{-rk}(A^-_{n_0+1})$. Then $p\text{-rk}(A^-_n) = p\text{-rk}(A^-_{n_0}) = \lambda^-$ for all $n > n_0$.*

Remark 2. The above application of Fukuda's Theorem requires $\mu = 0$; it is known that in this case the p-rank of A_n must stabilize, but here it is shown that it must stabilize after the first time this rank stops growing from A_n to A_{n+1}. We have restricted the result to the minus part, which is of interest in our context. Note that the condition $\mu = 0$ can be easily eliminated, by considering the module $(\mathbf{A}^-)^{p^m}$ for some $m > \mu$.

The following elementary, technical lemma will allow us to draw additional information from Lemma 4.

Lemma 5. *Let A and B be finitely generated abelian p-groups denoted additively, and let $N : B \to A$, $\iota : A \to B$ two $\mathbb{Z}_p$ – linear maps such that:*

1. N is surjective and ι is injective[4]*;*
2. The p-ranks of A and B are both equal to r and $|B|/|A| = p^r$.
3. $N(\iota(a)) = pa, \forall a \in A$ and ι is rank preserving, so $p\text{–rk}(\iota(A)) = p\text{–rk}(A)$;

Then $\iota(A) = pB$ and $\text{ord}(x) = p \cdot \text{ord}(Nx)$ for all $x \in B$.

Proof. The condition 3. is certainly fulfilled when ι is injective, as we assume, but it also follows from $\text{sexp}(A) > p$, even for lift maps that are not injective. We start by noting that for any finite abelian p – group A of p – rank r and any pair α_i, β_i; $i = 1, 2, \ldots, r$ of minimal systems of generators there is a matrix $E \in \text{Mat}(r, \mathbb{Z}_p)$, which is invertible over $\mathbb{Z}_p$, such that

$$\beta = E\alpha. \tag{6}$$

This can be verified directly by extending the map $\alpha_i \mapsto \beta_i$ linearly to A and, since $(\beta_i)_{i=1}^r$ is also a minimal system of generators, deducing that the map is invertible, thus regular. It represents a unimodular change of base in the vector space $A \otimes_{\mathbb{Z}_p} \mathbb{Q}_p$.

The maps ι and N induce maps

$$\overline{\iota} : A/pA \to B/pB, \ \overline{N} : B/pB \to A/pA.$$

From 1, we see $\overline{N}$ is surjective and since, by (2), it is a map between finite sets of the same cardinality, it is actually an isomorphism. But 3. implies that $\overline{N} \circ \overline{\iota} : A/pA \to A/pA$ is the trivial map and since $\overline{N}$ is an isomorphism, $\overline{\iota}$ must be the trivial map, hence $\iota(A) \subset pB$.

Since ι is injective, it is rank preserving, i.e. $p\text{–rk}(A) = p\text{–rk}(\iota(A))$. Let b_i, $i = 1, 2, \ldots, r$ be a minimal set of generators of B: thus, the images $\overline{b}_i$ of b_i in B/pB form an $\mathbb{F}_p$ – base of this algebra. Let $a_i = N(b_i)$; since $p\text{–rk}(B/pB) = p\text{–rk}(A/pA)$, the set $(a_i)_i$ also forms a minimal set of generators for A. We claim that $|B/\iota(A)| = p^r$.

Pending the proof of this equality, we show that $\iota(A) = pB$. Indeed, we have the equality of p-ranks:

$$|B/pB| = |A/pA| = |B/\iota(A)| = p^r,$$

implying that $|pB| = |\iota(A)|$; since $\iota(A) \subset pB$ and the p – ranks are equal, the two groups are equal, which is the first claim. The second claim will be proved after showing that $|B/\iota(A)| = p^r$.

---

[4]The same results can be proved if the injectivity assumption is replaced by the assumption that $\text{sexp}(A) > p$ – injectivity then follows. In our context the injectivity is however part of the premises, so we give here the proof of the simpler variant of the lemma.

Let $S(X)$ denote the socle of the finite abelian p – group X. There is the obvious inclusion $S(\iota(A)) \subset S(B) \subset B$ and since ι is rank preserving, $p\text{–rk}(A) = p\text{–rk}(S(A)) = p\text{–rk}(B) = p\text{–rk}(S(B)) = p\text{–rk}(S(\iota(A)))$, thus $S(B) = S(\iota(A))$. Let $(a_i)_{i=1}^r$ be a minimal set of generators for A and $a'_i = \iota(a_i) \in B, i = 1,2,\ldots,r$; the $(a'_i)_{i=1}^r$ form a minimal set of generators for $\iota(A) \subset B$. We choose in B two systems of generators in relation to a'_i and the matrix E will map these systems according to (6).

First, let $b_i \in B$ be such that $p^{e_i} b_i = a'_i$ and $e_i > 0$ is maximal among all possible choices of b_i. From the equality of socles and p-ranks, one verifies that the set $(b_i)_{i=1}^r$ spans B as a $\mathbb{Z}_p$-module; moreover, $\iota(A) \subset pB$ implies $e_i \geq 1$. On the other hand, the norm being surjective, there is a minimal set of generators $b'_i \in B,\ i = 1,2,\ldots,r$ such that $N(b'_i) = a_i$. Since b_i, b'_i span the same finite $\mathbb{Z}_p$-module B, (6) in which $\alpha = \mathbf{b}$ and $\beta - \mathbf{b}'$ defines a matrix with $\mathbf{b} = E \cdot \mathbf{b}'$. On the other hand,

$$\iota(\mathbf{a}) = \mathbf{a}' = \mathbf{Diag}(p^{e_i})\mathbf{b} = \mathbf{Diag}(p_i^{e_i})E \cdot \mathbf{b}',$$

The linear map $N : B \to A$ acts component-wise on vectors $\mathbf{x} \in B^r$. Therefore,

$$\begin{aligned} N\mathbf{b} = \mathbf{Nb_i} = N(E\mathbf{b}') &= N\left(\left(\prod_j b'^{\sum_j e_{i,j}}_j\right)_{i=1}^r\right) \\ &= \left(\prod_j (Nb'_j)^{\sum_j e_{i,j}}\right)_{i=1}^r = \left(\prod_j (a_j)^{\sum_j e_{i,j}}\right)_{i=1}^r \\ &= E(\mathbf{a}). \end{aligned}$$

Using the fact that the subexponent is not p, we obtain thus two expressions for $N\mathbf{a}'$ as follows:

$$\begin{aligned} \mathbf{Na}' &= p\mathbf{a} = pI \cdot \mathbf{a} \\ &= N\left(\mathbf{Diag}(p^{e_i})\mathbf{b}\right) = \mathbf{Diag}(p^{e_i}) \cdot N(\mathbf{b}) = \mathbf{Diag}(p^{e_i}) \cdot E\mathbf{a}, \quad \text{so} \\ \mathbf{a} &= \mathbf{Diag}(p^{e_i - 1}) \cdot E\mathbf{a}. \end{aligned}$$

The a_j form a minimal system of generators and E is regular over $\mathbb{Z}_p$; therefore, $(\alpha) := (\alpha_j)_{j=1}^r = E\mathbf{a}$ is also minimal system of generators of A and the last identity above becomes

$$\mathbf{a} = \mathbf{Diag}(p^{e_i - 1}) \cdot \alpha.$$

If $e_i > 1$ for some $i \leq r$, then the right-hand side is not a generating system of A while the left side is: it follows that $e_i = 1$ for all i. Therefore, $|B/\iota(A)| = p^R$ and we have shown above that this implies the injectivity of ι.

Finally, let $x \in B$ and $q = \text{ord}(Nx) \geq p$. Then $qN(x) = 1 = N(qx)$, and since $qx \in \iota(A)$, it follows that $N(qx) = pqx = 1$ and thus pq annihilates x. Conversely, if

$\text{ord}(x) = pq$, then $pqx = 1 = N(qx) = qN(x)$, and $\text{ord}(Nx) = q$. Thus, $\text{ord}(x) = p \cdot \text{ord}(Nx)$ for all $x \in B$ with $\text{ord}(x) > p$. If $\text{ord}(x) = p$, then $x \in S(B) = S(\iota(A) \subset \iota(A)$ and $Nx = px = 1$, so the last claim holds in general.

One may identify the modules A, B in the lemma with subsequent levels A_n^-, thus obtaining:

Proposition 1. *Let $\mathbb{K}$ be a CM field, let $\mathbf{A}^- = \varprojlim_n A_n^-$ and assume that $\mu(\mathbf{A}) = 0$. Let $n_0 \in \mathbb{N}$ be the bound proved in Lemma 4, such that for all $n \geq n_0$ we have $p\text{-rk}(A_n^-) = \mathbb{Z}_p\text{-}rk(\mathbf{A}^-) = \lambda^-$. Then the following facts hold for $x = (x_n)_{n\in\mathbb{N}} \in \mathbf{A}^-$ with $x_n \neq 1$ for some $n > n_0$:*

$$\begin{aligned} px_{n+1} &= \iota(N_{n+1,n}(x_n)), \quad \iota(A_n^-) = pA_{n+1}^-, \\ \omega_n x_{n+1} &\in \iota_{n,n+1}(A_n^-[p]). \end{aligned} \tag{7}$$

In particular,

$$\nu_{n+1,n}(x_{n+1}) = px_{n+1} = \iota_{n,n+1}(x_n). \tag{8}$$

Proof. We let $n > n_0$. Since $\mathbf{A}^-$ is $\mathbb{Z}_p$-torsion free, we may also assume that $\text{sexp}(A_n^-) > p$. We use the notations from Lemma 5 and let $\iota = \iota_{n,n+1}, N = N_{n+1,n}$ and $N' = \nu_{n+1,n}$.

For proving (8), thus $px = \iota(N(x)) = N'(x)$, we consider the development $t := \omega_n = (T+1)^{p^n} - 1$ and

$$N' = p + t \cdot v = p + t\left(\binom{p}{2} + tw)\right), \quad v, w \in \mathbb{Z}[t],$$

as follows from the binomial development of $N' = \frac{(t+1)^p - 1}{t}$. By definition, t annihilates A_n^- and a fortiori $\iota(A_n^-) \subset A_{n+1}^-$; therefore, for arbitrary $x \in A_{n+1}^-$ we have $(pt)x = t(px) = t\iota(x_1) = 0$, where the existence of x_1 with $px = \iota(x_1), x_1 \in A_n^-$ follows from Lemma 5. Since ι is injective and thus rank preserving, we deduce that $tx \in A_{n+1}^-[p] = \iota(A_n^-[p])$, which is the first claim in (7). Then

$$t^2x = t \cdot (tx) = tx_2 = 0, \qquad \text{since } x_2 = tx \in \iota(A_n^-).$$

Using $t^2x = ptx = 0$, the above development for N' plainly yields $N'x = px$, as claimed. Injectivity of the lift map then leads to (7). Indeed, for $a = (a_n)_{n\in\mathbb{N}}$ and $n > n_0$ we have

$$\begin{aligned} \text{ord}(a_n) &= \text{ord}(\iota_{n+1,n}(a_n)) = \text{ord}(\iota_{n+1,n} \circ N_{n+1,n}(a_{n+1})) \\ &= \text{ord}(\nu_{n+1,n}a_{n+1}) = \text{ord}(pa_{n+1}) = \text{ord}(a_{n+1})/p. \end{aligned}$$

This completes the proof.

Remark 3. The restriction to the minus part $\mathbf{A}^-$ is perfectly compatible with the context of this paper. However, we note that Lemma 5 holds as soon as $\mathrm{sexp}(A) > p$. As a consequence, all the facts in Proposition 1 hold true for arbitrary cyclic modules Λa with $\mathrm{ord}(a) = \infty$. The proof being algebraic, it is not even necessary to assume that $\mathbb{K}_\infty$ is the cyclotomic Λ-extension of $\mathbb{K}$, it may be any $\mathbb{Z}_p$-extension and $\mathbf{A} = \varprojlim_n A_n$ is defined with respect to the p-Sylow groups of the class groups in the intermediate levels of $\mathbb{K}_\infty$. The field $\mathbb{K}$ does not need to be CM either. The Proposition 1 is suited for applications in Kummer theory, and we shall see some in the chapter 4. This remark shows that the applications reach beyond the frame imposed in this paper.

As a consequence, we have the following elegant description of the growth of orders of elements in A_n^-:

Lemma 6. *Let $\mathbb{K}$ be a CM field and $A_n, \mathbf{A}$ be defined as above, with $\mu(\mathbf{A}) = 0$. Then there exists an $n_0 > 0$ which only depends on $\mathbb{K}$, such that:*

1. $p-\mathrm{rk}(A_n^-) = p-\mathrm{rk}(A_{n_0}^-) = \lambda^-$ *for* $n \geq n_0$,
2. *For all* $a = (a_n)_{n\in\mathbb{N}} \in \mathbf{A}^-$ *there is a* $z = z(a) \in \mathbb{Z}$ *such that, for all* $n \geq n_0$ *(3) holds.*

Proof. The existence of n_0 follows from Lemma 4 and relation (7) implies that $\mathrm{ord}(a_n) = p^{n-n_0}\mathrm{ord}(a_{n_0})$ for all $n \geq n_0$, hence the definition of z. This proves point 2 and (3).

The above identities show that the structure of $\mathscr{A}$ is completely described by $\mathscr{A}_{n_0}$: both the rank and the annihilator $f_a(T)$ of $\mathscr{A}$ are equal to rank and annihilator of $\mathscr{A}_{n_0}$. Although $\mathscr{A}_{n_0}$ is a finite module and thus its annihilator ideal is not necessarily principal, since it also contains ω_{n_0+1} and p^{n+z}, the polynomial $f_a(T)$ is a distinguished polynomial of least degree, contained in this ideal. Its coefficients may be normed by choosing minimal representatives modulo p^{n+z}. It appears that the full information about $\mathscr{A}$ is contained in the *critical section* $\{\mathscr{A}_n : n \leq n_0\}$.

2.4 *The Case of Increasing Ranks*

In this section, we shall give some generic results similar to Lemma 5, for the case when the groups A and B have distinct ranks. Additionally, we assume that the groups A and B are endowed with a common group action, which is reminiscent from the action of Λ on the groups $\mathscr{A}_n$ of interest.

The assumptions about the groups A, B will be loaded with additional premises, which are related to the case $A = \mathscr{A}_n, B = \mathscr{A}_{n+1}$. We define:

Definition 3. A pair of finite abelian p-groups A, B is called a *conic transition*, if the following hold:

1. A, B are abelian p-groups written additively and $N : B \to A$ and $\iota : A \to B$ are linear maps which are surjective, respectively injective. Moreover $N \circ \iota = p$ as a map $B \to B$. The ranks are $r = p\text{–rk}(A) \leq r' = p\text{–rk}(B)$. Note that for $r = r'$ we are in the case of Lemma 5, so this will be considered as a *stable* case.
2. There is a finite cyclic p-group $G = \mathbb{Z}_p \tau$ acting on A and B, making B into cyclic $\mathbb{Z}_p[\tau]$ – modules. We let $T = \tau - 1$.
3. We assume that there is a polynomial $\omega(T) \in \mathbf{R} := \mathbb{Z}/(p^N \cdot \mathbb{Z})[T]$, for $N > 2\exp(B)$, with

$$N = \frac{(\omega(T)+1)^p - 1}{\omega} \in \mathbf{R},$$

$$\omega \equiv T^{\deg(\omega)} \bmod p\mathbb{Z}_p[T], \quad \text{and} \quad \omega \equiv 0 \bmod T.$$

In particular, (4) holds; we write $d = \deg(\omega(T)) \geq 1$. We also assume that $\omega A = 0$.
4. The kernel $K :=$ Ker $(N : B \to A) \subset B$ is assumed to verify $K = \omega B$ and if $x \in B$ verifies $\omega x = 0$, then $x \in \iota(A)$.
5. There is an $a \in A$ such that $a_i = T^i a, i = 0, \ldots, r-1$ form a $\mathbb{Z}_p$-base of A, and $a_0 = a$.

The transition is *regular* if $r' = pd$; it is regular *flat*, if $\text{sexp}(B) = \exp(B)$ and it is regular wild, if it is regular and $\exp(B) > \text{sexp}(B)$. It is *initial* is $r = d = 1$ and it is *terminal* if $r' < pd$. If $r = r'$, the transition is called *stable*. The module associated with the transition (A, B) is the *transition module* $\mathscr{T} = B/\iota(A)$. We shall write $\nu = \iota \circ N : B \to B$. Then $\nu = \nu(T)$ is a polynomial of degree $\deg(\nu) = (p-1)d$ and $\omega\nu$ annihilates B.

We introduce some notions for the study of socles. Let $\varpi : B \to \mathbb{N}$ be the map $x \to \text{ord}(x)/p$ and $\psi : B \to S(B)$ be given by $x \mapsto \varpi(x) \cdot x$, a $\mathbb{Z}_p$-linear map. Let $\Omega(b) = \{q_i := \varpi(T^i b) \ : \ i = 0, 1, \ldots, r'-1\}$. Then $q_0 \geq q_1 \geq \cdots q_{r'-1}$. The *jumps* of $\Omega(b)$ are the set

$$J := \{i \ : \ q_i > q_{i+1}\} \subset \{0, 1, \ldots, r'-2\}.$$

We shall write

$$B_j := \sum_{i=0}^{j} \mathbb{Z}_p T^i b \subset B, \quad 0 \leq j < r'.$$

We consider in the sequel only transitions that are not stable, thus we assume that $r < r'$. We show below that point 4 of the definition reflects the specific properties of conic modules, while the remaining ones are of general nature and apply to transitions in arbitrary cyclic Λ-modules. Throughout this chapter, a and b are generators of A and B as $\mathbb{Z}_p[T]$-modules. Any other generators differ from a and b by units.

We start with an elementary fact, which holds for finite cyclic $\mathbb{Z}_p[T]$-modules X, such as the elements of conic transitions.

Lemma 7. *If (A,B) be a conic transition. If $y,z \in B \setminus pB$ are such that $y-z \in pB$, then they differ by a unit:*

$$y,z \notin pB, \quad y-z \in pB \quad \Rightarrow \quad \exists\, v(T) \in (\mathbb{Z}_p[T]^\times),\ z = v(T)y. \tag{9}$$

Moreover, if $S(A), S(B)$ are $\mathbb{F}_p[T]$-cyclic and $y \in B \setminus \{0\}$ is such that $Ty \in TS(B)$, then either $y \in S(B)$ or there are $a' \in \iota(a)$ and $z \in S(B)$ such that

$$y = z + a', \quad Ta' = 0, \quad \mathrm{ord}(a') > p. \tag{10}$$

Proof. Let $b \in B$ generate this cyclic $\mathbb{Z}_p[T]$ module. Then $R(B) = B/pB$ is a cyclic $\mathbb{F}_p[T]$ module; since $\tau^{p^M} = 1$ for some $M > 0$, it follows that $(T+1)^{p^M} - 1 = T^{p^M} = 0 \in \mathbb{F}_p[T]$, so the element T is nilpotent.

Let $y \in B$ with image $0 \neq \overline{y} \in R(B)$. Then there is a $k \geq 0$ such that $T^k b\mathbb{F}_p[T]R(B) = y\mathbb{F}_p[T]R(B)$: consider the annihilator ideal of the image $b' \in B/(pB,y)$. Since b is a generator, we also have $y = g(T)b, g \in \mathbb{Z}_p[T]$. The above shows that $g(T) \equiv 0 \bmod T^k$, so let $g(T) = T^k g_1(T)$ with $g_1(T) = \sum_{j \geq 0} c_j T^j$, $c_j \in \mathbb{Z}_p$. The above equality of ideals in B/pB implies that $c_0 \in \mathbb{Z}_p^\times$, since otherwise $T^k B\mathbb{F}_p[T]R(B) \supsetneq y\mathbb{F}_p[T]R(B)$. Therefore, $g(T) \in (\mathbb{Z}_p[T])^\times$. Applying the same fact to y,z, we obtain (9) by transitivity.

Finally suppose that $0 \neq Ty \in \ S(B) = B[p]$. Then $y \neq 0$; if $\mathrm{ord}(y) = p$ then $y \in S(B)$ and we are done. Suppose thus that $\mathrm{ord}(y) = p^e, e > 1$ and let $y' = p^{e-1}y \in S(B)$. The socle $S(B)$ is $\mathbb{F}_p[T]$ cyclic, so there is a $z \in S(B)$ such that $Ty = Tz \in TS(B)$. Then $T(y-z) = 0$ and $y - z \in \iota(A)$ by point 4 of the definition 3; therefore $y = z + a', a' \in \iota(a)$. Moreover, $\mathrm{ord}(y) = \mathrm{ord}(a') > p$ while $Ty = Tz + Ta'$, thus $Ta' = 0$. This confirms (10).

2.5 *Transition Modules and Socles*

The following lemmata refer to conic transitions. We start with several results of general nature, which will then be used in the next section for a case-by-case analysis of transitions and minimal polynomials.

Lemma 8. *The following facts hold in conic transitions:*

(i) Suppose that $S(A)$ is $\mathbb{F}_p[T]$-cyclic; then the socle $S(B)$ is also a cyclic $\mathbb{F}_p[T]$-cyclic module.

(ii) Let $x^\top = \{t \in \mathbb{Z}_p[T] \ : \ tx = 0\}$ be the annihilator ideal of x and $\overline{\omega} \in \mathbb{Z}_p[T]$ be a representant of the class $(\omega \bmod b^\top) \in B/b^\top$. We have

$$\iota(S(A)) \subset S(K), \quad K \cap \iota(A) = \iota(S(A)), \quad \textit{and} \tag{11}$$

$$K = \overline{\omega}B = a^\top B. \tag{12}$$

Proof. The point (i) follows from Lemma 3. Indeed, $S(B) \supseteq \iota(S(A))$ are elementary p-groups by definition. If $x \in S(B)[T]$, then $Tx = 0$ and point 3 of the definition of conic transitions implies that $x \in \iota(A) \cap S(B) = \iota(S(A))$. Thus, $S(B)[T] \subseteq \iota(S(A))[T]$ and since $S(A)$ is $\mathbb{F}_p[T]$ cyclic, we know that $p\text{–rk}(\iota(S(A))[T]) = p\text{–rk}(S(B)[T]) = 1$. The Lemma 3 implies that the T-rank of $S(B)$ is one and $S(B)$ is cyclic as an $\mathbb{F}_p[T]$-module.

Let now $x \in \iota(S(A))$, so $\omega x = px = 0$. Then $Nx = (pu + \omega^{p-1})x = 0$, and thus $x \in S(K)$. If $x' \in \iota(A) \cap K$, then $Nx' = px' = 0$ and thus $x' \in S(K) \cap \iota(A) = \iota(S(A))$, showing that (11) is true.

By point (4) of the definition of conic transitions, we have $K = \omega B = \omega \mathbb{Z}_p[T]b$ and since ω acts on b via its image modulo the annihilator of this generator, it follows that $K = \overline{\omega} b$ for any representant of this image in $\mathbb{Z}_p[T]$. For $t \in a^\top$, we have $N(tb) = tN(b) = ta = 0$; conversely, if $x = t'b \in K$, then $N(t'b) = t'N(b) = t'a = 0$ and thus $t' \in a^\top$. We thus have $K = a^\top B$, which confirms (12) and completes the proof of (ii).

An important consequence of the structure of the kernel of the norm is

Corollary 2. *Let (A,B) be a transition with $r < d$. Then $r' = r$.*

Proof. Let $\theta = T^r + pg(T) \in a^\top$ be a minimal annihilator polynomial of a. Then $\theta B \subset K = \overline{\omega} B$ so there is a $y \in \mathbb{Z}_p[T]$ such that $\theta b = \omega y b$ and since $\theta b \notin pB$, it follows that $y \notin p\mathbb{Z}_p[T]$. Let thus $y = cT^j + O(p, T^{j+1})$ with $j \geq 0$ and $(c,p) = 1$. Then

$$\alpha = T^r + pg(T) - \omega \cdot y = T^r + T^{d+j} + O(p, T^{d+j+1}) \in b^\top.$$

Using $d > r$, Weierstrass Preparation implies that there is a distinguished polynomial $h(T)$ of degree r and a unit $v(T)$ such that $\alpha = hv$. Since v is a unit, $h \in b^\top$. But then $T^r b \in P$ and thus $B = P$ and $p\text{–rk}(B) = r$, which confirms the statement of the Lemma.

The corollary explains the choice of the signification of flat and terminal transitions: a terminal transition can only be followed by a stable one.

We analyze in the next lemma the transition module $\mathscr{T}$ in detail.

Lemma 9. *Let (A,B) be a conic transition and $\mathscr{T} = B/\iota(A)$ be its transition module. Then*

1. *The module $\mathscr{T}$ is $\mathbb{Z}_p[T]$-cyclic, annihilated by ν, and*

$$r' \leq r + (p-1)d.$$

Moreover,

$$\exp(B) \leq p \exp(A), \tag{13}$$

and there is an $\ell(B)$ with $\operatorname{ord}(T^{\ell-1}b) = \exp(B) = p \cdot \operatorname{ord}(T^\ell B)$.

2. *If $S(A) \subset TA$ and $S(B)$ is folded, then $r' = r$.*

Proof. Since $\nu(B) = \iota(A)$, it follows that $\nu(\mathcal{T}) = 0$, showing that

$$r' = p\text{–rk}(B) \leq p\text{–rk}(A) + p\text{–rk}(\mathcal{T}) \leq p\text{–rk}(A) + \deg(\nu) = r + (p-1)d,$$

which confirms the first claim in point 1.

Let now $q = \exp(A)$, so $q\iota(A) = 0$ and $qB \supseteq \iota(A)$. Thus $\mathcal{T}^{\top} \supseteq (B/qB)$. We let $\ell(B) = p\text{–rk}(B/qB)$ and prove the claims of the lemma. We have

$$pqu(T)b = -qT^{p-1}b + q\iota(a) = -qT^{p-1}b,$$

Assuming that $qT^{p-1}b \neq 0$, we obtain $p\text{–rk}(\mathcal{T}) \leq (p-1)d < p\text{–rk}(B/qB)$, in contradiction with the fact that B/qB is a quotient of $\mathcal{T}$. Therefore, $qT^{p-1}b = 0$ and thus $pqu(T)b = 0$, so $\exp(B) = pq$. Therefore, the module $qB \subset S(B)$ and it has rank $\ell(B)$. Let $s' \in S(B)$ be a generator. Comparing ranks in the $\mathbb{F}_p[T]$-cyclic module $S(B)$, we see that $T^{r'-\ell}s = qbv(T)$, $v(T) \in (\mathbb{F}_p[T])^{\times}$.

Suppose now that $S(B)$ is folded; then $\iota(S(A)) = T^k S(B), k = r' - r$. If $s = Tg(T)\iota(a)$ is a generator of $\iota(S(A))$, then $T^{r'-r}s' = v(T)\iota(s), v(T) \in (\mathbb{F}_p[T])^{\times}$. Thus,

$$T(T^{r'-r-1}s' - g(T)v\iota(a)) = 0,$$

and by point 4 of the definition of conic transitions, $T^{r'-r-1}s' \in \iota(S(A))$. But then

$$r' = p\text{–rk}(S(B)) \leq p\text{–rk}(\iota(S(A)) + r' - (r+1) = r + r' - (r+1) = r' - 1.$$

This is a contradiction which implies that $r = r'$ and (A, B) is in this case a stable transition.

In view of the previous lemma, we shall say that the transition (A, B) is *wild* if $r' = pd$ and $S(B)$ is folded. The flat transitions are described by:

Lemma 10. *Let (A, B) be a conic transition. The following conditions are equivalent:*

(i) The exact sequence

$$0 \to \iota(A) \to B \to \mathcal{T} \to 0, \tag{14}$$

is split.

(ii) The jump-set $J(B) = \emptyset$,

(iii) The socle $S(B)$ is straight,

(iv) $\text{sexp}(B) = \exp(B)$,

Moreover, if (A, B) is a transition verifying the above conditions and $\exp(A) = q$, *then* $\exp(B) = q$.

Proof. The conditions (ii) and (iv) are obviously equivalent: if $q = \text{sexp}(B) = \exp(B)$, then the exponent $q_i = \text{ord}(T^i b) = q$ are all equal, and conversely, if these exponents are equal, then $\mathrm{sexp}(B) = \exp(B)$: to see this, consider $x \in B \setminus pB$ such that $\mathrm{ord}(x) = \mathrm{sexp}(B)$. Since $0 \neq \overline{x} \in R(B)$, Lemma 7 shows that $x = T^k v(T)$, $k \geq 0, v \in (\mathbb{Z}_p[T])^\times$. Therefore $\mathrm{ord}(x) = q_k = q$, as claimed.

Suppose that $S(B)$ is straight, so $\psi(b)$ generates $S(B)$. Then $T^j\psi(b) \neq 0$ for all $0 \leq j < r'$ and thus $T^j\varpi(b) \cdot b = \varpi(b)(T^jb) \neq 0$. Since $\mathrm{ord}(\varpi(b)T^jb) \leq p$, it follows that the order is p and $\varpi(b) = \varpi(T^jb)$, so $\mathrm{ord}(b) = \mathrm{ord}(T^jb)$ for all j, and thus $\mathrm{sexp}(B) = \exp(B)$. Hence, $(iii) \Rightarrow (ii), (iv)$. Conversely, suppose that the socle is folded. Then let $\psi(T^kb)$ be a generator of the socle, $k > 0$. The same argument as above shows that $\mathrm{ord}(T^{k-1}b) > \mathrm{ord}(T^kb)$ and thus $J(B) \neq \emptyset$. Therefore, (ii)–(iv) are equivalent.

We show that (14) is split if (ii)–(iv) hold. Suppose that $\iota(a) \notin pB$; then $\iota(a) = vb$ has non trivial image in $R(B)$ and thus $\mathrm{ord}(a) = \mathrm{ord}(b)$. If (14) is not split, then $\psi(\iota(a)) \in \sum_{i=0}^{\deg(v)-1} \mathbb{Z}_pT^ib$, in contradiction to $S(B)$ being straight. Thus, $J(b) = \emptyset$ implies (14) being split.

Conversely, we show that if (14) is split, then $S(B)$ is straight and $\mathrm{sexp}(B) = \exp(B)$. We have $B = \iota(A) \oplus B_{r'-r-1}$ and $S(B) = S(B_{r'-r-1}) \oplus \iota(S(A))$. On the other hand, $\iota(S(A)) = \iota(A) \cap K \neq \emptyset$; therefore, $S(B_{r-1}) \cap S(K) = \emptyset$, and it follows that $S(B)$ is straight. This completes the proof of the equivalence of (i)–(iv).

If $\exp(A) = q$ and the above conditions hold, then $\exp(B) = \mathrm{sexp}(B)$ and thus $\ell(B) = p\text{–}\mathrm{rk}(B) = r'$. We prove by induction that $r' = pd$: Assume thus that $p\text{–}\mathrm{rk}(A) = d$ and let $s' = (q/p)b \in S(B)$, a generator. Then $s := Ns' = (q/b)\iota(a) \in \iota(S(A))$ will be a generator of $\iota(S(A))$. A rank comparison then yields

$$r' = p\text{–}\mathrm{rk}(S(B)) = p\text{–}\mathrm{rk}(S(A)) + (p-1)d = pd.$$

From $\iota(a) = pbu(\omega) + \omega^{p-1}b$, we gather that $\mathrm{ord}(a) \geq \max(p\,\mathrm{ord}(b), \mathrm{ord}(\omega^{p-1}b))$. Since $r' = p\text{–}\mathrm{rk}(B) = pd$ and $\ell(B) = r'$, it follows that $\mathrm{ord}(\omega^{p-1}b) = \mathrm{ord}(b) = \mathrm{ord}(a) = q$. The claim follows by induction on the rank of A.

We have seen in the previous lemma that regular flat transitions can be iterated indefinitely: this is the situation for instance in Λ-modules of unbounded rank: note that upon iteration, the exponent remains equal to the exponent of the first module and this may be any power of p. The regular wild transitions will be considered below, after the next lemma that generalizes Lemma 5 to the case of increasing ranks, and gives conditions for a large class of terminal transitions.

Lemma 11. *Suppose that $q' := \mathrm{ord}(a) > p$, $r' > r$ and $\iota(a) \in pB$. For $b \in B$ with $Nb = a$, we let the module $C = C(b) := \sum_{i=0}^{r-1} \mathbb{Z}_pT^ib$. Then*

1.

$$C \supset \iota(A) \quad \text{and} \quad \iota(A) = pC. \tag{15}$$

2. The element b spans B as a cyclic $\mathbb{Z}_p[T]$-module and

$$K = S(B). \tag{16}$$

Moreover, $r' \leq (p-1)d$ and the transition (A,B) is terminal.

Proof. Let $a' = \iota(a)$ and $c \in B$ be maximal and such that $p^e c = a'$, thus $T^i p^e c = T^i a'$. Let $C = \sum_{i=0}^{r-1} \mathbb{Z}_p T^i c$. Since $T^i a' = p^e T^i c$, we have $\iota(A) \subset C$ and thus $p\text{–rk}(C) \geq p\text{–rk}(A)$; on the other hand, the generators of C yield a base for C/pC, so the reverse inequality $p\text{–rk}(C) \leq p\text{–rk}(A)$ follows; the two ranks are thus equal.

We show that $N : C \to A$ is surjective. We may then apply the lemma 5 to the couple of modules A, C. Let $x \in \mathbb{Z}_p[T]$ be such that $N(c) = xa$. If $x \in (\mathbb{Z}_p[T])^\times$, then $N(x^{-1}c) = a$ and surjectivity follows.

Assume thus that $x \in \mathfrak{M} = (p, T)$. We have an expansion

$$N(c) = h(T)a = \left(h_0 + \sum_{i=1}^{r-1} h_i T^i \right) a, \quad h_i \in \mathbb{Z}_p,$$

and we assume, after possibly modifying h_0 by a p-adic unit, that $h_0 = p^k$ for some $k \in \mathbb{N}$. If $k = 0$, then $h(T) \in (\mathbb{Z}_p[T])^\times$, so we are in the preceding case, so $k > 0$. We rewrite the previous expansion as

$$N(c) = (p^k + Tg(T))a, \tag{17}$$

with $g(T) \in \mathbb{Z}_p[T]$ of degree $< r-1$. Let $f = e + k - 1$; from $p^e c = a$, we deduce:

$$p^f c = p^{k-1} \cdot (p^e c) = p^{k-1} a \quad \text{and} \quad N(p^f c) = N(p^{k-1} a) = p^k a.$$

By dividing the last two relations, we obtain $(1 - p^f)N(c) = Tg(T)a$. Since B is finite, we may choose $M > 0$ such that $p^{Mf} c = 0$. By multiplying the last expression with $(1 - p^{Mf})/(1 - p^f)$, we obtain

$$N(c) = Tg(T)(1 + p^f + \cdots)a.$$

We compare this with (17), finding $Tg(T)(p^f + p^{2f} + \cdots)a = p^k a$.

Since $\iota(a) \in pB$, we have $e > 0$. It follows that

$$p^k \cdot \left(1 - p^{e-1} Tg(T)(1 + p^f + \cdots)\right) a = 0,$$

so $p^k a = 0$ – since the expression in the brackets is a unit. Introducing this in (17), yields: $N(c) = Tg(T)a$. From $p^e c = a$, we then deduce $N(p^e c) = pa = p^e Tg(T)a$, and this yields $p(1 - p^{e-1}Tg(T))a = 0$. It follows from $e > 0$ that $pa = 0$, in contradiction with the hypothesis that $\text{ord}(a) > p$. We showed thus that if $\iota(a) \in pB$ and $\text{ord}(a) > p$, the norm $N : C \to A$ is surjective and we may apply the lemma 5. Thus $pC = A = N(C)$ and $pc = a$.

The module B is $\mathbb{Z}_p[T]$-cyclic, so let b be a generator with $Nb = a$ and let $\tilde{C} = \sum_{i=0}^{r'-1} T^i c$. We claim that $\tilde{C} = B$. For this, we compare $R(B)$ to $R(\tilde{C})$; we obviously have $R(\tilde{C}) \subseteq R(B)$. If we show that this is an equality, the claim follows from Nakayama's Lemma 2. The module $R(B)$ if $\mathbb{F}_p[T]$ cyclic, so there is an integer $k \geq 0$ with $\overline{c} \in T^k R(B)$. But then $N(\tilde{C}) \subset T^k N(B)$ and since $N : C \rightarrow A$ is surjective and $C \subset \tilde{C}$, we must have $k = 0$, which confirms the claim and completes the proof of point 1.

Note that $\iota(S(A)) \subset A = pC$, thus $C \cap K \supseteq \iota(S(A))$. Conversely, if $x \in C \cap K$, then $x \in pC$, since $T^i c \notin \iota(A)$ for $0 \leq i < r$, from the assumption $a \in pB$. Therefore, $x \in pC \cap K = \iota(A) \cap K = \iota(S(A))$, as shown in (11), and

$$C \cap K = \iota(S(A)).$$

If $r' > r$, then $p\text{--rk}(K) = r' - r$; if $r = r'$, the transition is stable and $K \subset C$.

We now prove (16). Let $x = gb \in K, g \in a^{\top} \subset \mathbb{Z}_p[T]$. Since $pb \in A$, we have $px = gpb \in gA = 0$. Thus $K \subset S(B)$; conversely,

$$\begin{aligned} p\text{--rk}(S(B)) &= p\text{--rk}(S(A)) + r' - r = p\text{--rk}(S(A)) + (p\text{--rk}(B) - p\text{--rk}(C)) \\ &= p\text{--rk}(S(A)) + p\text{--rk}(K) \end{aligned}$$

and since $S(A) \subset K$ and $S(B)$ is cyclic, it follows that $S(B) = K$, which confirms (16) and assertion 2.

Finally, note that $p\text{--rk}(S(B)) = r'$ and since $\kappa = \omega b$ generates the socle and

$$0 = ((\omega + 1)^p - 1)b = \omega^{p-1}\kappa = T^{(p-1)d}\kappa,$$

it follows that $r' \leq (p-1)d$, as claimed.

We now investigate regular wild transitions and show that not more than two such consecutive transitions are possible.

Lemma 12. *Let (X,A) be a wild transition with $p\text{--rk}(X) = 1$ and (A,B) be a consecutive transition. Then*

1. *$S(A) \not\subset K(A)$ and there is an $x' \in X$ with $\mathrm{ord}(x') = p^2$ together with $g = Tf(T)a \in K(A)$ such that $s = \iota(x') + g$ is a generator of $S(A)$ and $\ell(A) = 2$.*
2. *The rank $r' \leq (p-1)d$ and B is terminal, allowing an annihilator $f_B(T) = T^{r'} - q/pw(T)$.*

Proof. In this lemma, we consider two consecutive transitions, so we write $T = \tau - 1$, acting on A, B and $\omega = (T+1)^p - 1$ annihilating A and acting on B. We shall also need the norm

$$\begin{aligned} \mathscr{N} = N_{B/X} &= \frac{1}{T}\left((T+1)^{p^2} - 1\right) = p^2 U(T) + pT^p V(T), \\ U, V &= 1 + O(T) \in (\mathbb{Z}_p[T])^{\times} \end{aligned}$$

We let $q = \exp(A), q/p = \exp(X)$ and $qp = \exp(B)$. In the wild transition (A,B), the socle has length p and if $s \in B$ is a generator, then $0 \neq Ns = T^{p-1}s \in S(X)$; it follows that $s \not\in K(A)$. Let $\iota(x) = Ns \in \iota(S(X))$; if $x \not\in pX$, there is a $c \in \mathbb{Z}_p^\times$ with $T^{p-1}s = cN(a) = cpu(T)a + cT^{p-1}a$ and thus $T^{p-1}(s - ca)u^{-1}(T) = pa$; then $pa \in K(A) \cap \iota(S(X))$, and thus $\mathrm{ord}(a) = p^2$ and Lemma 11 implies that $p{-}\mathrm{rk}(A) < p$, which is a contradiction to our choice. Therefore, $\exp(X) > p$ and there is an $x' \in X$ of order p^2 such that $p\iota(x') = N(\iota(x')) = N(s)$, so we conclude that $N(s - x') = 0$ and $x' - s \in K(A)$, which implies the first part in claim 1. We show now that $\ell(A) = 2$; indeed, $s - q/p^3\iota(x) \in K(A)$, so there is a power p^k such that $s - q/p^3\iota(x) = Tp^k w(T)b$ and $w \in \mathbb{Z}_p[T] \setminus p\mathbb{Z}_p[T]$. From $Ts = T^2p^kw(T)b$, and since $p{-}\mathrm{rk}(Ts\mathbb{F}_p[T]) = p - 1$, we conclude that $w \in (\mathbb{F}_p[T])^\times$. Moreover, the above identities in the socle imply

$$q/p^k \geq p^2\mathrm{ord}(Tp^ka) > p = \mathrm{ord}(T^2p^ka) \geq \mathrm{ord}(T^{p-1}p^ka) = q/p^{k+1}.$$

Consequently, $p^k = q/p^2$ and $\mathrm{ord}(Ta) = q$ while $\mathrm{ord}(T^2a) = q/p$, thus $\ell(A) = 2$.

For claim 2 we apply point 1. in Lemma 9. Let $q = \exp(A) = p\exp(X) > p^2$ and $\ell' = \ell(B) = p{-}\mathrm{rk}(qB)$, $\ell = \ell(A) = p{-}\mathrm{rk}(q/pA)$. From the cyclicity of the socle, we have

$$\begin{aligned} S(B)[T] &= q/p^2x\mathbb{F}_p = q/p^2\mathcal{N}(b)\mathbb{F}_p, \quad \text{hence } \exists c \in \mathbb{F}_p^\times, \\ T^{\ell'-1}qb &= cq/p^2\mathcal{N}(b) = (cqU(T) + cq/pT^pV(T))b. \end{aligned} \tag{18}$$

Assuming that $\ell' > 1$, then $qb(1 - O(T)) = q/pT^pV(T)b$ and thus $q/pT^pb = qbV_1(T)$. Then $\mathrm{ord}(T^pb) = q$ and thus $\ell' < p$.
Moreover, $q/pT^{p+\ell'}b = qT^{\ell'}pV_1(T)b = 0$, thus $\mathrm{ord}(T^{p+\ell'}b) \leq q/p$ and a fortiori

$$\mathrm{ord}(T^{2p-1}b) \leq q/p, \quad \mathrm{ord}(T^{p-1}b) \leq q. \tag{19}$$

We now apply the norm of the transition (A, B), which may be expressed in ω as $N_{B/A} = pu(\omega) + \omega^{p-1}$. Note that $u(\omega) = v(T) \in (\mathbb{Z}_p[T])^\times$, for some v depending on u. Also

$$\begin{aligned} \omega &= T^p + pTu(T) = T(T^{p-1} + pu(T)) \\ \omega^{p-1} &= T^{p(p-1)} + T^{(p-1)^2}p(p-1)u(T) + O(p^2). \end{aligned}$$

Since $p \geq 3$, (19) implies $\mathrm{ord}(T^{p(p-1)}b) \leq q/p$ and $\mathrm{ord}(T^{(p-1)^2}b) \leq q$. We thus obtain:

$$\begin{aligned} \iota(a) &= pv(T)b + (T^{p(p-1)} + pu_1(T)T^{(p-1)^2} + O(p^2))b, \quad \text{hence} \\ q/p\iota(a) &= qbv(T), \end{aligned}$$

and it follows in this case that $1 < \ell = \ell' = 2 < p$. Consider now the module $Q = \iota(A)/(\iota(A) \cap pB)$. Since $p\iota(A) \subset (\iota(A) \cap pB)$, this is an $\mathbb{F}_p[T]$ module; let T^i be its minimal annihilator. Then $T^i\iota(a)v_1(T) = pb, v_1 \in (\mathbb{Z}_p[T])^\times$; but $q/pv^{-1}(T)\iota(a) = qb$, and thus $q/p(T^i - v_2(T))\iota(a) = 0, v_2 \in (\mathbb{Z}_p[T])^\times$. If $i > 0$, then $v_2(T) - T^i \in (\mathbb{Z}_p[T])^\times$ and this would imply $q/p\iota(a) = 0$, in contradiction with the definition $q = \mathrm{ord}(a)$. Consequently $i = 0$ and $\iota(a) \in pB$. We may thus apply the Lemma 11 to the transition (A,B). It implies that the transition is terminal and $r' < (p-1)d$.

Finally, we have to consider the case when $\ell' = 1$, so the relation (18) becomes

$$q(1/c - U(T))b = q/pT^pV(T))b.$$

If $c \neq 1$, then $q/pT^pb = qw(T), w \in (\mathbb{F}_p[T])^\times$ and the proof continues like in the case $\ell' > 1$. If $c = 1$, then we see from the development of $U(T) = 1 + T\binom{p^2}{2} + O(T)^2$ that there is a unit $d = \frac{p^2-1}{2} \in \mathbb{Z}_p^\times$ such that

$$qdTU_1(T)b = q/pT^pV(T)b = 0,$$

since $\ell = 1$ and thus $Tqb = 0$. We may deduce in this case also that $q/p\iota(a) = qb \cdot c_1,\ c_1 \in \mathbb{Z}_p^\times$ and complete the proof like in the previous cases. The annihilator polynomial of B is easily deduced from Lemma 11: $T^{r'}b \in \iota(S(A))$ is a generator of the last socle, so $T^{r'}b = Tq/p^2\iota(a) + cq/p^3\iota(x)$ and some algebraic transformations lead to

$$T^{r'}bw(T) = q/pb, \quad w \in (\mathbb{Z}_p[T])^\times,$$

which is the desired shape of the minimal polynomial. Note that $q/p = \exp(X)$; also, the polynomial is valid in the case when (A,B) is stable. We could not directly obtain a simple annihilating polynomial for A, but now it arises by restriction.

The previous lemma shows that an initial wild regular transition cannot be followed by a second one. Thus, growth is possible over longer sequences of transitions only if all modules are regular flat. The following lemma considers the possibility of a wild transition following flat ones.

Lemma 13. *Suppose that (A,B) is a transition in which A is a regular flat module of rank p–$\mathrm{rk}(A) \geq p$ and $\exp(B) > \exp(A)$. Then B is terminal and $d < r' < (p-1)d$. Moreover, there is a binomial $f_B(T) = \omega T^{r'-d} - qw(T) \in b^\top; w \in (\mathbb{F}_p[T])^\times$.*

Proof. Since $\exp(B) > \exp(A)$, the transition is not flat. Assuming that B is not terminal, then it is regular wild. Let $q = \exp(A) = \mathrm{sexp}(A)$ and $s' \in S(B)$ be a generator of the socle of B. By comparing ranks, we have $T^{(p-1)d}s' = \omega^{p-1}s' = q/p\iota(a)v(T), v \in (\mathbb{F}_p[T])^\times$. If $q = p$, then

$$T^{(p-1)d}s' = vv_1(T)b \quad \Rightarrow \quad vs' = pu(\omega)s' + \omega^{p-1}s' = vv_1(T)b,$$

and thus $s' - v_1(T)b \in K$. Since $K = \overline{\omega}B$, we have $(\overline{\omega}x + v_1(T))b \in S$. The factor $\overline{\omega}x + v_1(T) \in (\mathbb{Z}_p[T])^{\times}$ and it follows that $b \in S(B)$, which contradicts the assumption $\exp(B) > \exp(A)$, thus confirming the claim in this case. If $q > p$, then the previous identity yields $s' - q/p^2\iota(a) \in K$ and thus $s' = q/p^2\iota(a) + \omega xb$; we let $x = p^k v(T)$ with $v(0) \not\equiv 0 \bmod p$.

We assumed that $p\text{–rk}(S(B)) = pd$, so it thus follows that $v(T) \in (\mathbb{Z}_p[T])^{\times}$. Recall that $\text{ord}(b) = qp$ as a consequence of $\exp(B) > \exp(A)$ and (13); the norm shows that

$$qu(\omega)b + (q/p)\omega^{p-1}b = (q/p)\iota(a) \neq 0.$$

If $(q/p)\omega^{p-1}b = 0$, then $qu(\omega) = (q/p)\iota(a)$, which implies that the annihilator of $Q = \iota(A)/(pB \cap \iota(A))$ is trivial and $\iota(A) \subset pB$. We are in the premises of Lemma 11, which implies that B is terminal.

It remains that $\text{ord}(\omega^{p-1}b) = q$. We introduce this in the expression for the generator of the socle:

$$s' = \omega p^k v(T)b - q/pu(\omega)b - q/p^2\omega^{p-1}b \in S(B).$$

We have

$$q/p^{k-1} = \text{ord}(p^k b) \geq p^2 = \text{ord}(p^k \omega b) \geq \text{ord}(p^k \omega^{p-1} b) = q/p^k,$$

and thus $q/p^2 \leq p^k \leq q/p$. Note that

$$\omega s' = \omega^2 p^k v(T)b + \omega q/p^2\iota(a) = \omega^2 p^k v(T)b \in S(B);$$

from $\omega^{p-1}(q/p)b \in S(B)$, it follows that $p^k = q/p$ and $\text{ord}(\omega b) = qp$ while $\text{ord}(\omega^2 b) = q = \text{ord}(\omega^{p-1}b)$. Let $i > 0$ be the least integer with $q/p\omega T^i b \in S(B)$. From the definition of s', we see that T^k is also the annihilator of $(q/p^2)\iota(a)$ in $\iota(A)/(S(\iota(A)))$, so $i = d$, since A is flat. It follows that $\ell(B) = p\text{–rk}(B/qB) = 2d$ and the cyclicity of the socle implies that

$$T^{(p-2)d}s' = qbv_1(T) = T^{(p-2)d}s' = \frac{q}{p}\omega^{p-1}v(T)b + \frac{q}{p^2}\omega^{p-2}\iota(a).$$

Hence, there is a unit $v_2(T) \in (\mathbb{Z}_p[T])^{\times}$ such that $\frac{q}{p}(p - \omega^{p-1}v_2(T))b = 0$. By comparing this with the norm identity $\frac{q}{p}(p + \omega^{p-1}u^{-1}(T))b = \frac{q}{p}\iota(a)$ we obtain, after elimination of $\omega^{p-1}b$, that $\frac{q}{p}(pbv_3(T) + \iota(a)) = 0$ and the reasonment used in the previous case implies that $\iota(a) \in pB$ so Lemma 11 implies that B is terminal.

We now show that $\iota(A) \subset pB$. Otherwise, $r' \geq (p-1)d$ and $T^{r'-1}s' = cT^{d-1}q/p\iota(a)$, so by cyclicity of the socle, $T^{r'-d}s' = v(T)q/p\iota(a)$ while $\nu s' = T^{pd-r'-1}\iota(a) \neq 0$. A similar estimation like before yields also in this case

$s' = q/p\omega\nu(T)b + q/p^2T^j\iota(a), j = pd - r' - 1$. Then $\omega^{p-2}s' = q/p\omega^{p-1}\nu(T)b \in S(B)$. Let $i > 0$ be the smallest integer with $pb \in \iota(a)$. Then $qb = T^i q/p\iota(a)\nu_1(T)$ and we find a unit $\nu_2(T)$ such that

$$q/p\big(\omega^{p-1}T^{i+r'-(p-1)d-1} - p\nu_2(T)\big)b = 0.$$

This implies by a similar argument as above, that $\iota(a) \in pB$. Therefore, we must have $r' < (p-1)d$ and $S(B) = K(B) \supset q/p\iota(A)$. Since $qb = q/p\iota(a) = \omega b T^{r'-d}\nu(T)$, we obtain an annihilator polynomial $f_B(T) = T^{r'-d} - q\nu^{-1}(T)$, which completes the proof of the lemma.

We finally apply the Lemma 10 to a sequence of flat transitions. This is the only case, which allows arbitrarily large growth of the rank, while the value of the exponent is fixed to q.

Lemma 14. *Suppose that $A_1, A_2, \ldots, A_n$ are a sequence of cyclic $\mathbb{Z}_p[T]$ modules such that (A_i, A_{i+1}) are conic non-stable transitions with respect to some $\omega_i \in \mathbb{Z}_p[T]$ and $p\text{–rk}(A_1) = 1$. If $n > 3$, then A_i are regular flat for $1 \le i < n$.*

Proof. If $n = 2$ there is only one, initial transition: this case will be considered in detail below. Assuming that $n > 2$, the transitions (A_i, A_{i+1}) are not stable; if (A_1, A_2) is wild, then Lemma 12 implies $n \le 3$. The regular transitions being by definition the only ones which are not terminal, it follows that $A_k, k = 1, 2, \ldots, n-1$ are flat. Lemma 13 shows in fact that A_2, which must be flat, can only be followed by either a regular flat or a terminal transition. The claim follows by induction.

2.6 *Case Distinctions for the Rank Growth*

We have gathered above a series of important building blocks for analyzing transitions. First, we have shown in point 2. of Lemma 9 that all transitions that are not flat are terminal. Thus for the cases of interest that allow successive growths of ranks, we must have $r = d, r' = pd$. The Lemma 10 shows that these reduce to $\exp(A) = q = \text{sexp}(A)$.

We start with an auxiliary result which will be applied in both remaining cases:

Lemma 15. *Let (A, B) be a transition with $\exp(A) = p$ and $r = d < r' < pd$. Then $S(B) \supseteq K$ with equality for $r' \le (p-1)d$. If $S(B) \ne K$, then $s = T^{pd-r'}$ is a generator of $S(B)$.*

Proof. Since $r = d = \deg(\omega)$, it follows that $N(T^i b)) = T^i\iota(a) \ne 0$ for $i = 0, 1, \ldots, r-1$ while $N(T^d b) = T^d\iota(a) = 0$, since $\exp(A) = p$. Therefore $K = \omega B$ and $R(K) = T^d R(B)$.

We first show that $S(B) \subset K$: let $s \in S(B)$ be a generator. Cyclicity of the socle implies that $T^{r'-1}s = c_0 T^{d-1}\iota(a) \ne 0, c_0 \in \mathbb{F}_p^\times$ and $T^{r'}s = 0$. We have

$N(s) = (pu(\omega) + \omega^{p-1})s = \omega^{p-1}s = T^{(p-1)d}s$. If $r' \leq (p-1)d$, then $T^{(p-1)d}s = 0$ and thus $s \in K$ and $S(B) = S(K)$. Otherwise, $T^{(p-1)d}\overline{b} = \overline{\iota(a)} \in R(B)$ and a fortiori $N(s) \in \iota(S(A)) = \mathbb{F}_p[T]\iota(a)$. Let $0 \leq k < d$ be such that $N(s) = T^k\iota(a)$: we may discard an implicit unit by accordingly modifying s. Then $T^{d-(k+1)}N(s) \neq 0$ and $T^{d-k}N(s) = 0$. Therefore, $T^{d-(k+1)}s \notin K = \omega B$ and $s \notin pB$. It follows that $\overline{s}R(B) = T^kR(B)$ and $s = T^k v(T)b$, by lemma 7. By comparing ranks, we see that $k = pd - r'$ and $s = T^{pd-r'}$ is a generator of $S(B)$.

For initial transitions we have:

Lemma 16. *Let (A,B) be a conic transition and suppose that $r = 1$ and the transition is terminal. If $r' < p$, then B has a monic annihilator polynomial $f_B(T) = T^{r'} - qw(T)$ with $q = \mathrm{ord}(a)$ and $w \in (\mathbb{Z}_p[T])^\times$.*

Proof. We let $q = \mathrm{ord}(a)$ throughout this proof. Assume first that $r' < p-1$, so $T^{p-1}b = T^{p-1-r'}(T^{r'}b) = 0$, since $T^{r'}a \in \iota(A)$ by definition of the rank. Then $\iota(a) = (pu(T) + T^{p-1})b = pu(T)b$ and $pb = u(T)^{-1}\iota(a) = \iota(a)$, since $u(T) \equiv 1 \bmod T$. We have thus shown that $\iota(a) \in pB$ and, for $q > p$, we may apply Lemma 11. It implies that $S(B) = K = TB$ and $T^{r'-1}(Tb) = cq/p\iota(a) = cqb, c \in \mathbb{Z}_p^\times$. This yields the desired result for this case. If $q = p$, the previous computation shows that $\iota(a) \in pB$, but Lemma 11 does not apply here. We can apply Lemma 15, and since, in the notation of the lemma, $d = 1$ and the transition is assumed not to be stable, we are in the case $1 < r' \leq p-d$ and thus $S(B) = K$ too. The existence of the minimal polynomial $f_B(T) = T^{r'} - pc \in b^\top$ follows from this point like in the case previously discussed.

If $r' = p-1$, then $T^{p-1}b = \iota(a) - pu(T)b$ and thus $T^pb = pTu(T)b = 0$ and $Tb \in S(B)$. Since the socle is cyclic and $K = TB$ it follows that $K = S(B)$. In particular, there is a $c \in \mathbb{Z}_p^\times$ such that

$$\begin{aligned} T^{p-1}b &= \frac{cq}{p}\iota(a) = \frac{cq}{p}(pu(T) + T^{p-1})b \\ &= cqu(T)b + \frac{cq}{p}u(T)T^{p-1}b, \quad \text{hence} \end{aligned}$$

$$T^{p-1}\left(1 - \frac{cq}{p}u(T)\right)b = cqu(T)b.$$

If $q > p$, then $1 - \frac{cq}{p}u(T) \in (\mathbb{Z}_p[T])^\times$; if $q = p$, it must also be a unit: otherwise, $1 - cu(T) \equiv 0 \bmod T$ and thus $T^pb = cqu(T)b = 0$, in contradiction with the fact that $qb = q/p\iota(a) \neq 0$. In both cases, we thus obtain an annihilator polynomial of the shape claimed.

Finally, in the case $r' = p$ and the transition is wild. We refer to Lemma 12 in which treats this case in detail.

Remark 4. Conic Λ-modules are particularly simple modules. The following example is constructed using Thaine's method used in the proof of his celebrated theorem [15]. Let $\mathbb{F}_1 \subset \mathbb{Q}[\zeta_{73}]$ be the subfield of degree 3 over $\mathbb{Q}$ and $\mathbb{K}_1 = \mathbb{F}_1 \cdot \mathbb{Q}[\sqrt{-23}]$. Then $A_1 := (\mathscr{C}(\mathbb{K}_1))_3 = C_9$ is a cyclic group with 9 elements. If $\mathbb{K}_2$ is the next level in the cyclotomic $\mathbb{Z}_3$-extension of $\mathbb{K}_1$, then

$$A_2 := (\mathscr{C}(K_2))_3 = C_{27} \times C_9 \times C_9 \times C_3 \times C_3 \times C_3.$$

The prime $p = 3$ is totally split in $\mathbb{K}_1$ and the classes of its factors have orders coprime to p. Although A_1 is $\mathbb{Z}_p$-cyclic, already A_2 has p-rank $2p$. Thus, $\mathbf{A}$ cannot be conic, and it is not even a cyclic Λ-module.

It is worth investigating, whether the result of this paper can extent to the case when socles are not cyclic and conicity is not satisfied, in one or more of its conditions. Can these tools serve to the understanding of Λ-modules as the one above?

3 Transitions and the Critical Section

We return here to the context of Λ modules and conic elements, and use the notation defined in the introduction, so $\mathscr{A}_n = \Lambda a_n$ are the intermediate levels of the conic Λ-module $\Lambda a \subset \mathbf{A}^-$. We apply the results of the previous chapter to the transitions $C_n = (\mathscr{A}_n, \mathscr{A}_{n+1})$ for $n < n_0$. By a slight abuse of notation, we keep the additive notation for the ideal class groups that occur in these concrete transitions. The first result proves the consistency of the models:

Lemma 17. *Let the notations be like in the introduction and $a = (a_n)_{n\in\mathbb{N}} \in \mathbf{A}^-$ a conic element, $\mathscr{A} = \Lambda a$ and $\mathscr{A}_n = \Lambda a_n \subset A_n^-$. Then the transitions $(\mathscr{A}_n, \mathscr{A}_{n+1})$ are conic in the sense of Definition 3, for all $n > 0$.*

Proof. Let $A = \mathscr{A}_n, B = \mathscr{A}_{n+1}$ and $N = \mathbf{N}_{\mathbb{K}_{n+1},\mathbb{K}_n}, \iota = \iota_{n,n+1}$ be the norms of fields and the ideal lift map, which is injective since $a \in \mathbf{A}^-$. We let $T = \tau - 1$ with τ the restriction of the topological generator of Γ to $\mathbb{K}_{n+1}$ and $\omega = \omega_n = (T+1)^{p^{n-1}} - 1$. Then a fortiori $\omega A = 0$, and all the properties (1)–(3) of conic transitions follow easily. Point (5) is a notation. We show that the important additional property (4) follows from the conicity of a. The direction $\omega A \subset K$ follows from $Y_1 = TX$ in Theorem 2. The inverse inclusion is a consequence of point (1) of the definition of conic elements. Conversely, if $x \in K$, we may regard $x = x_{n+1} \in \mathscr{A}_{n+1}$ as projection of a norm coherent sequence $y = (x_m)_{m\in\mathbb{N}} \in \mathscr{A}$: for this, we explicitly use point (3) of the definition of conic elements. Since $x = y_{n+1} = 1$ we have by point (2) of the same definition, $y \in \omega_n \cdot \mathscr{A}$. This implies $y_n = Nx = 1$; this is the required property (4) of Definition 3

The next lemma relates $v_p(a_1)$ to the minimal polynomials $f_a(T)$:

Lemma 18. *Let $a \in \mathbf{A}^-$ be conic and $m = v_p(a_1)$. Then $v_p(f_a(0)) = m$. In particular, if $v_p(a_1) = 1$, then $f_a(T)$ is an Eisenstein polynomial.*

Proof. Let $q = p^m$ and $b = qa \in \Lambda a$. Then $b_1 = 0$ and, by conicity, it follows that $qa = b = Tg(T)a$. It follows that $Tg(T) - q$ annihilates a. We may choose g such that $\deg(g(T)T) = \deg(f_a(T))$, so there is a constant $c \in \mathbb{Z}_p$ such that $Tg(T) - q = cf_a(T)$. Indeed, if c is the leading coefficient of $Tg(T)$, the polynomial $D(T) = Tg(T) - q - cf_a(T)$ annihilates a and has degree less than $\deg(f_a)$. Since f_a is minimal, either $D(T) = 0$, in which case $c = 1$ and $f_a(T) = Tg(T) - q$, which confirms the claim, or $D(T) \in p\mathbb{Z}_p[T]$ and $c \equiv 1 \bmod p$. Since c is a unit in this case, we may replace b by $c^{-1}b = Tg_1(T)a$ and the polynomial $Tg(T)$ is now monic. The previous argument implies that $f_a(T) = Tg_1(T) - c^{-1}q$, which completes the proof. Since $f_a(T)$ is distinguished, we have $f_a(T) \equiv T^d \bmod p$ and if $m = 1$, then $p^2 \nmid f_a(0)$, so $f_a(T)$ is Eisenstein. The converse is also true.

If $m > 1$, we have seen in the previous chapter that there are minimal polynomials of $\mathscr{A}_{n_0}$, which are essentially binomials; in particular, they are square free. It would be interesting to derive from this fact a similar conclusion about $f_a(T)$. We found no counterexamples in the tables in [6]; however, the coefficients of $f_{n_0}(T)$ are perturbed in the stable growth too, and there is no direct consequence that we may derive in the present setting. The next lemma describes the perturbation of minimal polynomials in stable growth:

Lemma 19. *Let $g_n(T) = T^r - p\tilde{g}_n(T)$ be a minimal polynomial of $\mathscr{A}_n$. If $n \geq n_0$, then*

$$g_n(T) \equiv f_a(T) \bmod \mathscr{A}_{n-1}^{\top}. \tag{20}$$

Proof. The exact annihilator $f_a(T)$ of $\mathscr{A}$ also annihilates all finite level modules $\mathscr{A}_n$. In particular, for $n \geq n_0$ we have $\deg(g_n(T)) = \deg(f_a)$ for all minimal polynomials g_n of $\mathscr{A}_n$, and thus $\deg(g_n - f_a) < \lambda(a)$. We note that $g_n - f_a = p\delta_n(T) \in p\mathbb{Z}_p[T]$ with $\deg(\delta_n) < r$. It follows that

$$0 = p\delta_n(T)a_n = \delta_n(T)\iota_{n-1,n}(a_{n-1}),$$

and since ι is injective, it follows that $\delta_n(T) \in \mathscr{A}_{n-1}^{\top}$, as claimed.

It is worthwhile noting that if a is conic and $f_a(T) = \prod_{i=1}^{k} f_i^{e_i}(T)$ with distinct prime polynomials $f_i(T)$, then $b_i := f_a(T)/f_i(T)a$ have also conic transitions, but the modules Λb_i are of course not complementable as Λ-modules.

3.1 *Proof of Theorem 1*

With this, we shall apply the results on conic transitions and prove the Theorem 1.

Proof. Let $a \in \mathbf{A}^- \setminus p\mathbf{A}^-$ be conic, let n_0 be its stabilization index and $\mathscr{A}_n = \Lambda a_n, n \geq 0$ be the intermediate levels of $\mathscr{A} = \Lambda a$. Proposition 1 implies the statement of point 1. in the theorem. Statement 2 is a consequence of Corollary 2. The final condition $\text{ord}(a_{n+1}) \leq p\,\text{ord}(a_n)$ for all n was proved in (13) of Lemma 9. This completes the proof of the theorem.

3.2 Some Examples

We shall discuss here briefly some examples[5] drawn from the paper of Ernvall and Metsänkylä [6] and the tables in its supplement. The authors consider the primes $p = 3$, $\rho = \zeta_p$ and base fields $\mathbb{K} = \mathbb{K}(m) = \mathbb{Q}[\sqrt{m}, \rho]$. They have calculated the annihilator polynomials of $f_a(T)$ for a large choice of cyclic $A(\mathbb{K}(m))^-$. Here, are some examples:

Example 1. In the case $m = 2732$, $A^-(\mathbb{K}_1(m)) \cong C_{p^2}$ and $A^-(\mathbb{K}_2(m)) \cong C_{p^3} \times C_p$. The growth stabilizes and the polynomial $f_a(T)$ has degree 2; the annihilator $f_2(T)$ is a binomial, but not $f_a(T)$, so the binomial shape is in general obstructed by the term $f_a(T) = f_2(T) + O(p)$.

Example 2. In the case $m = 3{,}512$, we have $A^-(\mathbb{K}_1(m)) \cong C_{p^2}$ and $A^-(\mathbb{K}_2(m)) \cong C_{p^3} \times C_p \times C_p$. The polynomial $f_a(T)$ has degree 3 and for $B = A(\mathbb{K}_2(m))^-$ and $A = A(\mathbb{K}_1(m))^-$. This is a wild transition, which is initial and terminal simultaneously. We did not derive a precise structure for such transitions in Lemma 12.

Example 3. In the case $m = -1{,}541$, the authors have found $\lambda = 4$. Unfortunately, the group $A(\mathbb{K}_3(m))$ cannot be computed with PARI, so our verification restricts to the structure of the transition $(A, B) = (\mathscr{A}_1, \mathscr{A}_2)$. This is the most interesting case found in the tables of [6] and the only one displaying a wild initial transition. The Lemma 12 readily implies that the transition $(\mathscr{A}_2, \mathscr{A}_3)$ must be terminal and $\lambda < (p-1)p = 6$, which is in accordance with the data. The structure is $\mathscr{A}_2 = C_{p^3} \times C_{p^3} \times C_p$ and with respect to this group decomposition we have the following decomposition of individual elements in $A = \mathscr{A}_1$ and $B = \mathscr{A}_2$:

$$\begin{array}{lll} b = (1,0,0) & Tb = (0,10,1) & T^2 b = (-6,9,1) \\ T^3 b = (18,-3,0) & T^4 b = (18,9,0) & T^5 b = (0,9,0) \\ a = (0,12,1) & 3a = (0,9,0) & 9b = (9,0,0). \end{array}$$

Some of the particularities of this examples are: $S(B)$ is generated by $s' = T^3 b - 2a$ and it is $\mathbb{F}_p[T]$-cyclic, as predicted. Moreover, $s' \in K + \iota(A)$ but $s \notin K$ and $pb \notin \iota(a)$, while $qb \in S(B)[T] = S(A)[T]$, both facts that were proved in the Lemma 12.

[5] I am grateful to an anonymous referee for having pointed out some very useful examples related to the present topic.

Example 4. In all further examples with $\lambda \geq 3$, the fields $\mathbb{K}(m)$ have more than one prime above p and $\mathbf{A}^-(m)$ is not conic. For instance, for $m = 2,516$, we also have $A^-(\mathbb{K}_1(m)) \cong C_{p^2}$ and $A^-(\mathbb{K}_2(m)) \cong C_{p^3} \times C_p \times C_p$, but $T^3 b = 0$, for b a generator of $A(\mathbb{K}_2(m))^-$. The module is thus obviously not conic. This examples indicates a phenomenon that was verified in more cases, such as our example in Remark 4: an obstruction to conicity arises from the presence of *floating* elements $b \in \mathbf{A}^-$. These are defined as sequences $b = (b_n)_{n \in \mathbb{N}} \in \mathbf{A}^- \setminus (p,T)\mathbf{A}^-$ having $b_1 = 0$. When such elements are intertwined in the structure of Λa, one encounters floating elements. It is an interesting question to verify whether the converse also holds: $a \in \mathbf{A}^- \setminus (p,T)\mathbf{A}^-$ is conic if it contains not floating elements. Certainly, the analysis of transitions in presence of floating elements is obstructed by the fact that the implication $Tx = 0 \Rightarrow x \in A$ is in general false. However, the obstruction set is well defined by the submodule of floating elements, which indicates a possible extension of the concepts developed in this paper. The analysis of floating elements is beyond the scope of this paper and will be undertaken in subsequent research.

Example 5. Let $\mathbb{K} = \mathbb{Q}[\sqrt{-31}]$ with $A(\mathbb{K}) = C_3$ and only one prime above $p = 3$. A PARI computation shows that $A(\mathbb{K}_2) = C_{p^2}$, so Fukuda's Theorem implies that $\mathbf{A}$ is Λ-cyclic with linear annihilator. Let $\mathbb{L}/\mathbb{K}$ be the cyclic unramified extension of degree p. There are three primes above p in $\mathbb{L}$ and $A(\mathbb{L}) = \{1\}$, a fact which can be easily proved and needs no verification. Let $\mathbb{L}_n = \mathbb{L} \cdot \mathbb{K}_n$ be the cyclotomic $\mathbb{Z}_p$-extension of $\mathbb{L}$. One can also prove that $A(\mathbb{L}_n) \cong (A_n(\mathbb{K}))^p$, so $\mathbf{A}(\mathbb{L})$ is also Λ cyclic with the same linear annihilator polynomial as $\mathbf{A}(\mathbb{K})$. Let $b \in \mathbf{A}(\mathbb{L})$ be a generator of the Λ-module. The above shows that b is a floating class.

The extension $\mathbb{L}/\mathbb{Q}$ in this example is galois but not CM and p splits in $\mathbb{L}/\mathbb{K}$ in three principal primes. If $\nu \in \mathrm{Gal}(\mathbb{L}/\mathbb{K})$ is a generator, it lifts in $\mathrm{Gal}(\mathbb{H}_\infty/\mathbb{K})$ to an automorphism $\tilde{\nu}$ that acts non-trivially on $\mathrm{Gal}(\mathbb{H}_\infty/\mathbb{L}_\infty)$.

Let $\mathbb{B}_\infty$ be the $\mathbb{Z}_p$-extension of $\mathbb{Q}$ and $\mathbb{H}_\infty$ be the maximal p-abelian unramified extension of $\mathbb{K}_\infty$ and of $\mathbb{L}_\infty$ (the two coincide in this case); then the sequence

$$0 \to \mathrm{Gal}(\mathbb{L}_\infty/\mathbb{B}_\infty) \to \mathrm{Gal}(\mathbb{H}_\infty/\mathbb{B}_\infty) \to \mathrm{Gal}(\mathbb{H}_\infty/\mathbb{L}_\infty) \to 0 \tag{21}$$

is not split in the above example, and this explains why $\tilde{\nu}$ lifts to a generator of $X' := \mathrm{Gal}(\mathbb{H}_\infty/\mathbb{K}_\infty)$.

Let $\mathfrak{p}, \nu\mathfrak{p}, \nu^2(\mathfrak{p}) \subset \mathbb{L}$ be the primes above p and $I_0, I_1, I_2 \subset \mathrm{Gal}(\mathbb{H}_\infty/\mathbb{L})$ be their inertia groups: then $I_1 = I_0^{\tilde{\nu}}, I_2 = I_0^{\tilde{\nu}^2}$. Let $I \subset \mathrm{Gal}(\mathbb{H}_\infty/\mathbb{K})$ be the inertia of the unique prime above p and $\tau \in \mathrm{Gal}(\mathbb{H}_\infty/\mathbb{K})$ be a generator of this inertia. We fix τ' as a lift of the topological generator of Γ: it acts, in particular, also on $\mathbb{L}$. Let τ be a generator of I_0 and $a \in X = \mathrm{Gal}(\mathbb{H}_\infty/\mathbb{L}_\infty)$ such that $\tau_1 = a\tau$ is a generator of I_1. We assume that both τ, τ_1 restrict to a fixed topological generator of $\Gamma = \mathrm{Gal}(\mathbb{L}_\infty/\mathbb{K}_\infty)$. Then

$$\tau_1 = a\tau = \tau^{\tilde{\nu}} = \tilde{\nu}^{-1}\tau\tilde{\nu} \quad \Rightarrow \quad a = \tilde{\nu}^{-1}\tau\tilde{\nu}\tau^{-1}.$$

Since τ acts by restriction as a generator of $\Gamma' = \mathrm{Gal}(\mathbb{K}_\infty/\mathbb{K})$ and $\tilde{\nu}$ generates X', the above computation implies that $a \in (\mathrm{Gal}(\mathbb{H}_\infty/\mathbb{K}))' = TX' = pX' = X$. In particular, a is a generator of $X \cong \mathbf{A}(\mathbb{L})$.

In this case, we have seen that the primes above p are principal, the module $\mathbf{A}(\mathbb{L})$ is floating and it is generated by $a = \tau_1 \tau^{-1} \notin TX$. Thus, $Y_1 = \Lambda a = \mathbb{Z}_p a$ and $[Y_1 : TX] = p$. Since TX is the commutator, there must be a cyclic extension $\mathbb{L}'/\mathbb{L}$ of degree p, which is p-ramified but becomes unramified at infinity. It arises as follows: let $\mathbb{H}_2$ be the Hilbert class field of $\mathbb{K}_2$. Then $\mathbb{H}_2/\mathbb{L}_2$ is cyclic of degree p and $\text{Gal}(\mathbb{H}_2/\mathbb{K}) =< \varphi(a_2) >$, with $a_2 \in A(\mathbb{K}_2)$ a generator. Thus $(T - cp)a_2 = 0$ for some $c \in \mathbb{Z}_p^\times$ and $\text{Gal}(\mathbb{H}_2/\mathbb{L}_2) = p < \varphi(a_2) >=< \varphi(Ta_2) >$. Since $T^2 a_2 = c^2 p^2 a_2$, it follows that $T\text{Gal}(\mathbb{H}_2/\mathbb{L}_2) = 0$ and thus $\mathbb{H}_2/\mathbb{L}_1$ is abelian. This induces a cyclic extension $\mathbb{L}_1'/\mathbb{L}_1$, which is p-ramified, but becomes unramified already over $\mathbb{L}_2$.

It also explains the role of the sequence (21) in Theorem 3. Phenomena in this context will be investigated together with the question about floating classes in a subsequent paper.

The prime $p = 3$ is interesting since it immediately displays the more delicate cases $r' = p - 1$ and $r = p$ in Lemma 16. We found no examples with $\lambda > p$, which require an intermediate flat transition according to the above facts.

4 The Ramification Module

In this section, we prove the theorems stated in Sect. 1.2. The terms and notations are those introduced in that introductory section. Note that the choice of $\mathbb{K}$ as a galois CM extension containing the pth roots of unity is useful for the simplicity of proofs. If $\mathbb{K}$ is an arbitrary totally real or CM extension, one can always take its normal closure and adjoin the roots of unity: in the process, no infinite modules can vanish, so facts which are true in our setting are also true for subextensions of $\mathbb{K}$ verifying our assumptions.

Let us first introduce some notations: $\mathbb{H}_1$ is the p-part of the Hilbert class field of $\mathbb{K}$ and $\overline{\mathbb{H}}_1 = \mathbb{H}_1 \cdot \mathbb{K}_\infty$; $\Omega/\mathbb{K}$ is the maximal p-abelian p-ramified extension of $\mathbb{K}$. It contains in particular $\mathbb{K}_\infty$ and $\mathbb{Z}_p\text{-rk}(\Omega/\mathbb{H}_1) = r_2 + 1 + \mathscr{D}(\mathbb{K})$, where $\mathscr{D}(\mathbb{K})$ is the Leopoldt defect. Since $\mathbb{K}$ is CM, complex multiplication acts naturally on $\text{Gal}(\Omega/\mathbb{K}_\infty)$ and induces a decomposition

$$\text{Gal}(\Omega/\mathbb{K}_\infty) = \text{Gal}(\Omega/\mathbb{K}_\infty)^+ \oplus \text{Gal}(\Omega/\mathbb{K}_\infty)^-;$$

this allows us to define

$$\begin{aligned} \Omega^- &= \Omega^{\text{Gal}(\Omega/\mathbb{K}_\infty)^+} \\ \Omega^+ &= \Omega^{\text{Gal}(\Omega/\mathbb{K}_\infty)^-}, \end{aligned} \tag{22}$$

two extensions of $\mathbb{K}_\infty$.

We shall review Kummer radicals below and derive a strong property of galois groups, which are Λ-modules with annihilator a power of some polynomial: the order reversal property. Combined with an investigation of the galois group of $\Omega^-/\mathbb{H}_1$ by means of class field theory, this leads to the proof of Theorem 3.

4.1 Kummer Theory, Radicals and the Order Reversal

Let $\mathbf{K}$ be a galois extension of $\mathbb{Q}$ which contains the pth roots of unity and $\mathbb{L}/\mathbf{K}$ be a finite Kummer extension of exponent $q = p^m, m \leq n$. Its classical Kummer radical $\mathrm{rad}(\mathbb{L}/\mathbf{K}) \subset \mathbf{K}^\times$ is a multiplicative group containing $(\mathbf{K}^\times)^q$ such that $\mathbb{L} = \mathbf{K}[\mathrm{rad}(\mathbb{L})^{1/q}]$ (e.g. [14], Chap. VIII, Sect. 8). Following Albu [1], we define the *cogalois* radical

$$\mathrm{Rad}(\mathbb{L}/\mathbf{K}) = \left([\mathrm{rad}(\mathbb{L}/\mathbf{K})^{1/q}]_{\mathbf{K}^\times}\right)/\mathbf{K}^\times, \tag{23}$$

where $[\mathrm{rad}(\mathbb{L}/\mathbf{K})^{1/q}]_{\mathbf{K}^\times}$ is the multiplicative $\mathbf{K}^\times$-module spanned by the roots in $\mathrm{rad}(\mathbb{L}/\mathbf{K})^{1/q}$ and the quotient is one of multiplicative groups. Then $\mathrm{Rad}(\mathbb{L}/\mathbf{K})$ has the useful property of being a finite multiplicative group isomorphic to $\mathrm{Gal}(\mathbb{L}/\mathbf{K})$. For $\rho \in \mathrm{Rad}(\mathbb{L}/\mathbf{K})$ we have $\rho^q \in \mathrm{rad}(\mathbb{L}/\mathbf{K})$; therefore, the Kummer pairing is naturally defined on $\mathrm{Gal}(\mathbb{L}/\mathbf{K}) \times \mathrm{Rad}(\mathbb{L}/\mathbf{K})$ by

$$\langle \sigma, \rho \rangle_{\mathrm{Rad}(\mathbb{L}/\mathbf{K})} = \langle \sigma, \rho^q \rangle_{\mathrm{rad}(\mathbb{L}/\mathbf{K})}.$$

Kummer duality induces a twisted isomorphism of $\mathrm{Gal}(\mathbf{K}/\mathbb{Q})$-modules $\mathrm{Rad}(\mathbb{L}/\mathbf{K})^\bullet \cong \mathrm{Gal}(\mathbb{L}/\mathbf{K})$. Here, $g \in \mathrm{Gal}(\mathbf{K}/\mathbb{Q})$ acts via conjugation on $\mathrm{Gal}(\mathbb{L}/\mathbf{K})$ and via $g^* := \chi(g)g^{-1}$ on the twisted module $\mathrm{Rad}(\mathbb{L}/\mathbf{K})^\bullet$; we denote this twist the *Leopoldt involution*. It reduces on $\mathrm{Gal}(\mathbf{K}/\mathbb{K})$ to the classical Iwasawa involution (e.g. [13], p. 150).

We now apply the definition of cogalois radicals in the setting of Hilbert class fields. Let $\mathbb{K}$ be like before, a CM galois extension of $\mathbb{Q}$ containing the pth roots of unity and we assume that, for sufficiently large n, the p^nth roots are not contained in $\mathbb{K}_{n-1}$, but they are in $\mathbb{K}_n$. Let $\mathbb{L} \subset \mathbb{H}_\infty$ be a subextension with galois group $\mathrm{Gal}(\mathbb{L}/\mathbb{K}_\infty) = \varphi(M)|_{\mathbb{L}}$, with $M \subset \mathbf{A}$ a Λ-submodule, which is $\mathbb{Z}_p$-free. Let $\mathbb{L}_n = \mathbb{L} \cap \mathbb{H}_n$ be the finite levels of this extension and let $z \in \mathbb{Z}$ be such that $\exp(M_n) = p^{n+z}$ in accordance with (7). If $z < 0$, we may take $z = \max(z, 0)$. We define $\mathbb{L}'_n = \mathbb{L}_n \cdot \mathbb{K}_{n+z}$, so that $\mathbb{L}'_n/\mathbb{K}_{n+z}$ is a Kummer extension and let $R_n = \mathrm{Rad}(\mathbb{L}'_n/\mathbb{K}_{n+z}$ and $B_n \cong R_n^{p^{n+z}} \subset \mathbb{K}_n^\times/(\mathbb{K}_n^\times)^{p^{n+z}}$. Then (7) implies, by duality, that $R_{n+1}^p = R_n$, for $n > n_0$; the radicals form a norm coherent sequence with respect to both the dual norm $N^*_{m,n}$ and to the simpler p-map. Since $\mathbb{L} = \cup_n \mathbb{L}'_n$, we may define $\mathrm{Rad}(\mathbb{L}/\mathbb{K}_\infty) = \varprojlim_n R_n$. The construction holds in full generality for infinite abelian extensions of some field containing $\mathbb{Q}[\mu_{p^\infty}]$, with galois groups, which are $\mathbb{Z}_p$-free Λ-modules and projective limits of finite abelian p-groups. But we shall not load notation here for presenting the details. Also, the extension $\mathbb{L}$ needs not be unramified, and we shall apply the same construction below for p-ramified extensions.

We gather the above mentioned facts for future reference in

Lemma 20. *Let $z \in \mathbb{N}$ be such that* $\mathrm{ord}(a_n) \leq p^{n+z}$ *for all n and $\mathbb{K}'_n = \mathbb{K}_{n+z}, \mathbb{L}'_n = \mathbb{L}_n \cdot \mathbb{K}_{n+z}$. Then $\mathbb{L}'_n/\mathbb{K}'_n$ are abelian Kummer extensions with galois groups*

$\mathrm{Gal}(\mathbb{L}'_n/\mathbb{K}_{n+z}) \cong \varphi(M_n)$, *galois over* $\mathbb{K}$ *and with radicals* $R_n = \mathrm{Rad}(\mathbb{L}'_n/\mathbb{K}_{n+z}) \cong (\mathrm{Gal}(\mathbb{L}'_n/\mathbb{K}_{n+z}))^\bullet$, *as* Λ*-modules. Moreover, if* $M = \Lambda c$ *is a cyclic* Λ*-module, then there is a* $\nu^*_{n+1,n}$*-compatible system of generators* $\rho_n \in R_n$ *such that* $R_n^\bullet = \Lambda\rho_n$ *and, for* n *sufficiently large,* $\rho_{n+1}^p = \rho_n$. *The system* R_n *is projective and the limit is* $R = \varprojlim_n R_n$. *We define*

$$\mathbb{K}_\infty[R] = \cup_n \mathbb{K}_{n+z}[R_n] = \mathbb{L}.$$

Note that the extension by the projective limit of the radicals R is a convention, the natural structure would be here an injective limit. However, this convention is useful for treating radicals of infinite extensions as stiff objects, dual to the galois group which is a projective limit. Alternatively, one can of course restrict to the consideration of the finite levels.

The order reversal is a phenomenon reminiscent of the inverse galois correspondence; if M is cyclic annihilated by $f^n(T)$, with f a distinguished polynomial, then there is an inverse correspondence between the f-submodules of M and the f^* submodules of the radical R. The result is the following:

Lemma 21. *Let* $f \in \mathbb{Z}_p[T]$ *be a distinguished polynomial and* $a \in A^- \setminus A^p$ *have characteristic polynomial* f^m *for* $m > 1$ *and let* $\mathscr{A}_n = \Lambda a_n, \mathscr{A} = \Lambda$. *Assume that* $\mathbb{L} \subset \mathbb{H}_\infty$ *has galois group* $\Delta = \mathrm{Gal}(\mathbb{L}/\mathbb{K}_\infty) \cong \mathscr{A}$ *and let* $R = \mathrm{Rad}(\mathbb{L}/\mathbb{K}_\infty)$. *At finite levels, we have* $\mathrm{Gal}(\mathbb{L}_n/\mathbb{K}_n) \cong \mathscr{A}_n$ *and* $R_n = \mathrm{Rad}(\mathbb{L}'_n/\mathbb{K}_{n+z})$, *with* $R_n = \Lambda\rho_n$. *Then*

$$\left\langle \varphi(a_n)^{f^k}, \rho_n^{(f^*)^j} \right\rangle_{\mathbb{L}'_n/\mathbb{K}_{n+z}} = 1 \quad \text{for } k+j \geq m. \tag{24}$$

Proof. Let $g = \varphi(a_n) \in \Delta_n$ be a generator and $\rho \in R_n$ generate the radical. The equivariance of Kummer pairing implies

$$\left\langle g^{f^k}, \rho^{(f^*)^j} \right\rangle_{\mathbb{L}'_n/\mathbb{K}_{n+z}} = \left\langle g, \rho^{(f^*)^{j+k}} \right\rangle = \left\langle g^{f^{j+k}}, \rho \right\rangle.$$

By hypothesis, $a_n^{f^m} = 1$, and using also duality, $g^{f^m} = \rho^{(f^*)^m} = 1$. Therefore, the Kummer pairing is trivial for $k + j \geq m$, which confirms (24) and completes the proof.

It will be useful to give a translation of (24) in terms of projective limits: under the same premises like above, writing $\rho = \varprojlim_n \rho_n$ for a generator of the radical $R = \mathrm{Rad}(\mathbb{L}/\mathbb{K}_\infty)$, we have

$$\left\langle \varphi(a)^{f^k}, \rho^{(f^*)^j} \right\rangle_{\mathbb{L}/\mathbb{K}} = 1 \quad \text{for } k+j \geq m. \tag{25}$$

We shall also use the following simple result:

Lemma 22. *Let* $\mathbb{K}$ *be a CM galois extension of* $\mathbb{Q}$ *and suppose that* $(\mathbf{A}')^-(T) \neq 0$. *Then* $\mathrm{ord}_T(\mathbf{A}^-(T)) > 1$.

Proof. Assuming that $(\mathbf{A}')^-(T) \neq 0$, there is some $a = (a_n)_{n\in\mathbb{N}} \in \mathbf{A}^-$ with image $a' \in (\mathbf{A}')^-[T]$. We show that $\mathrm{ord}_T(a) = 2$. Let $\mathfrak{Q}_n \in a_n$ be a prime and n sufficiently large; then $\mathrm{ord}(a_n) = p^{n+z}$ for some $z \in \mathbb{Z}$ depending only on a and not on n. Let $(\alpha_0) = \mathfrak{Q}^{p^{n+z}}$ and $\alpha = \alpha_0/\overline{\alpha_0}$; since $a' \in (\mathbf{A}')^-[T]$ it also follows that $a_n^T \in \mathbf{B}^-$ and thus $\mathfrak{Q}^T = \mathfrak{R}_n$ wit $b_n := [\mathfrak{R}_n] \in \mathbf{B}_n$. If $b_n \neq 1$, then $\mathrm{ord}_T(a) = 1 + \mathrm{ord}_T(a') = 2$, and we are done.

We thus assume that $b_n = 1$ and draw a contradiction. In this case, $\mathfrak{R}_n^{1-j} = (\rho_n)$ is a p-unit and $(\alpha^T) = \left(\rho_n^{p^{n+z}}\right)$, so

$$\alpha^T = \delta \rho_n^{p^{n+z}}, \quad \delta \in \mu_{p^n}.$$

Taking the norm $N = N_{\mathbb{K}_n/\mathbb{K}}$ we obtain $1 = N(\delta)N(\rho_n)^{p^{n+z}}$. The unit $N(\delta) \in \mu(\mathbb{K}) =< \zeta_{p_k} >$ – we must allow here, in general, that $\mathbb{K}$ contains the p^kth roots of unity, for some maximal $k > 0$. It follows that $\rho_1 := N(\rho_n)$ verifies $\rho_1^{p^{n+z}} = \delta_1$, and since $\delta_1 \not\in E(\mathbb{K})^{p^{k+1}}$, it follows that $\rho_1^{p^k} = \pm 1$ and by Hilbert 90 we deduce that $\rho_n^{p^k} = \pm x^T, x \in \mathbb{K}_n^\times$. In terms of ideals, we have then

$$\mathfrak{Q}^{(1-j)Tp^{n+z}} = (\alpha^T) = \left(x^{Tp^{n+z-k}}\right), \quad \text{hence}$$

$$\left(\mathfrak{Q}^{(1-j)p^k}/(x)\right)^{Tp^{n+z-k}} = (1) \quad \Rightarrow (\mathfrak{Q}^{(1-j)p^k}/(x))^T = (1).$$

But $\mathfrak{Q}$ is by definition not a ramified prime, so the above implies that a_n has order bounded by p^k, which is impossible since $a_n \in A_n^-$. This contradiction confirms the claim and completes the proof of the lemma.

4.2 Units and the Radical of Ω

The extension $\Omega/\mathbb{K}$ is an infinite extension and $\mathbb{Z}_p$-rk$(\mathrm{Gal}(\Omega/\mathbb{K})) = \mathscr{D}(\mathbb{K}) + r_2(\mathbb{K}_n) + 1$. Here, $r_2(\mathbb{K}_n)$ is the number of pairs of conjugate complex embedding and the 1 stands for the extension $\mathbb{K}_\infty/\mathbb{K}$. Let $\wp \subset \mathbb{K}$ be a prime above p, let $D(\wp) \subset \Delta$ be its decomposition group, and $C = \Delta/D(\wp)$ be a set of coset representatives in Δ. We let $s = |C|$ be the number of primes above p in $\mathbb{K}$. Moreover,

$$\mathbb{Z}_p\text{-rk}\left(\mathrm{Gal}((\Omega^-/\mathbb{K}_\infty)\right) = r_2(\mathbb{K}_n).$$

It is a folklore fact, which we shall prove constructively below that the *regular* part $r_2(\mathbb{K}_n)$ in the above rank stems from $\Omega^- \subset \Omega_E$, where $\Omega_E = \cup_n \mathbb{K}_n[E_n^{1/p^n}]$. The radical is described precisely by:

Lemma 23. *Notations being like above, we define for* $n > 1$: $\mathscr{E}_n' = \{e^{v_{n,1}^*} : e \in E_n\}$ *and* $\mathscr{E}_n = \mathscr{E}_n' \cdot (E_n)^{p^n}$. *Then*

$$\Omega^- = \mathbb{H}_1 \cdot \cup_n \mathbb{K}_n[\mathscr{E}_n^{1/p^n}] \times \mathbb{T}_1, \tag{26}$$

where $\mathbb{T}_1/\mathbb{K}_1$ *is an extension which shall be described in the proof. It has group* $\mathrm{Gal}(\mathbb{T}_1/\mathbb{K}_1) \cong (\mathbb{Z}/(p\cdot\mathbb{Z}))^{s-1}$.

Proof. We show that the subgroups $\mathscr{E}_m$ give an explicit construction of Ω^-, as radicals. The proof uses reflection, class field theory and some technical, but strait forward estimations of ranks.

Let $U = \mathscr{O}(\mathbb{K} \otimes_{\mathbb{Q}} \mathbb{Q}_p)$ and $U^{(1)}$ be the units congruent to one modulo an uniformizor in each completion of $\mathbb{K}$ at a prime above p. The global units $E_1 = E(\mathbb{K}_1)$ embed diagonally in U and we denote by $\overline{E}$ the completion of this embedding, raised to some power coprime to p, so that $\overline{E} \subset U^{(1)}$. A classical result from class field theory [13] p. 140, says that

$$\mathrm{Gal}(\Omega/\mathbb{H}_1) \cong U^{(1)}/\overline{E}.$$

Since $(U^{(1)})^- \cap \overline{E} = \mu_p$, it follows that $\mathrm{Gal}(\Omega^-/\mathbb{H}_1^-) = (U^{(1)})^- \times \mathscr{T}(U^-)/\mu_p$, where the torsion part $\mathscr{T}(U^-) = \prod_{v \in C} \mu_p$ is[6] the product of the images of the pth roots of unity in the single completions, factored by the diagonal embedding of the global units.

For the proof, we need to verify that ranks are equal on both sides of (26). Let $\pi_v \in \mathbb{K}_n$ be a list of integers such that $(\pi_v) = \wp^{vh}$ for h the order of the class of $\wp^v$ in the ideal class group $\mathscr{C}(\mathbb{K})$. Then we identify immediately $\mathbb{T}_1 = \prod_{v \in \mathbb{C}} \mathbb{K}[\pi_v^{1/p}]$ as a p-ramified extension with group $\mathrm{Gal}(\mathbb{T}_1/\mathbb{K}) = \mathscr{T}(U^-)/\mu_p \subset \mathrm{Gal}(\Omega^-/\mathbb{H}^-)$.

A straightforward computation in the group ring yields that $T^*x \equiv 0 \bmod (\omega_n, p^n)\Lambda$ iff $x \in v_{n,1}^*\Lambda$. On the other hand, suppose that $x \in \mathrm{rad}(\Omega^-/\mathbb{K}_n) \cap E_n$; note that here the extensions can be defined as Kummer extensions of exact exponent p^n, so there is no need of an index shift as in the case of the unramified extensions treated above. This observation and Kummer theory imply that $x^{T^*} \in E_n^{p^n}$, and thus $x \in \mathscr{E}_n$. We denote as usual $\Omega_E = \cup_n \mathbb{K}_n[E_n^{1/p^n}]$. We found that $\cup_m \mathbb{K}_m[\mathscr{E}_m^{1/p^m}] = \Omega^- \cap \Omega_E$; by comparing ranks, we see that if $\Omega^- \neq \mathbb{T}_n \cdot \mathbb{H}_1 \cdot (\Omega^- \cap \Omega_E)$, then there is an extension $\Omega^- \supset \Omega'' \supsetneq (\Omega^- \cap \Omega_E)$, such that

$$\mathbb{Z}_p\text{-rk}(\mathrm{Gal}(\Omega''/\mathbb{K}_\infty)) = r_2(\mathbb{K}) = \mathbb{Z}_p\text{-rk}(\mathrm{Gal}(\Omega^- \cap \Omega_E)).$$

---

[6]We have assumed for simplicity that $\mathbb{K}$ does not contain the p^2th roots of unity. The construction can be easily generalized to the case when $\mathbb{K}$ contains the p^kth but not the p^{k+1}th roots of unity.

Since $\Omega_E \subset \overline{\Omega}$, where $\overline{\Omega}$ is the maximal p-abelian p-ramified extension of $\mathbb{K}_\infty$, it follows that $\mathrm{Gal}((\Omega^- \cap \Omega_E)/\mathbb{K}_\infty)$ is a factor of $\mathrm{Gal}(\Omega^-/\mathbb{K}_\infty)$ and also of $\mathrm{Gal}(\Omega''/\mathbb{K}_\infty)$.

The index $[\mathrm{Gal}(\Omega'' : \mathbb{K}_\infty) : \mathrm{Gal}((\Omega^- \cap \Omega_E)/\mathbb{K}_\infty)] < \infty$ and since $\mathrm{Gal}(\Omega''/\mathbb{K}_\infty)$ is a free $\mathbb{Z}_p$-module and thus has no finite compact subgroups, it follows from infinite galois theory that $\Omega'' = \Omega^- \cap \Omega_E$, which completes the proof.

We note that for $\Omega_n \supset \mathbb{K}_n$, the maximal p-abelian p-ramified extension of $\mathbb{K}_n$, the same arguments lead to a proof of

$$\Omega_n^- = \cup_{m \geq n} \mathbb{K}_m \left[E(\mathbb{K}_m)^{N^*_{m,n}/p^m} \right]. \tag{27}$$

4.3 Construction of Auxiliary Extensions and Order Reversal

On minus parts we have $\mathbb{Z}_p\text{-rk}(\Omega^-/\mathbb{K}^-) = r_2 + 1$ and the rank $\mathbb{Z}_p\text{-rk}(\Omega^-/\overline{\mathbb{H}_1}^-) = r_2$ does not depend on Leopoldt's conjecture. We let $G = \mathrm{Gal}(\mathbb{H}_\infty/\mathbb{K})$ and $X = \varphi(\mathbf{A}) = \mathrm{Gal}(\mathbb{H}_\infty/\mathbb{K}_\infty)$, following the notation in [16], Lemma 13.15. The commutator is $G' = TX$ and the fixed field $\mathbb{L} = \mathbb{H}_\infty^{TX}$ is herewith the maximal abelian extension of $\mathbb{K}$ contained in $\mathbb{H}_\infty$. From the definition of Ω, it follows that $\mathbb{L} = \Omega \cap \mathbb{H}_\infty$ (see also [11], p. 257). Consequently, $\mathrm{Gal}(\mathbb{L}/\overline{\mathbb{H}_1}) \cong X/TX$. Let $F(T) = T^m G(T)$ be the annihilator polynomial of $p^M \mathbf{A}$, with p^M an annihilator of the $\mathbb{Z}_p$-torsion (finite and infinite) of $\mathbf{A}$. If $\mathbf{A}^\circ$ is this $\mathbb{Z}_p$-torsion, then $\mathbf{A} \sim \mathbf{A}(T) + \mathbf{A}(G(T)) + \mathbf{A}^\circ$.

From the exact sequences

$$\begin{array}{ccccccccc}
0 & \longrightarrow & K_1 & \longrightarrow & p^\mu \mathbf{A}^- & \longrightarrow & p^\mu \mathbf{A}^-(T) + \mathbf{A}^-(G) & \longrightarrow & K_2 & \longrightarrow & 0 \\
 & & \downarrow & & \downarrow & & \downarrow & & \downarrow & & \\
0 & \longrightarrow & 0 & \longrightarrow & p^M \mathbf{A}^- & \longrightarrow & p^M \mathbf{A}^-(T) \oplus \mathbf{A}^-(G) & \longrightarrow & 0 & \longrightarrow & 0
\end{array}$$

in which $M \geq \mu$ is such that annihilates the finite kernel and cokernel K_1, K_2 and the vertical arrows are multiplication by $p^{M-\mu}$, we see that it is possible to construct a submodule of $\mathbf{A}^-$, which is a direct sum of G and T-parts. We may choose M sufficiently large, so that the following conditions also hold: $p^M \mathbf{A}^-(T)$ is a direct sum of cyclic Λ-modules and if a prime above p is inert in some $\mathbb{Z}_p$-subextension of $\mathbb{H}_\infty/\mathbb{H}_\infty^{p^M \varphi(\mathbf{A}^-)}$, then it is totally inert. Let $\tilde{\mathbb{K}} = \mathbb{H}_\infty^{p^M \varphi(\mathbf{A}^-)}$ for some M large enough to verify all the above conditions. Let $\mathbb{K}_T = \mathbb{H}_\infty^{p^M \mathbf{A}^-(G)}$; by construction, $\tilde{X}_T := \mathrm{Gal}(\mathbb{K}_T/\tilde{\mathbb{K}}) \sim \mathbf{A}^-(T)$ and it is a direct sum of cyclic Λ-modules. Let $a_1, a_2, \ldots, a_t \in p^M \mathbf{A}^-(T) = \varphi^{-1}(\tilde{X}_T)$ be such that

$$\tilde{X}_T = \bigoplus_{j=1}^{t} \varphi(\Lambda a_i). \tag{28}$$

From the definition $\mathbb{K}_{\overline{B}}^- = \Omega^- \cap \mathbb{H}_\infty \subset \mathbb{K}_T$ and Lemma 22 implies that $\mathrm{Gal}(\mathbb{H}_\infty/\mathbb{K}_{\overline{B}}^-) \sim \tilde{X}_T[T]$. Let now $a \in p^M \mathbf{A}^-(T) \setminus (p,T)p^M \mathbf{A}^-(T)$ – for instance $a = a_1$ and let $\mathscr{A} = \Lambda a$ while $\mathscr{C} \subset p^M \mathbf{A}^-(T)$ is a Λ-module with $\mathscr{A} \oplus \mathscr{C} = p^M \mathbf{A}^-(T)$. We assume that $m = \mathrm{ord}_T(a)$ and let $b = T^{m-1} a \in \mathbf{A}^-[T]$. We define $\mathbb{K}_a = \mathbb{K}_T^{\varphi(\mathscr{C})}$, an extension with $\mathrm{Gal}(\mathbb{K}_a/\tilde{\mathbb{K}}) \cong \mathscr{A}$. At finite levels, we let $\mathbb{K}_{a,n} := \mathbb{K}_a \cap \mathbb{H}_n$ and let z be a positive integer such that $\mathbb{K}'_{a,n} := \mathbb{K}_{n+z}\mathbb{K}_{a,n}$ is a Kummer extension of $\mathbb{K}'_n := \mathbb{K}_{n+z}$, for all sufficiently large n – we may assume that M is chosen such that the condition $n > M$ suffices. The duals of the galois groups $\varphi(\mathscr{A}_n)$ are radicals $R_n = \mathrm{Rad}(\mathbb{K}_{a,n}/\tilde{\mathbb{K}})$, which are cyclic Λ-modules too (see also the following section for a detailed discussion of radicals), under the action of Λ, twisted by the Iwasawa involution. We let $\rho_n \in R_n$ be generators which are dual to a_n and form a norm coherent sequence with respect to the p-map, as was shown above, since $n > M > n_0$; by construction, $\rho_n^{p^{n+z}} \in \mathbb{K}'_n$. We gather the details of this construction in

Lemma 24. *Notations being like above, there is an integer $M > 0$, such that the following hold:*

1. *The extension $\tilde{\mathbb{K}} := \mathbb{H}_\infty^{p^M X}$ has group $\tilde{X} := \mathrm{Gal}(\mathbb{H}_\infty/\tilde{\mathbb{K}}) = \tilde{X}(T) \oplus \tilde{X}(G)$ below $\mathbb{H}_\infty$.*
2. *The extension $\mathbb{K}_T := \mathbb{H}_\infty^{\tilde{X}(G)}$ has group $\tilde{X}_T = \bigoplus_{i=1}^t \Lambda\varphi(a_i)$.*
3. *For $a \in p^M \mathbf{A}^-(T) \setminus (p,T)p^M \mathbf{A}^-(T)$, we define $\mathscr{A} = \Lambda a$ and let $\mathscr{C} \subset p^M \mathbf{A}^-(T)$ be a direct complement. We define $\mathbb{K}_a = \mathbb{H}_T^{\varphi(\mathscr{C})}$, so $\mathrm{Gal}(\mathbb{K}_a/\tilde{\mathbb{K}}) = \varphi(\mathscr{A})$ and let $\mathbb{K}_{a,n} = \mathbb{K}_a \cap \mathbb{H}_n$.*
4. *There is a positive integer z such that for all $n > M$,*
$$\mathbb{K}'_{a,n} = \mathbb{K}_{n+z} \cdot \mathbb{K}_{a,n} \subset \mathbb{H}_{n+z}$$
is a Kummer extension of $\mathbb{K}'_n := \mathbb{K}_{n+z}$.
5. *For $\mathbb{K}_{\overline{B}}^- = \Omega^- \cap \mathbb{H}_\infty$, we have $\mathbb{K}_{\overline{B}}^- \subset \mathbb{K}_T$ and $\mathrm{Gal}(\mathbb{H}_\infty/\mathbb{K}_{\overline{B}}^-) \sim \tilde{X}_T[T]$.*
6. *The radical $R_n = \mathrm{Rad}(\mathbb{K}_{a,n}/\tilde{\mathbb{K}}) \cong \mathscr{A}_n^\bullet$ and we let $\rho_n \in R_n$ generate this radical as a Λ^*-cyclic module, so that $\rho_n^{(T^*)^i}, i = 0, 1, \ldots, m-1$ form a dual base to the base $a_n^{T^i}, i = 0, 1, \ldots, m-1$ of $\mathscr{A}_n$. We have $\rho_n^{p^{n+z}} \in \mathbb{K}'_n$.*

We may apply the order reversal to the finite Kummer extensions $\mathbb{K}_{a,n}/\tilde{\mathbb{K}}_n$ defined in Lemma 24. In the notation of this lemma, we assume that $m = \mathrm{ord}_T(a) > 1$. We deduce from 24 that

$$\left\langle \varphi(a_n)^{T^i}, \rho_n^{(T^*)^{m-1-i}} \right\rangle_{\mathbb{K}_{a,n}/\tilde{\mathbb{K}}} = \zeta_{p^v},$$
$$v \geq n + z - M, \; i = 0, 1, \ldots, m-1. \qquad (29)$$

This fact is a direct consequence of (24) for $i = 0$ and it follows by induction on i, using the following fact. Let $\mathbb{F}_i = \tilde{\mathbb{K}}\left[\rho_n^{(T^*)^{m-1-i}}\right]$; then $\overline{\mathbb{F}}_i = \prod_{j=0}^i \mathbb{F}_i$ are galois extensions of $\tilde{\mathbb{K}}_1$ and in particular their galois groups are Λ-modules. In particular, $\mathrm{Gal}(\mathbb{F}_{m-1}/\tilde{\mathbb{K}}) \cong \mathscr{A}_n^{T^{m-1}} = \mathscr{A}_n[T]$. From Lemma 22, we know that $\mathscr{A}_n[T] \subset \mathbf{B}_n^-$, so at least one prime $\mathfrak{p} \subset \tilde{\mathbb{K}}$ above p is inert in $\mathbb{F}_{m-1}$, and the choice of M in Lemma 24 implies that it is totally inert in $\mathbb{F}_{m-1}/\tilde{\mathbb{K}}_n$. Let $\wp \subset \mathbb{K}$ be a prime below $\mathfrak{p}$. It follows in addition $\overline{\mathbb{F}}_{m-2} \subset \mathbb{H}'_\infty \cdot \tilde{\mathbb{K}}$ and all the primes above p are split in $\overline{\mathbb{F}}_{m-2}$: this is because

$$\mathrm{Gal}(\overline{\mathbb{F}}_{m-2}/\tilde{\mathbb{K}}) \cong \mathscr{A}_n/\mathscr{A}_n[T] = \mathscr{A}_n/(\mathscr{A}_n \cap \mathbf{B}_n) \subset A'_n.$$

Let now $\mathbb{K}_a$ be like above and $\mathbb{K}_b = \Omega^- \cap \mathbb{K}_a$, so $\mathbb{K}_b/\tilde{\mathbb{K}}$ is a $\mathbb{Z}_p$-extension. Moreover, we assume that $\mathbb{K}_b \not\subset \mathbb{H}'_\infty$, so not all primes above p are totally split. By choice of M, we may assume that there is at least on prime $\wp \subset \mathbb{K}$ above p, such that the primes $\tilde{K} \supset \mathfrak{p} \supset \wp$ are inert in $\mathbb{K}_b$. By the construction of Ω^- in the previous section, we have $T^*\mathrm{Rad}(\mathbb{K}_b/\tilde{\mathbb{K}}) = 0$. The order reversal lemma implies then that $\mathrm{Gal}(\mathbb{K}_b/\tilde{\mathbb{K}}) \cong \mathscr{A}/(T\mathscr{A})$. Assuming now that $m = \mathrm{ord}_T(a) \geq 1$, the Lemma 22 implies that $T^{m-1}a \in \mathbf{B}^-$ and the subextension of $\mathbb{K}_a$, which does not split all the primes above p is the fixed field of $T^{m-1}\mathscr{A}$; but then order reversal requires that $\mathrm{Rad}(\mathbb{K}_b/\tilde{\mathbb{K}})$ is cyclic, generated by ρ, which is at the same time a generator of $\mathrm{Rad}(\mathbb{K}_a/\tilde{\mathbb{K}})$ as a Λ-module. Since we have seen that $T^*\mathrm{Rad}(\mathbb{K}_b/\tilde{\mathbb{K}}) = 0$, we conclude that $T^*\mathrm{Rad}(\mathbb{K}_a/\tilde{\mathbb{K}}) = 0$, and by duality, $Ta = 0$. This holds for all $a \in p^M\mathbf{A}^-(T)$, so we have proved:

Lemma 25. *Let $\mathbb{H}_B^- = \Omega^- \cap \mathbb{H}_\infty$. If $[\mathbb{K}_B^- \cap \mathbb{H}'_\infty : \mathbb{K}_\infty] < \infty$, then $\mathbf{A}^-(T) = \mathbf{B}^-$.*

4.4 The Contribution of Class Field Theory

We need to develop more details from local class field theory in order to understand the extension $\mathbb{H}_B^- = \Omega^- \cap \mathbb{H}_\infty$. This is an unramified extension of $\mathbb{K}_\infty$, which is abelian over $\mathbb{H}_1$. We wish to determine the $\mathbb{Z}_p$-rank of this group and decide whether the extensions in $\mathbb{H}_B^-$ split the primes above p or not.

Let $\wp \subset \mathbb{K}$ be a prime over p and $\wp^+ \subset \mathbb{K}^+$ be the real prime below it. If $\wp^+$ is not split in $\mathbb{K}/\mathbb{K}^+$, then $\mathbf{B}^- = \{1\}$ and it is also known that $(\mathbf{A}')^-(T) = \{1\}$ in this case – this follows also from the Lemma 22. The case of interest is thus when $\wp$ is split in $\mathbb{K}/\mathbb{K}^+$. Let $D(\wp) \subset \Delta$ and C, s be defined like above and let $\mathfrak{p} \subset \Omega$ be a prime above $\wp$.

Local class theory provides the isomorphism $\mathrm{Gal}(\Omega/\mathbb{H}_1) \cong U^{(1)}/\overline{E}$ via the global Artin symbol (e.g. [13]). We have the canonic, continuous embedding

$$\mathbb{K} \hookrightarrow \mathbb{K} \otimes_{\mathbb{Q}} \mathbb{Q}_p \cong \prod_{\nu \in C} \mathbb{K}_{\nu\wp},$$

and $U^{(1)} = \prod_{\nu \in C} U^{(1)}_{\nu\wp}$, where $U^{(1)}_{\mathfrak{p}}$ are the one-units in the completion at the prime $\mathfrak{p}$. The ring $U^{(1)}$ is a galois algebra and $\Delta = \mathrm{Gal}(\mathbb{K}/\mathbb{Q}) \hookrightarrow \mathrm{Gal}(U^{(1)}/\mathbb{Q}_p)$. Thus, complex conjugation acts on $U^{(1)}$ via the embedding of $\mathbb{K}$, and if $u \in U^{(1)}$ has $\iota_{\wp}(u) = x, \iota_{\overline{\wp}}(u) = y$, then ju verifies

$$\iota_{\wp}(ju) = \overline{y}, \quad \iota_{\overline{\wp}}(ju) = \overline{x}.$$

Moreover, $u \in U^-$ iff $u = v^{1-j}, v \in U$. Thus, if $\iota_{\wp}(v) = v_1$ and $\iota_{\overline{\wp}}(v) = v_2$, then

$$\iota_{\wp}(u) = v_1/\overline{v_2}, \quad \iota_{\overline{\wp}}(u) = v_2/\overline{v_1} = 1/\overline{\iota_{\wp}(u)}. \tag{30}$$

One can analyze U^+ in a similar way. Note that $\mathbb{Z}_p$ embeds diagonally in U^+; this is the preimage of $\mathrm{Gal}(\mathbb{K}_\infty/\mathbb{K})$, under the global Artin symbol.

Next we shall construct by means of the Artin map some subextension of Ω^-, which are defined uniquely by some pair of complex conjugate primes $\wp, \overline{\wp} \supset (p)$ and intersect $\mathbb{H}_\infty$ is a $\mathbb{Z}_p$-extension. Since $U^{(1)}$ is an algebra, there exists for each pair of conjugate primes $\wp, \overline{\wp}$ with fixed primes $\mathfrak{P}, \mathfrak{P}^{\tilde{j}} \subset \Omega$ above $(\wp, \overline{\wp})$, a subalgebra

$$V_{\wp} = \left\{ u \in U^{(1)} \ : \ \iota_{\mathfrak{P}}(u) = 1/\overline{\iota_{\mathfrak{P}^j}(u)}; \iota_{\nu\wp} = 1, \ \forall \nu \in C \setminus \{1, j\} \right\}. \tag{31}$$

Accordingly, there is an extension $\mathbb{M}_{\wp} \subset \Omega^-$ such that

$$\varphi^{-1}(\mathrm{Gal}(\mathbb{M}_{\wp}/\mathbb{K}_\infty)) = V_{\wp}.$$

By construction, all the primes above p above $\wp, \overline{\wp}$ are totally split in $\mathbb{M}_{\wp}$. Since $\mathrm{Gal}(\mathbb{K}_{\wp}/\mathbb{Q}_p) = D(\wp)$ and $U_{\wp}$ is a pseudocyclic $\mathbb{Z}_p$-module, pseudoisomorphic to $\mathbb{Z}_p[D(\wp)]$ (e.g. [13], p. 140–141), it follows that there is exactly one $\mathbb{Z}_p$-subextension $\mathbb{U}_{\wp} \subset \mathbb{M}_{\wp}$ with galois group fixed by the augmentation of $D(\wp)$. Since the augmentation and the norm yield a direct sum decomposition of $\mathbb{Z}_p[D(\wp)]$, this extension and its galois group are canonic – up to possible finite quotients. Locally, the completion of $\mathbf{U}/\mathbb{Q}_p$ of $\mathbb{U}_{\wp}$ at the primes above $\wp$ is a $\mathbb{Z}_p$-extension of $\mathbb{Q}_p$, since its galois group is fixed $D(\wp)$. It follows by a usual argument that $\mathbb{U}_{\wp}/\mathbb{K}_\infty$ is unramified at all primes above p, so $U_{\wp} \subset \mathbb{H}$. One has by construction that $\mathbb{U}_{\wp}^- \subset \Omega^-$, so we have proved:

Lemma 26. *Let $\mathbb{K}$ be a CM extension like above and assume that the primes $\wp^+ \subset \mathbb{K}^+$ split in $\mathbb{K}/\mathbb{K}^+$. For each prime $\wp \subset \mathbb{K}$, there is a canonic (up to finite subextensions) $\mathbb{Z}_p$-extension $\mathbb{U}_{\wp} \subset \Omega^- \cap \mathbb{H}_\infty$ such that $\mathrm{Gal}(\mathbb{U}_{\wp}/\mathbb{K}_\infty) = \varphi\left(V_{\wp}^{\mathfrak{A}(\mathbb{Z}_p[D_{\wp}])}\right)$, where $\mathfrak{A}(\mathbb{Z}_p[D_{\wp}]$ is the augmentation ideal of this group ring and $V_{\wp}$ is defined by (31). In particular, Ω^- contains exactly $s' = |C|/2$ unramified extensions.*

Our initial question boils down to the following: is $\mathbb{U}_{\wp} \subset \mathbb{H}'_\infty$?

The following example perfectly illustrates the question:

Example 6. Let $\mathbb{K}/\mathbb{Q}$ be an imaginary quadratic extension of $\mathbb{Q}$ in which p is split. Then $U^{(1)}(\mathbb{K}) = (\mathbb{Z}_p^{(1)})^2$ and $\Omega = \mathbb{K}_\infty \cdot \Omega^-$ is the product of two $\mathbb{Z}_p$-cyclotomic extensions; we may assume that $\mathbb{H}_1 = \mathbb{K}$, so $\mathrm{Gal}(\Omega/\mathbb{K}) = \varphi(U^{(1)}(\mathbb{K}))$. One may take the second $\mathbb{Z}_p$-extension in Ω also as being the anticyclotomic extension. In analyzing a similar example, Greenberg makes in [10] the following simple observation: since $\mathbb{Q}_p$ has only two $\mathbb{Z}_p$-extensions and $\mathbb{K}_\infty$ contains the cyclotomic ramified one, it remains that locally $\Omega^-/\mathbb{K}_\infty$ is either trivial or an unramified $\mathbb{Z}_p$-extension. In both cases, $\Omega^- \subset \mathbb{H}_\infty$ is a global, totally unramified $\mathbb{Z}_p$-extension – we have used the same argument above in showing that $\mathbb{U}_{\wp}/\mathbb{K}_\infty$ is unramified. The remark settles the question of ramification, but does not address the question of our concern, namely splitting. However, in this case we know more. In the paper [9] published by Greenberg in the same year, he proves that for abelian extensions of $\mathbb{Q}$, thus in particular for quadratic ones, $(\mathbf{A}')^-(T) = \{1\}$. Therefore in this example, Ω^- cannot possibly split the primes above p.

How can this fact be explained by class field theory?

We give here a proof of Greenberg's theorem [9] for imaginary quadratic extensions, and thus an answer to the question raised in the last example; we use the notations introduced there:

Proof. We shall write $\mathbb{L} = \mathbb{K}_\infty \cdot \mathbb{H}_1$; we have seen above that $\Omega/\mathbb{L}$ must be an unramified extension. Let $\mathfrak{P} \in \Omega$ be a prime above $\wp$, let $\tilde{j} \in \mathrm{Gal}(\Omega/\mathbb{H}_1)$ be a lift of complex conjugation and let $\tau \in \mathrm{Gal}(\Omega/\mathbb{H}_1)$ be a generator of the inertia group $I(\mathfrak{P})$: since $\Omega_{\mathfrak{P}}/\mathbb{K}_{\wp}$ is a product of $\mathbb{Z}_p$-extensions of $\mathbb{Q}_p$ and $\mathbb{Q}_p$ has no two independent ramified $\mathbb{Z}_p$-extensions, it follows that $I(\mathfrak{P}) \cong \mathbb{Z}_p$ is cyclic, so τ can be chosen as a topological generator. Then $\tau^j = j \cdot \tau \cdot j$ generates $I(\mathfrak{P}^{\tilde{j}}) \cong \mathbb{Z}_p$. Iwasawa's argument used in the proof of Thereom 2 holds also for $\Omega/\mathbb{H}_1$: there is a class $a \in A_n$ with $\tau^j = \tau\varphi(a)$, where the Artin symbol refers to the unramified extension $\Omega/\mathbb{L}$. Thus,

$$j \cdot \tau \cdot j \cdot \tau^{-1} = \tau^{j-1} = \varphi(a).$$

The inertia groups $I(\mathfrak{P}) \neq I(\mathfrak{P}^{\tilde{j}})$: otherwise, their common fixed field would be an unramified $\mathbb{Z}_p$-extension of the finite galois field $\mathbb{H}_1/\mathbb{Q}$, which is impossible: thus $\tau^{j-1} = \varphi(a) \neq 1$ generates a group isomorphic to $\mathbb{Z}_p$. Let now $\mathfrak{p} = \mathfrak{P} \cap \mathbb{L}$; the primes $\mathfrak{p}, \mathfrak{p}^{\tilde{j}}$ are unramified in $\Omega_n/\mathbb{L}$, so τ restricts to an Artin symbol in this extension. The previous identity implies

$$\left(\frac{\Omega/\mathbb{L}}{a}\right) = \left(\frac{\Omega/\mathbb{L}}{\mathfrak{p}^{j-1}}\right);$$

Since the Artin symbol is a class symbol, we conclude that the primes in the coherent sequence of classes $b = [\mathfrak{p}^{j-1}] \in \mathbf{B}^-$ generate $\mathrm{Gal}(\Omega/\Omega^{\varphi(a)})$ and $a = b$, which completes the proof.

4.5 Proof of Theorems 3 and 1

We can turn the discussion of the example above into a proof of Theorem 3 with its consequence, the Corollary 1. The proof generalizes the one given above for imaginary quadratic extensions, by using the construction of the extensions $\mathbb{U}_{\wp}$ defined above.

Proof. Let $\mathbb{L} = \mathbb{H}_1 \cdot \mathbb{K}_\infty$, like in the previous proof. Let $\wp \subset \mathbb{K}$ be a prime above p and $\mathbb{U}$ be the maximal unramified extension of $\mathbb{L}$ contained in $\mathbb{U}_{\wp}$, the extension defined in Lemma 26, and let $\tilde{j}$ be a lift of complex conjugation to $\mathrm{Gal}(\mathbb{U}/\mathbb{Q})$. Since $\Omega/\mathbb{H}_1$ is abelian, the extension $\mathbb{U}/\mathbb{H}_1$ is also galois and abelian.

Let $\mathfrak{P} \subset \mathbb{U}$ be a fixed prime above $\wp$ and $\tilde{j} \in \mathrm{Gal}(\mathbb{U}/\mathbb{H}_1)$ be a lift of complex conjugation. Consider the inertia groups $I(\mathfrak{P}), I(\mathfrak{P}^{\tilde{j}}) \subset \mathrm{Gal}(\mathbb{U}/\mathbb{H}_1)$ be the inertia groups of the two conjugate primes. Like in the example above, $\mathrm{Gal}(\mathbb{U}/\mathbb{H}_1) \cong \mathbb{Z}_p^2$ and $\mathbb{U}_{\mathfrak{P}}/\mathbb{K}_{\wp}$ is a product of at most two $\mathbb{Z}_p$ extensions of $\mathbb{Q}_p$. It follows that the inertia groups are isomorphic to $\mathbb{Z}_p$ and disinct: otherwise, the common fixed field in $\mathbb{U}$ would be an uramified $\mathbb{Z}_p$-extension of $\mathbb{H}_1$.

For $\nu \in C \setminus \{1, j\}$, the primes above $\nu\wp$ are totally split in $\mathbb{U}_{\wp}/\mathbb{K}_\infty$, so a fortiori in $\mathbb{U}$. Let $\tilde{\tau} \in \mathrm{Gal}(\mathbb{U}/\mathbb{H}_1)$ generate the inertia group $I(\mathfrak{P})$; then $\tilde{\tau}^{\tilde{j}} \in \mathrm{Gal}(\mathbb{U}/\mathbb{H}_1)$ is a generator of $I(\mathfrak{P}^{\tilde{j}})$. Since $\mathbb{U}/\mathbb{L}$ is an unramified extension, there is an $a \in \mathbf{A}^-$ such that

$$\tilde{\tau}^j = j\tilde{\tau}j = \left(\frac{\mathbb{U}/\mathbb{L}}{a}\right) \cdot \tilde{\tau}.$$

Thus,

$$\varphi(a) = j\tilde{\tau}j\tilde{\tau}^{-1}. \tag{32}$$

Like in the previous proof, we let $\mathfrak{p} = \mathfrak{P} \cap \mathbb{L}$ and note that since $\mathfrak{p}$ does not ramify in $\mathbb{U}/\mathbb{L}$, the automorphism $\tilde{\tau}$ acts like the Artin symbol $\left(\frac{\mathbb{U}/\mathbb{L}}{\mathfrak{p}}\right)$. The relation (32) implies:

$$\left(\frac{\mathbb{U}/\mathbb{L}}{a}\right) = \left(\frac{\mathbb{U}/\mathbb{L}}{\mathfrak{p}^{j-1}}\right).$$

In particular, the primes in the coherent sequence of classes $b = [\mathfrak{p}^{j-1}] \in \mathbf{B}^-$ generate $\mathrm{Gal}(\mathbb{U}/\mathbb{L})$ and $\mathbb{U}$ does not split all the primes above p. This happens for all $\wp$ and by Lemma 26 we have $\mathbb{H}_B^- = \prod_{\nu \in C/\{1,j\}} U_{\nu\wp}$, so it is spanned by $\mathbb{Z}_p$-extensions that do not split the primes above p and consequently

$$[\mathbb{H}'_\infty \cap \mathbb{H}_B] < \infty.$$

We may now apply Lemma 25 which implies that $\mathbf{A}^-(T) = \mathbf{B}^-$. This completes the proof of Theorem 3. The corollary 1 is a direct consequence: since $\mathbf{A}^-(T) = \mathbf{A}^-[T] = \mathbf{B}^-$, it follows directly from the definitions that $(\mathbf{A}')^-(T) = \{1\}$.

Remark 5. The above proof is intimately related to the case when $\mathbb{K}$ is CM and $\mathbb{K}_\infty$ is the $\mathbb{Z}_p$-cyclotomic extension of $\mathbb{K}$. The methods cannot be extended without additional ingredients to non-CM fields, and certainly not other $\mathbb{Z}_p$-extensions than the cyclotomic. In fact, Carroll and Kisilevsky have given in [3] examples of $\mathbb{Z}_p$-extensions in which $\mathbf{A}'(T) \neq \{1\}$.

A useful consequence of the Theorem 3 is the fact that the $\mathbb{Z}_p$-torsion of X/TX is finite. As a consequence, if $M = \mathbf{A}[p^\mu], \mu = \mu(\mathbb{K})$, then $Y_1 \cap M^- \subset TX$. In particular, if $a \in M^-$ has $a_1 = 1$, then $a \in TM^-$. We shall give in a separate paper a proof of $\mu = 0$ for CM extensions, which is based upon this remark. Note that the finite torsion of X/TX is responsible for phenomena such as the one presented in the example (5) above.

5 Conclusions

Iwasawa's Theorem 6 reveals distinctive properties of the main module $\mathbf{A}$ of Iwasawa Theory, and these are properties that are not shared by general Noeterian Λ-torsion modules, although these are sometimes also called "Iwasawa modules". In this paper, we have investigated some consequences of this theorem in two directions. The first was motivated by previous results of Fukuda: it is to be expected that the growth of specific cyclic Λ-submodules, which preserve the overall properties of $\mathbf{A}$ in Iwasawa's Theorem, at a cyclic scale, will be constrained by some obstructions. Our analysis has revealed some interesting phenomena, such as

1. The growth in rank of the modules $\mathscr{A}_n$ stops as soon as this rank is not maximal (i.e., in our case, p^{n-1} for some n.
2. The growth in the exponent can occur at most twice before rank stabilization.
3. The most *generous* rank increase is possible for regular flat module, when all the group $\mathscr{A}_n$ have a fixed exponent and subexponent, until rank stabilization, and the exponent is already determined by $\mathscr{A}_n$. It is an interesting fact that we did not encounter any example of such modules in the lists of Ernvall and Metsänkylä.

Although these obstruction are quite strong, there is no direct upper bound either on ranks or on exponents that could be derived from these analysis.

Turning to infinite modules, we have analyzed in chapter 4 the structure of the complement of TX in Iwasawa's module Y_1^- in the case of CM extentions. This was revealed to be $\mathbf{B}^-$, a fact which confirms the conjecture of Gross-Kuz'min in this case.

The methods introduced here suggest the interest in pursuing the investigation of consequences of Iwasawa's Theorem. Interesting open topics are the occurrence of floating elements and their relation to the splitting in the sequence (21) and possible intersections of Λ-maximal modules. It is conceivable that a better understanding of these facts may allow to extend our methods to the study of arbitrary Λ-cyclic submodules of $\mathbf{A}$. It will probably be also a matter of taste to estimate whether the detail of the work, that such generalizations may require, can be expected to be compensated by sufficiently simple and structured final results.

Acknowledgements The material of this paper grew, was simplified and matured over a long period of time, first as a central target, then as a byproduct of deeper investigations in Iwasawa's theory. I thank all my colleagues at the Mathematical Institute of the University of Göttingen for their support in this time. I thank Victor Vuletescu who actively helped the development of the ideas during the time of the Volkswagen Foundation grant and Tobias Bembom and Gabriele Ranieri for their critical reading and active help with the completion of this paper.

Last but not least, the questions discussed here were the first building block of a series of Seminar lectures held together with S. J. Patterson in the years 2007–2008, the notes of which will appear in the sequel in the series entitled SNOQIT: *Seminar Notes on Open Questions in Iwasawa Theory*. I thank Paddy for the inspirational discussions we had since that time. The first results of this paper were developed before the year 2008, while the author was sponsored by a grant of the Volkswagen Foundation. Some of the central questions were first considered during the month of February 2010, spent as a guest professor in Caen, at the invitation of Bruno Anglès: these led also to the more recent developments.

References

1. T. Albu. *Cogalois theory*. Number 252 in Monographs and textbooks in pure and applied mathematics. Marcel Dekker Inc., 2003.
2. M. Bhargava. The density of discriminants of quadratic rings and fields. *Ann. Math. (2)*, 162:1031–1063, 2005.
3. J. Carroll and H. Kisilevsky. On the iwasawa invariants of certain $\mathbb{Z}_p$-extensions. *Compositio Mathematica*, 49(2):217–229, 1983.
4. H. Cohen and H. W. Lenstra Jr. Heuristics on class groups of number fields. In Springer, *Number Theory, Noordwijkerhout*, volume 1068 of *Lecture Notes in Math.*, pages 33–62, 1984.
5. H. Cohen and J. Martinet. Étude heuristique des groupes de classes des corps de nombres. *J. reine und angew. Math.*, 404:39–76, 1990.
6. R. Ernvall and T. Metsänkylä. Computation of the Zeros of p-Adic L-Functions. *Math. Comp.*, 58(198):815–830, 1992.
7. L.J. Federer and B.H. Gross. Regulators and Iwasawa modules. *Invent. Math.*, 62(3):443–457, 1981.
8. T. Fukuda. Remarks on $\mathbb{Z}_p$-extensions of number fields. *Proc. Japan Acad. Ser. A Math. Sci.*, 70(8):264–266, 1994.
9. R. Greenberg. On a ceratin ℓ-adic representation. *Invent. Math.*, 21:117–124, 1973.
10. R. Greenberg. On the Iwasawa invariants of totally real fields. *American Journal of Mathematics*, 98:263–284, 1973.
11. K. Iwasawa. On $\mathbb{Z}_\ell$ - extensions of number fields. *Ann. Math. (2)*, 98:247–326, 1973.
12. L.V. Kuz'min. The Tate module for algebraic number fields. *Math. USSR Izvestija*, 6(2): 263–321, 1972.
13. S. Lang. *Cyclotomic fields I and II*, volume 121 of *Graduate Texts in Mathematics*. Springer, combined Second Edition edition, 1990.
14. S. Lang. *Algebra*. Addison Wesley, second Edition edition, 1994.
15. Francisco Thaine. On the ideal class groups of real abelian number fields. *Ann. of Math. (2)*, 128(1):1–18, 1988.
16. L. Washington. *Introduction to Cyclotomic Fields*, volume 83 of *Graduate Texts in Mathematics*. Springer, 1996.